VOLUME FIVE

Developments in NEUROETHICS AND BIOETHICS

Neurodivergence and Architecture

Serial Editor

Judy Illes, CM, PHD
Distinguished University Scholar and UBC Distinguished Scholar in Neuroethics
Division of Neurology, Department of Medicine
Neuroethics Canada
The University of British Columbia
2211 Wesbrook Mall, Koerner S124
Vancouver BC V6T 2B5
Canada

Katherine Bassil, Editorial Assistant to Judy Illes

VOLUME FIVE

DEVELOPMENTS IN NEUROETHICS AND BIOETHICS

Neurodivergence and Architecture

Edited by

ANTHONY CLARKE
BLOXAS, Melbourne, VIC, Australia

JOS BOYS
Learning Environments Equality Diversity and Inclusion Centre (LEEDIC), The Bartlett Faculty of the Built Environment, University College London;
The DisOrdinary Architecture Project, London, United Kingdom

JOHN GARDNER
School of Social Sciences, Monash University, Clayton, VIC, Australia

ELSEVIER

ACADEMIC PRESS
An imprint of Elsevier

Academic Press is an imprint of Elsevier
50 Hampshire Street, 5th Floor, Cambridge, MA 02139, United States
525 B Street, Suite 1650, San Diego, CA 92101, United States
The Boulevard, Langford Lane, Kidlington, Oxford OX5 1GB, United Kingdom
125 London Wall, London, EC2Y 5AS, United Kingdom

First edition 2022

ISBN: 978-0-12-824562-0
ISSN: 2589-2959

For information on all Academic Press publications
visit our website at https://www.elsevier.com/books-and-journals

Publisher: Zoe Kruze
Acquisitions Editor: Sam Mahfoudh
Developmental Editor: Cindy Angelita Gardose
Production Project Manager: Sudharshini Renganathan
Cover Designer: Victoria Pearson

Typeset by STRAIVE, India

Contents

Contributors

Clare Bond
Sarah Wigglesworth Architects, London, United Kingdom

Sonia Boue
Artist and Consultant for Neurodiversity in the Arts, Oxford, United Kingdom

Jos Boys
Learning Environments Equality Diversity and Inclusion Centre (LEEDIC), The Bartlett Faculty of the Built Environment, University College London; The DisOrdinary Architecture Project, London, United Kingdom

Anthony Clarke
BLOXAS, Melbourne, VIC, Australia

J.T. Eisenhauer Richardson
Department of Arts Administration, Education, and Policy, The Ohio State University, Columbus, OH, United States

John Gardner
School of Social Sciences, Monash University, Clayton, VIC, Australia

Marie K. Harder
Values & Sustainability Research Group, University of Brighton, Brighton, United Kingdom

Camille Y. Huang
Neuroethics Canada, Division of Neurology, Department of Medicine, University of British Columbia, Vancouver, BC, Canada

Judy Illes
Neuroethics Canada, Division of Neurology, Department of Medicine, University of British Columbia, Vancouver, BC, Canada

Allen Kong
Allen Kong Architect Pty LTD, North Melbourne, VIC, Australia

Yeoryia Manolopoulou
The Bartlett University College, London, United Kingdom

Molly Mattaini
Interdisciplinary Theater Studies, University of Wisconsin–Madison, Madison, WI, United States

Raquel Meseguer Zafe
Founder, Unchartered Collective and PMStudio Resident, Arnolfini Gallery Bristol 2021–2022, Narrow Quay, Bristol, England

Tanya Petrovich
Dementia Australia, Parkville, VIC, Australia

William Renel
Touretteshero CIC, London, United Kingdom

Jim Sinclair
Autism Network International (ANI), Syracuse, NY, United States

Jessica Thom
Touretteshero CIC, London, United Kingdom

Natasha M. Trotman
The Wellcome Collection Hub, The Wellcome Trust; The Helen Hamlyn Centre for Design, Royal College of Art; Kensington Gore, South Kensington, London, United Kingdom

Sarah Wigglesworth
Sarah Wigglesworth Architects, London, United Kingdom

Foreword

The Indus River Valley, which lies near the border of the modern states of India and Pakistan, was the site of one of the earliest human civilizations on the planet. Coursed by snow- and monsoon-fed rivers, this rich alluvial plain was originally home to an agrarian society, which grew to become a large and technically advanced bronze-age urban culture—a culture that helped define the subsequent growth of the Indian subcontinent. At its ancient peak, the Indus Valley civilization developed modern forms of governance, manufacturing, and trade and excelled in both fine and practical arts. Building design and construction, and urban planning, were among the strengths of this society, and the practical and philosophical aspects of these disciplines were captured in early Hindu Vedas as "Vaastu Shastra," the science of architecture.

Vaastu is a complex set of rules and formalisms for building design, much of which has origins in religious and mystical beliefs, but one of its guiding principles is that "forms, space, proportion, energy points and aesthetic features, can be put together in such a way that the entire experience created by the building would bring about an extraordinary response in the occupants" (Ananth, 1999). "Response" here refers to mental states—perceptual or cognitive—that influence decisions, beliefs, emotions, and actions. Perhaps, most importantly, this early Hindu approach was earnestly egalitarian, founded on the principle that each individual has unique spiritual and material connections to the universe. It naturally followed that a building can only function, can only achieve an "enhancement and enrichment of the users' consciousness," if the unique elements of the user are understood and incorporated in design (Ananth, 1999). The simple but fundamental principle here is that the building must be designed for the person who occupies it, with recognition that every individual has specific—and equally valid—requirements for fulfillment through the spaces they inhabit.

In modern cultures of the industrialized West, this person-centric design ethic is all but lost. Architects who attend to the individual and their relationship to the site, to the community, and to the natural environment must charge hefty fees for doing so. This has become an elitist practice such that fewer than 2% of residential housing in America is now designed by architects (Dickinson, 2016). Outside of the upscale residences of the wealthy and sometimes architecturally sophisticated, home design, today, is largely driven by profit and expedience motives of real estate developers. Within a few miles of my own home in Southern California lies a blight of tract houses, bounded by the freeway and extending as far east as the eye can see. Over a few short decades, large swaths of some of the most beautiful land on the planet was sold to developers, who leveled the natural warp and woof of the earth and built houses of mind-numbing sameness. All of this practice denies any need for real design or creativity and greatly reduces the cost of construction. Contrary to the architectural principles that guided the built environment of millennia past, the result is that our buildings today dictate who the occupant must be, rather than the person as the source of inspiration for the building.

The editors and contributors to this new volume have, quite rightly and usefully, cast architecture as an ethical challenge. When a system of practice discounts the manifold diversity of the human population, forcing all individuals to conform to the same ideal, due process fails and disparities of opportunities, access, and treatment are inevitable. Architects and observers of social history alike point to the patent and profound influence of the built environment on human behavior and cognitive function. But we shy away from frank scientific discussions of the psychological deprivation caused by the sameness of planned residential communities, the banality of our public-school classrooms, the sensory and motor inequities that result when profit drives the built environment, and the fact that design for people with genetic or acquired disorders of the brain is at risk of having the same status as an orphan drug.

The authors of the chapters in this collection bring these issues out into the open and, in doing so, contribute to an ongoing discussion of individual differences and fairness in the light of fast-growing knowledge of how the human brain develops and functions and how it gives rise to the uniqueness of individual human character and ability. The chapters in this collection define neuroethics and frame its relationship to the built environment (and indeed space more generally) and to modern neuroscience research. Perhaps, the most effective way to communicate these concepts and strategies is in the context of specific cases, including how the design needs of individuals who think differently from one another or who suffer from neurological and cognitive impairments are slowly but steadily being addressed. Several chapters in this collection adopt this approach with focus, for example, on dementia and autism. Advocacy and intervention are naturally critical elements of a path forward and are appropriately highlighted in this collection. Finally, and very compellingly, the collection includes contributions by individuals who, by virtue of their inability to fit inside the standard box, have suffered at the hands of a society inured of its diversity.

Many will continue to ask why people are not all content with the same three baths, elevated entryway, and shiny open-plan kitchen, with the convenience and pastel peace that it affords. The answer, well understood by the ancient Hindus and re-articulated herein, is that people are not all the same. Modern neuroscience and cognitive science have probed the depths of difference and revealed that structural and functional attributes of the brain are unique products of personal life and experiences of the environmental and social contexts in which each individual learns, decides, and acts. The cultural diversity of these experiences, expressed by humans through assorted personalities with distinctive tastes, colorfully eccentric design preferences, and varied strengths and weaknesses, are what defines the Manushyalaya—the "human temple" in Vaastu, the living space for "discovering and honing the individual spirit" (Ananth, 1999).

This person-centric and egalitarian approach to the built environment does not scale easily in a free market, particularly in the context of large disparities of wealth and power. Buttressed by a continuously growing knowledge of brain function, however, it is an approach that might lead back to an era in which the science of the human condition actually matters to architecture. Cultivating that approach, as this volume does, advances a vision and a promise of a new architecture for the human good.

Thomas D. Albright
The Salk Institute for Biological Studies
La Jolla, CA, United States
August 2022

References

Ananth, S. (1999). *The Penguin guide to Vaastu: The classical Indian science of architecture and design*. India: Penguin Books.

Dickinson, D. (2016). Architects design just 2% of all houses—Why?. Common Edge 7, 2016. https://commonedge.org/architects-design-just-2-of-all-houses-why/.

Introduction

John Gardner[a], Jos Boys[b], and Anthony Clarke[c]
[a]School of Social Sciences, Monash University, Clayton, VIC, Australia
[b]Learning Environments Equality Diversity and Inclusion Centre (LEEDIC), The Bartlett Faculty of the Built Environment, University College London; The DisOrdinary Architecture Project, London, United Kingdom
[c]BLOXAS, Melbourne, VIC, Australia

Architecture, as Amos Rapoport famously illustrated (1969), is a clear expression of social and cultural norms. The physical spaces of our everyday lives—our homes and workplaces, towns, cities, and transport infrastructures—reflect deeply-held assumptions about human behavior and interaction. Domestic spaces are obvious examples (Duncan & Lambert, 2004; Gorman-Murray, 2007; Percival & Hanson, 2009). Most newly built homes in western countries contain two or three bedrooms, reflecting the cultural intransigence of the nuclear family structure. The dominance of steps and staircases (instead of level accesses and ramps) assumes that individuals move bipedally. So too does the conventional width of door frames or the size of powder rooms and toileting spaces, both of which typically inhibit the use of walking aids and wheelchairs. The trend toward open-plan kitchen and dining areas in many countries reflects the increasing centrality of the kitchen and food preparation in social entertainment and, no doubt, the assumption that meal preparation and child-minding responsibilities are undertaken by the same individual simultaneously (Dowling & Power, 2012). Some behavioral norms have of course become encoded in domestic building regulations and guidelines (e.g., the need to separate toileting areas from food preparation), some are reinforced by economic dynamics (who can afford a house with three bedrooms, let alone four?), and many are the consequence of the habitual professional practices of developers, architects, designers, and builders (e.g., the near-universal preference for smooth and hard and largely featureless internal walls).

Architecture, however, is not simply the expression of social and cultural norms. It is a very powerful means by which such norms are imposed and perpetuated. Indeed, the ideal that very much underpins architecture as a profession is that better architecture will make genuine improvements to peoples' lives, and is, then, a public good (Lautner, 2011). There is also now a well-known and substantial body of evidence that the arrangement of our physical spaces—such as the simple proximity of windows with views—can make measurable improvements in peoples' health and well-being (Rice, 2019). The assumption that "we shape our buildings and thereafter they shape us" (Churchill, 1943) needs to be treated with care. Who exactly is this "we?" Who is being valued in the design of our built surroundings and who is being ignored, marginalized or mis-represented? Who has power over what gets financed and constructed and maintained? Who is left out of these processes? And how exactly does architecture work in relation to how we occupy it? As the feminist architecture collective Matrix wrote in 1984:

> *Buildings do not control our lives. They reflect the dominant values in our society, political and architectural views, people's demands and the constraints of finance, but we can live in them in different ways from those originally intended. Buildings only affect us insomuch as they contain ideas about [...] our 'proper place', about what is private and what is public activity, about which things should be kept separate and which put together (pp. 9–10)*

While matrix was thinking mainly about gender, many others have critiqued assumptions about architecture and its occupation. Disability advocates and academics from various disciplines have long demonstrated that the material fabric of our built environment is enabling for some people and disabling and alienating for others, reifying social divisions and stigma (Blunt & Dowling, 2006; Delitz, 2017; Duncan, 1996; Gleeson, 1964; Heylighen, Van Doren, & Vermeersch, 2013; Imrie, 1996; Jackson, 2018). More recently, disability studies scholars and activists have extended these debates, for example, by critically exploring histories of building access and inclusive design (Hamraie, 2017; Hendren, 2020; Titchkosky, 2003; Williamson, 2019) and building directly on disabled and neurodivergent people's own diverse perspectives and experiences of built space as included in academic scholarship, autobiography, memoir and other forms of narrative (Gadsby, 2022; Limburg, 2021; Mainspring Arts, 2021; Wong, 2020; Yergeau, 2017).

Architecture and the built environment, therefore, are *ethically* relevant. They are ethically relevant because they enact a generalized set of norms about how humans should conduct themselves, and in doing so, they constrain, enable, and transmute the actions of individuals and collectives in ways that significantly impact their well-being. Architecture and the built environment are also, as many scholars have argued, important targets for ethically relevant interventions (Fox, 2012). These might be permanent adjustments and reconfigurations of the built environment, aimed at directly mitigating inequalities and, ideally, ultimately enabling a diversity of peoples to flourish (Watchorn, Larkin, Hitch, & Ang, 2014). Or, they might include the creation of temporary installations, aimed at prompting a greater awareness and reflexivity among those who engage with them about the ethical relevance of the built environment (e.g., Boys, 2014; Cachia, 2017). Or they might be aimed at shifting both every day and disciplinary practices that—whether because of unconscious bias or deliberate intention—reproduce particular patterns of inequality in access to, and use of, space and resources.

Architecture as an ethical endeavor

Conceptualizing architecture as an ethical endeavor raises a number of important questions. In general terms, what processes will bring about spaces that genuinely support the flourishing of a diversity of people? More specifically, we might ask: what are the relationships between human variety, in all its neuro- and biological complexity, and the design and occupation of built spaces? To what extent should architecture engage with rapidly emerging neuroscientific knowledge about the nature of human difference? How might lived experiences and the accumulated knowledge of diverse individuals and groups be meaningfully incorporated? And how can people who consider themselves unproblematically normal pay attention to, and take account of, nonnormative and neurodivergent others that goes beyond commonplace tropes of an assumed abnormality or pathology?

The recognition and celebration of neurodiversity over the past two decades (see Kapp, 2019) has made these questions and the issues that underpin them especially relevant. Our homes and workplaces and built outdoor spaces are based on innumerable assumptions about the cognitive capabilities and emotional responses of typical humans. They make demands on our memories, our habits of reasoning, and our levels of energy. They assume shared sensitivities to some elements (e.g., air temperature, the height of door handles), and a general insensitivity or irrelevance to others (e.g., the feeling of stepping on the grout lines between tiles, the physical sensation of brick surfaces). For those who approximate a "typical human" in their cognitive capacities and emotional responses, the experience of inhabiting everyday built environments will be mostly unremarkable, aside from various levels of comfort and discomfort. For others, however, such environments can be deeply perplexing, alienating, and indeed traumatic. These atypical engagements may be a connected to learning disabilities, chronic illness, cognitive decline, result from injury, old age or neurodegenerative disease. They are also indicative of conditions that have acquired a diagnostic label such as autism spectrum disorder, posttraumatic stress disorder,

Tourettes syndrome or similar, some of which are now recognized and often celebrated as evidence of normal human neurodiversity in our many forms of responsiveness and sensitivities to our surroundings. These modes of existence and experience foreground what we can call the *neuroethical* relevance of architecture. This refers to the conventional capacity of the built environment to perpetuate and enable neurotypical (NT) modes of existence and expression, and to disable neurodivergent and bio-divergent people; but also due to its *potential* as a means of supporting and enabling other, neurodivergent modes of existence and expression.

So what might an architecture that also takes into account neurodivergent modes of existence and expression actually be like? How would the design of our buildings and infrastructure need to change? What processes, collaborations, tools, and sensitivities might meaningfully engage neurodivergent people and their allies in the design and evolution of built environments? What is the relevance of neuroscience and emerging knowledge of neural and cognitive mechanisms? And to what extent do educators and professionals—both those involved in neuroscience and those across the built environment sector—need to critically rethink their practices and attitudes? What role can disability art and theater play in both generating awareness among practitioners and communities about the relationships between architecture and neurodivergence; and in opening-up creative and provocative spaces that suggest the potential of built space to support all its diverse occupants?

Framing neurodivergence

The purpose of this edited collection—the fifth in the *Developments in Neuroethics and Bioethics* series—is to address these questions. Our aims here are to demonstrate the necessity for critical engagement with architecture and architectural processes as we recognize and celebrate neurodivergent modes of existence and expression, and to illustrate what has and is being done in this regard, by academic scholars, practitioners, and importantly, by neurodivergent people themselves. We have therefore brought together articles that range across academic disciplines including neuroethics, sociology, architectural theory and disability studies; that center neurodivergent people's own knowledge and experiences of built space and its associated design practices and processes; and that include contributions from architects and designers working with neurodivergent clients and users in a variety of ways. One consequence of this is that this volume is quite different in its form and format compared to others in the series, and indeed conventional collaborative academic texts more generally. Some contributions are indeed conventional in their academic style and tone. Others, however, employ many images; some use a more casual and widely-accessible tone and style; some are deliberately written outside of academic norms; and some contributions are significantly shorter than conventional academic texts. We have more to say about the importance of this pluralistic approach below.

It is also important to mention here that many contributions in this volume are written by autistic people, and/or are about autistic perceptions and experiences. "Neurodiversity" and "neurodivergence" are—as we discuss in more depth below—evolving concepts. In addition, neurodivergence is often currently used in the literature as a synonym for autism. It is not our place here to delineate what exactly constitutes neurodivergence, and we want to remain open to the prospect that it can be applied to many different modes of existence and expression, including but not restricted to learning disabilities, ADHD, dyslexia, trauma, brain injury or degeneration and the like. The fact that more chapters focus on autism in this volume than other modes of neurodivergence is a reflection of the pioneering work that has been undertaken by autistic scholars and their advocates, which, we believe, illustrates themes and concerns which are relevant to neurodivergence and architecture more generally. At the same time, we want to resist a tendency to try and delineate a fixed and final list of "neurodivergent" categories. We are more interested in exploring both the similarities and differences in the many neurodivergent ways of being intersect with built spaces.

Exploring complex intersections as this volume does makes certain demands. It demands a clarity about concepts and methods, so these can be shared effectively across multiple disciplinary perspectives and agendas, and it demands a criticality about conventional academic writing itself, to ensure that nonnormative voices are not excluded. While our intended readers span across people working in neuroscience, those working in built environment areas, and neurodivergent scholars, artists and others interested in architecture, the build environment and neurodivergent space more generally; we are particularly keen to make sure that various neurodivergent voices and perspectives are represented. In this space, it is essential that we recognize and value both alternative nonnormative forms of academic writing, and the centrality of visualization, illustrations and other forms of representations within architecture and other built environment disciplines.

This also requires that we first look at the terminology of neurodiversity, neurodivergence and neurotypicality, and its associated everyday practices. This is an area of activism, advocacy, and scholarship where the socio-ethical implications of language, terms, and concepts have been widely discussed.

Here we provide some general definitions for the key concepts and terms that are used by authors in this volume. In doing this, we also want to illustrate "what is at stake" in our choice of terms and phrases. Like the architecture of our built environments, the language we use to make sense of each other is an assertion of particular conventions and norms, and this can have stigmatizing and repressive consequences. We recognize the need for a shared understanding of key concepts and terms—hence the provisional definitions we provide here—but we also recognize the necessity that such concepts and terms evolve via contestation and negotiation.

Readers will note that we have chosen for the title of this book *Neurodivergence and Architecture*, rather than *Neurodiversity and Architecture*. Here, we have made use of the increasingly common distinction between neurodivergence and neurodiverse: the former refers to modes of being that *diverge from* the norm, and are thus *distinct from* so-called NT modes of being, whereas the latter refers to the full spectrum, including modes of being conventionally deemed normal or typical. Our decision here reflects our focus on architectural and other creative processes that engage with atypical participants. A focus on neurodiversity, in contrast, would imply the inclusion of architectural responses and processes that engage also with NT people, who by definition represent the largest and dominant segment of neurodiverse spectrum. Such a focus would of course be a redundant exercise given the everyday unnoticed privileging of NT people in architectural processes and accompanying scholarship, and it would undermine our objective to champion architectures and processes that celebrate modes of existence and expression.

What exactly, then, might be included within the term "neurodivergent?" As noted above, we do not make a firm distinction about what modes of being are—or should be—included within the term. It is not our place to make such a delineation, and it is likely that the application of the term will evolve. In general terms, however, we assume that the term can apply to a range of atypical modes of existence and expression. This of course includes atypical modes that are now recognized and celebrated by an increasing number of scholars and advocates as normal but atypical, such as autism or ADHD.

This is informed by a long history of challenges to the medical model of disability, that is, one that sees impairments as individual problems, predominantly articulated within a medical framework as diagnosable conditions characterized by abnormal brain functioning and a dysfunctional engagement with the world, and that therefore require medical treatment and—preferably—cure. Common terminology that is often used to make sense of autism, such as "high-" and "low-functioning" reflects this medical framework, and has been consistently resisted by many autistic activists and scholars (see Harder, this volume). The premise of this book is that such modes may indeed be enacted and experienced as dysfunctional, but that this is the product of engagement with inadequate built environments (as well as other stigmatizing and repressive social and cultural factors). From this perspective, disability needs to be understood through a social model that examines the interplay between discriminatory societal attitudes, social and material barriers, and impairment.

This in turn, has broadened our engagements with neuroethics, as shown by the range of selected contributions to this volume. If neuroethics within neuroscience focuses on the ongoing improvements in our ability to monitor and influence brain function, requiring "the examination of what is right and wrong, good and bad, about the treatment of, perfection of, or unwelcome invasion of and worrisome manipulation of, the human brain" (Marcus, 2002: p. 5), then a neuroethics that also engages with the social model and with our built surroundings, necessarily opens up the need for conceptual frameworks that connect how our brains give rise to the people that we are and the social and spatial structures that we inhabit and create. In different ways, from the personal through to the theoretical, many of our authors are concerned to connect neuroethics to human rights around access and inclusion; to equality and equity; and to social spatial and material justice—often called disability justice (Costanza-Chock, 2020; Mingus, 2011; Piepzna-Samarasinha, 2018; Sins Invalid, 2019).

Toward neurodivergent architecture and spaces?

In this regard, we take the position that many neurodivergent modes of existence and experience are reflections of natural and vital variation in neural and cognitive capabilities, that can and should be valued, and are not inherently pathological or inferior. We also note that many autistic activists and scholars see themselves as not disabled but as having an autistic way of being; often deliberately reversing the terminology that only names them as having a deficit, by using terms such as normate and NT in order to highlight how typical bodies and minds simultaneously value themselves over others, while assuming their own ways of being as nonproblematic. However, our use of "neurodivergent" also encompasses modes of existence which are subject to much debate regarding their status as neurological illnesses or disabilities, and others which perhaps have more stable definitions. The former may include Tourettes, schizophrenia, and psychological trauma, while the latter may include conditions arising from neurodegenerative disease and injury such as Alzheimers, learning disabilities or brain trauma. This is about whether particular "mental differences are forms of illness or disorder at all" as well as the difficulties of creating accurate diagnostic labels (Radden, 2019). This can be seen in the multiple and partial attempts through history (and across place) to define various forms of cognitive difference into clear and mutually exclusive categories (Silberman, 2015). In addition, the language used in these definitions has changed, away from the belief of mental illnesses as "evil," as "willful error," as "grave misfortune" or, at the very least, regarded as unwelcome (Radden, 2019).

At the same time, disabled people's perceptions and experiences are—of course—very diverse, intersected with other kinds of label (by gender or race or sexuality for example); so that attempts at impairment categorization, and/or generalizations about needs and preferences, are always potentially fraught. Jess Thom, a neurodivergent performer and writer (see also Chapter 15 this volume) recently created a diagram that explores and explains her own personal experiences of coming to identify positively and politically as a disabled person. Called "Acceptance and Identity: Seven Stages" (Thom, 2020), it starts from the social model of disability (already outlined in this introduction) by separating out people's various impairments (that is, the facts of their own body mind) from disability as the lived experience of attitudinal and material barriers, because of that impairment. Thom then explores the different stages through which people may go in understanding both how they position themselves as disabled, and how society positions them. These range from feelings of shock and grief through to acceptance; and of moving from the frustrating experience barriers to valuing the contribution that disabled people make. This is relevant here, because all too often disabled people are asked to participate in design processes as if they are authentic representatives for a whole impairment category. This is problematic for two reasons. First, it fails to recognize both how much impairments can vary within a disability category as well as across categories. Second, it does not take account of the multitude of ways disabled people place themselves in relationship to wider societal views (Fig. 1). Boys (2021) has also explored a version of this variation in positioning for nondisabled people, to try and draw out equivalent differences and journeys, that recognizes a considerable range of attitudes to disability—from feelings of pity to full engagement with enabling social, spatial and material justice (Fig. 2).

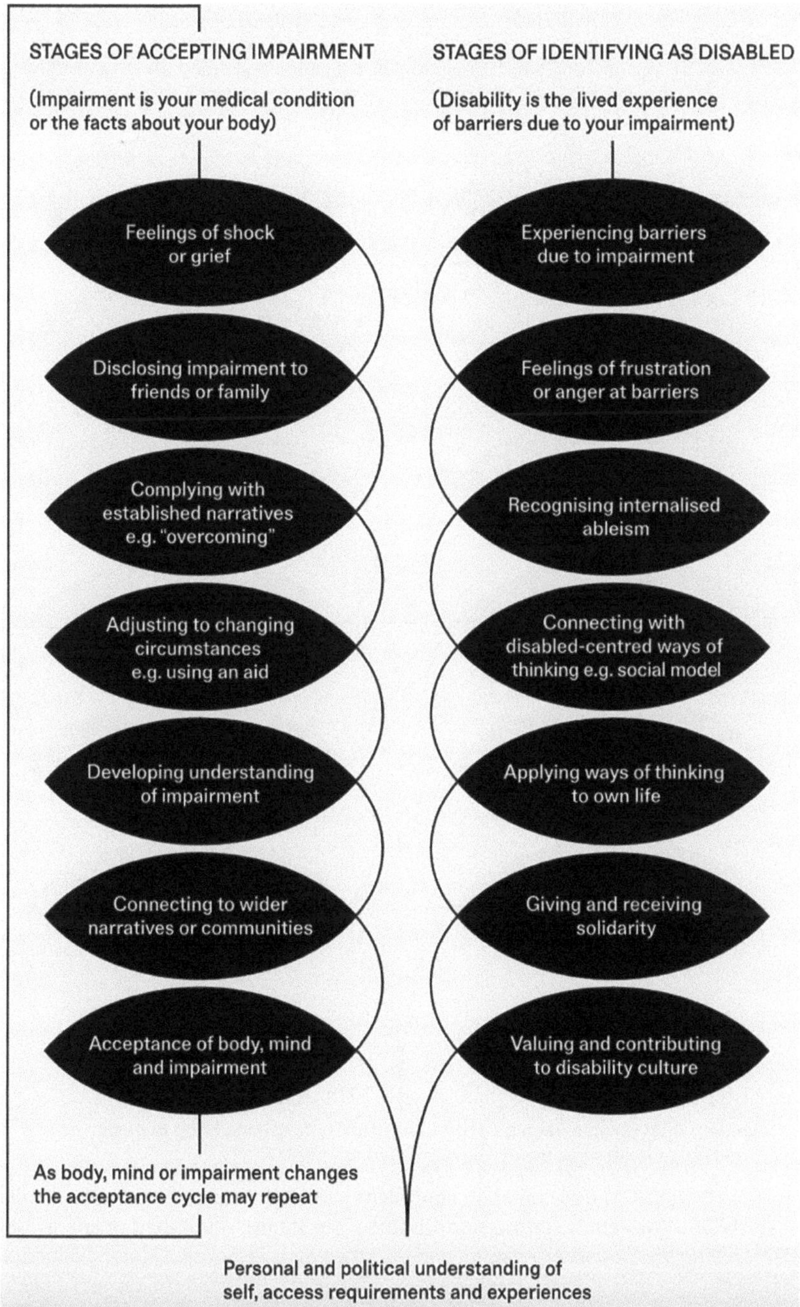

Fig. 1 Diagram adapted from Jess Thom's "Acceptance and Identity: Seven Stages," as published on the Touretteshero blog 2020 (3rd September).

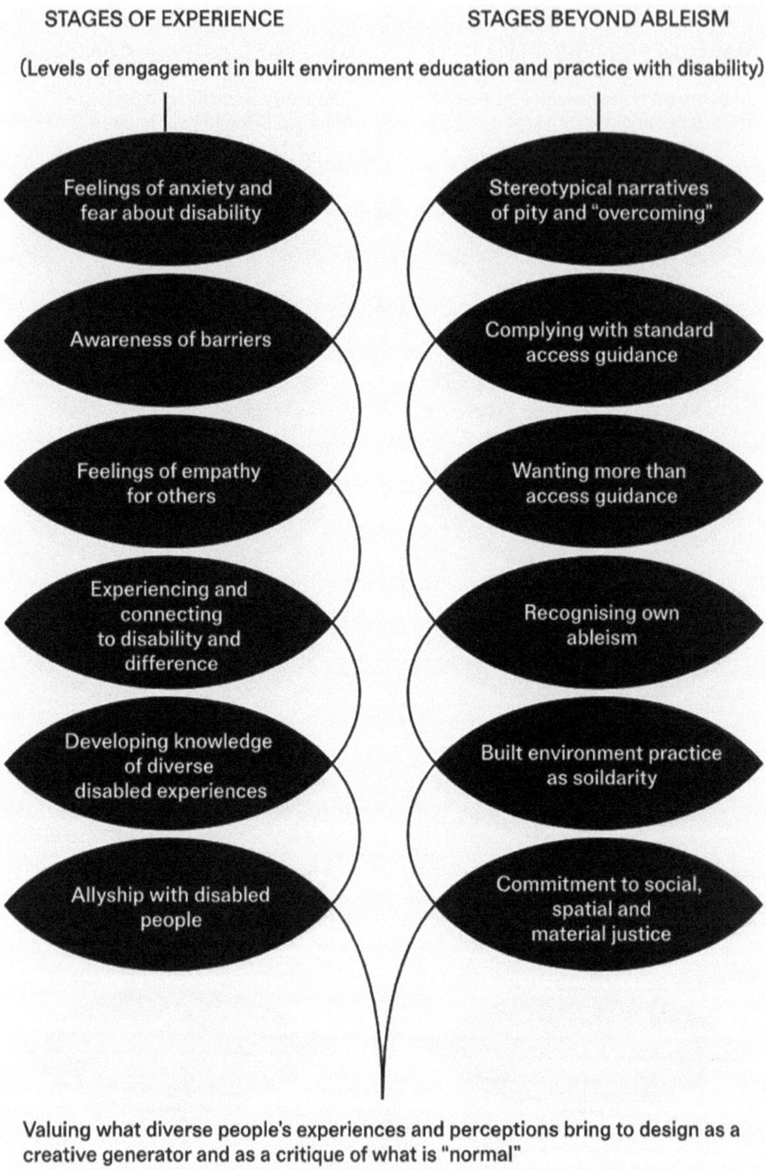

Fig. 2 Jos Boys (2021) diagram of an equivalent journey for nondisabled people called "Disability and Awareness for the Non-Disabled: Six Stages," published online as Boys (2021) *Shifting the Ground: an essay on re-thinking access and inclusion* Molonglo (online).

This acceptance of a fluidity of definitions of, expansiveness about, and contested views over, what it is to be disabled and/or neurodivergent, is therefore part of a broader interest in how diverse bodies and minds engage with built space. Here, similarities and differences in access needs are to be found (within and across impairments, perceptions and experiences) in what makes supportive built spaces. This is about tracing the emergent and dynamic patterns of both alignments and potential contradictions in our multiple ways of being in the world; and proposing that this can be critically and creatively learnt from (for all body minds). That is, we see neurodivergence as a creative place to start, as a potential generator of design ideas (Richard Rogers, the world-renowned architect was famously dyslexic), and as a challenge to normative assumptions built into our material surroundings that disable some and enable others.

As editors, we have a similar expansiveness in our definition of architecture. We have selected contributors who understand architecture as more than its built products. Authors have explored architecture and design as a process and practice; as a dynamic service that can meet the needs of diverse people; as a site for performance and other creative actions that can indicate new design possibilities; and as built space that supports nonnormative bodies and minds This focus on architecture as always ongoing and dynamic, rather than a static thing fits with much contemporary work in the built environment disciplines—understanding buildings and public spaces in terms of the processes and practices through which they are both created and occupied. This is particularly relevant when connected to neurodivergence, as it demands a critical engagement with normative and NT assumptions about how buildings should be designed; who and what is valued in its conventional processes and practices; what become framed as solutions; and what are examples of good practice for alternatives and improvement.

Chapter outlines

This volume is divided into three sections—Frameworks, Advocacy, and Practices. The first section presents theoretical and methodological frameworks for making sense of the intersections between neurodivergence, architecture and ethics. There are, as this section will argue, a range of relevant theories around conceptualizing relationships between minds, bodies and built space across neuroscience, neuro and bioethics, the social sciences, as well as some arts and humanities disciplines. It is important to represent a range of perspectives and agendas that engage with these issues, as well as illustrating the diverse scholarship that supports the various forms of advocacy and design practices and processes that are explored in subsequent sections.

The Advocacy section centers on disability-led approaches to neurodivergence and architecture. By advocacy, we mean scholarship and activism that starts from, and supports, a particular cause or policy; in this case, foregrounding the perceptions and experiences of multiple neurodivergent people as they engage with built space and its design. Practices, the third section, centers real-life examples—both design processes and built projects—to examine how neurodivergence as a concept, and neurodivergent people as a constituency across many variations, are positively impacting on architecture and the built environment.

Frameworks

We begin with examining neuroethics as a scholarly field in its own right. In "The Neuroethics of Architecture," Judy Illes (Editor-in-Chief) and Camille Huang argue that modern neuroscience, psychology, and behavioral studies provide tools that can significantly improve our understanding of human responses to the built environment. They argue that, together, these give rise to a science-based framework for understanding how architecture can be responsive to diverse neurological needs. Illes and Huang begin by outlining basic understandings of how the brain interprets the built environment, illustrating that neuroscience foregrounds the connection between the built environment and individuals' mental responses. They then go on to propose a neuroethical framework that reflects this scientific understanding of the brain-environment connection; and how this can inform design so that it is genuinely aligned with human health and well-being, and facilitative of human flourishing.

The second contribution is John Gardner's "Science and Technology Studies (STS) and the neuroethics of architecture." In this chapter, Gardner draws on the interdisciplinary field of Science and Technology Studies (STS) to conceptualize the relationship between architecture and neurodiversity, and to advocate for interventions aimed at creating built environments that accommodate and celebrate neurodivergent modes of existence and expression. An important advantage of an STS-informed approach, Gardner suggests, is that it encourages us to recognize the value of neuroscientific knowledge of human responses to the built environment while also avoiding the pitfalls of neuro-reductionism or neuro-essentialism. Gardner then argues that the built environment reflects social norms and values, including those that privilege NT modes of existence, and it therefore participates in enabling and supporting people who approximate "typical" individuals, while disabling and constraining those who do not. The chapter thus concludes by providing a description of general neuroethical interventions that can bring about environments that are more supportive and celebratory of neurodivergent modes of existence and expression.

In Chapter 3, "Disability studies, neurodivergence and architecture," Jos Boys first outlines some key aspects of disability studies, and then considers how this strongly emerging field is intersecting with scholarship and activism across many varieties of neurodivergence—as well as wider intersectionalities—to draw out some of this approach's underlying principles, as well as its various ethical challenges to everyday negative and ableist attitudes to disability. Boys then explores connections with architecture—both as a disciplinary practice, and as built space. She offers examples of projects that challenge the deficit model of disability and neurodivergence (which assumes being "normal" and NT is inherently a naturally preferred and superior way of being) by treating mental nonnormativity as a creative generator, rather than a problem to be solved. And she explores projects and processes that aim toward a future where starting from neurodivergence enables alternative ways of thinking and doing architecture.

In the final chapter in this section, "Autoethnographic reflections on architectural design for neurodivergence" Anthony Clarke uses selected case studies from his own architectural practice to highlight the importance of the physical environment to individual needs, desires, and expectations, using design processes that start from the mind and body differences of his clients, and questioning what effects this might have on designing for neurodivergence. Clarke uses an auto-ethnographic design methodology, which he argues is a valuable tool for architects and designers to appreciate the benefits of reflecting on, and working with, the emotive dimensions of design and the design process.

Advocacy

The first chapter in this section is by Jim Sinclair. As a cofounder of Autism Network International (ANI) in 1992 and as its coordinator since then, they have been a central figure in the development of autistic-led activism. In

"Cultural commentary: Being autistic together," written in 2010 with an extract reprinted here, Sinclair notes that autistic people are generally seen by NT people as lacking in ability to share common interests with others, disconnected from social participation and fellowship, and unable to understand or manage "normal" social behaviors and attitudes. They suggest that for autistic people, the situation is not so clear-cut, and that isolation and withdrawal is as much a result of the distress caused by normative demands on behavior. During the past 20 years, facilitated by the availability of the Internet, more and more autistic people have been reaching out and forming connections, creating community, and discovering their own styles of autistic togetherness. ANI was the probably the first autistic community to be created naturalistically by autistic people and remains the largest autistic-run organization to have regular physical gatherings of autistic people. In this article, Sinclair reports on the social and spatial culture that such meetings create; drawing on the experiences of the ANI and its retreat-style conferences, Autreat, to provide reflections and guidance on preparing shared autistic spaces.

In Chapter 6, Professor Marie Harder, an autistic academic who specializes in sustainable waste management and is based at the University of Fudan Shanghai, considers her own working environments. In "Self-made design notes for an autistic's office," she argues that meaningful work/occupation is a basic human need and that meaningful occupation for her requires place and space to immerse in tacit thinking processes. This is developed as a principle for examining the relationship of her neurodivergence to various built environment attributes. Harder concludes that valid design for neurodivergent persons such as herself needs a holistic and autoethnographic consideration, rather that checklist approaches and bimodal medical descriptions.

The third contribution to this section is by Natasha Trotman. "Equalities design: Toward post-normative equity" describes the approach used to inform and guide their practice as a designer. Trotman outlines their own creative development journey and the associated tools, methods and approaches developed, aimed at enabling the interrogation and queering of normative (including neuronormative) assumptions in order to create alternative body mind explorations and experimental outputs. Rather than simplistic design solutions that have been presented as inclusive, this work is about centering intersectionality and the profound challenges encountered by multiply-marginalized groups. The chapter argues for the necessity of such an approach and describes a Design Exosystem Framework that builds in equality, and thus ethical positioning, right from the start. The chapter also introduces examples of the implications of this approach, through Trotman's involvement with the Wellcome Trust's *Design and the Mind* initiative, focused on cocreation and engagement with neurodivergent, disabled and D/deaf people and groups, and which produced outputs such as toolkits and a public-facing 5-day program in the Wellcome Collection Hub.

In the final chapter for this section, "Dialogic drawing," architect Yeoryia Manolopoulou writes about the *Losing Myself* project, developed from the perspective of architects needing to know much more about dementia, in order to better inform the design process. Between 2015 and 2016, together with architect Niall McLaughlin, she conducted a series of interviews with experts in the health, cognitive and behavioral fields and with people and carers affected by dementia across the UK and Ireland. These were envisioned as a series of critical dialogues that sought to examine the experiences of people living with dementia and, more fundamentally, how the brain comprehends space and the implications of this for architecture. While most literature produced on dementia and architecture focuses on best practice guidance, these discussions questioned reductive design guidelines and asked: What spatial capacities do we have that we might lose because of dementia? What findings from neuroscience, anthropology, art, healthcare, and policy research can help architects design for people living with dementia and, more broadly, for all of us? Manolopoulou's chapter reflects on selected content from these dialogues and expands the discussion to newer research and policy initiatives on inclusive design and brain health. It also discusses Mikhail Bakhtin's theory of dialogue and a dialogic process of architectural drawing that conjoins divergent perspectives and imaginations to form better understandings of the human experience. She concludes by arguing that many of our limitations in understanding neurodivergence holistically and imaginatively, both as architects and as citizens, can be overcome by developing learning and communication processes that are truly dialogic.

Practices

The first contribution in the Practices section is by architect Allen Kong. It presents examples of inclusive architecture for atypical people, designed by Allen Kong Architects (AKA). The design philosophy of AKA is that the quality of life of the individual is first and foremost, and that this requires addressing their individual needs in a way that also recognizes their social, economic, and environmental contexts. The chapter summarizes this design approach and present three AKA projects in Australia: the Port Melbourne Hostel for particularly vulnerable elderly individuals who would otherwise be homeless; The Odyssey House Family Units for individuals receiving therapy for drug and alcohol addiction and their families; and the Wallara-Wintringham Potter Street Redevelopment, for individuals with disabilities and their elderly parent carers.

Chapter 10 is by autistic artist Sonia Boué. "Not all surfaces catch the light at the same time" presents work written during her Arts Council England (ACE) funded *Neither Use Nor Ornament* (NUNO) curated exhibition project. This chapter both reflects on, and presents, a curated selection of blog posts originally published on *The Other Side* (Boué's Wordpress blog). This both documents the development of the project and the evolution of her own thinking about enabling inclusive processes and projects. The aims of NUNO were twofold; to lead the project autistically, and level the playing field for the autistic artists on the project. Using a methodology of worked iteratively (with the support of mentors), Boué explored how to make access for all—regardless of neurological status—a priority above all other criteria, including project design and outcomes. This recognized that all participants had needs that required accommodation, and demonstrated that when access was centered, project outcomes took care of themselves. Both the reflections and the associated collection of writings provide insights into an outstandingly successful experiment in accessible project design, which created a peak positive experience for participants. Boué concludes by exploring questions around sustaining such initiatives, and how the spirit of NUNO might be replicated or translated to other contexts.

In the next chapter, by J.T. Eisenhauer Richardson, the author returns to an installation artwork created in May 2008 entitled Admission and a prior *Disability Studies Quarterly (DSQ)* (2009) article titled "Admission: Madness and (Be)coming Out Within and Through Spaces of Confinement." Invoking Karen Barad's "re-turning" as an iterative and diffractive practice, Richardson considers neuroqueer intimacy and neuroqueer touch in relation to both their original artwork and to experiences of confinement within psychiatric spaces. This centers experiences in architectures of madness, including relationships with nonhuman actants beyond constructs of insight and intentionality, through a neuroqueer response of "I don't know." Such not-knowing challenges the explanatory impulse of criticality through insight. Extending Ervin Goffman's framing of "admission" within psychiatric spaces as dispossession, Richardson argues that neuroqueer touch and neuroqueer intimacy is integral to a more complex understanding of psychiatric architectures of confinement beyond NT frames of agency and action.

This is followed by a contribution from Molly Mattaini, reprinted from *Youth Theater Journal* (2019) called "Creating autistic space in ability-inclusive sensory theater (AIST)". This is an emerging genre of Theater for Young Audiences which serves young people with autism and other cognitive disabilities. Distinct from sensory friendly or relaxed performances, these productions build the entire aesthetic experience to cater to an audience with sensory differences. The space created in these productions mirror the social concept of "autistic space" in which the needs of people on the autism spectrum inform the physical and social design of the environment. This chapter explores how AIST productions create autistic space through script development, audience engagement, immersive design, soft transitions, sensory objects, and an audience-centric dramaturgy.

In Chapter 13, Tanya Petrovich outlines "The virtual dementia experience" (VDE), developed in 2013. Its aim is to use game technology to provide learning experiences about dementia so as to enable carers, and others living with people with dementia, better understanding of how to provide enabling support. Creating a virtual world allowed developers to illustrate how the built environment may be interpreted differently by people living with dementia. Petrovich argues that the impact of the experience on learners has been very positive in relation to their care practices and that the VDE has also offered a powerful tool for understanding the built environment and the principles of dementia-friendly design.

In the following chapter, Sarah Wigglesworth and Clare Bond from Sarah Wigglesworth Architects (SWA) discuss some of the practice's projects. As with many of our other authors, "Designing with neurodiverse children and adults: Learning a different lesson with every engagement" take as a starting point the specificities and complexities of each design process. SWA uses a "learning through doing" method that combines project work with research and knowledge building. Their chapter explores this approach through two different projects, Classroom for the Future at Mossbrook Special School in Sheffield UK, a completed building; and a collaborative research project with Habinteg Housing Association. Both these projects are focused on inclusive and accessible design for people with neurodiversities; with the latter building on the principles of research and engagement which were embedded within the design of Mossbrook School.

Working with Habinteg has also required the translation of lessons from the education sector to the residential sector and beyond; thus enabling the updating and deepening of the practice's understanding of how adults as well as children with neurodiversities and other physical impairments navigate the built environment within everyday life.

Chapter 15 "Relax and resist: Reflections on the Touretteshero relaxed venue methodology" is by William Renel and Jessica Thom. In February 2020, Battersea Arts Centre (BAC) in London launched as the world's first Relaxed Venue, aiming to radically embed access and inclusivity across all of their activities by utilizing a new "Relaxed Venue" methodology. This methodology was developed by Touretteshero, a disabled and neurodivergent-led Community Interest Company and is underpinned by Touretteshero's mission to create an inclusive and socially just world for disabled and nondisabled people through cultural practice. This chapter outlines the process through which the Relaxed Venue methodology was created, focusing on moments of change at BAC and describing the three Relaxed Venue commitments: "No New Barriers," "Equality of Experience," and "Reduce Fuss." Renel and Thom argue that their Relaxed Venue methodology challenges the depoliticized understanding of accessibility as a check-list, an end goal or a "nice thing to do" and sounds a call to action for more environments that are designed to support a diversity of bodies and minds to take up space meaningfully and equitably together.

The final contribution to this section, by disabled dance theater practitioner Raquel Meseguer, is called "Chronic pain & chronic illness: A crash course in cloudspotting." In it, she asks what it is like to navigate our built environment in a body that is hurt or fatigued; and how we might reimagine the relationship between resting bodies and public space. Working from her creative and immersive project, *A crash course in cloudspotting*, Meseguer is able to draw on nearly 300 stories accumulated from people with chronic pain, chronic fatigue, and/or other disabilities and/or neurodivergence. This work aims to challenge the etiquette of our public space, to make horizontal bodies more visible, to re-frame and elevate acts of rest, and to convey something of the lived experience of a disability like chronic pain. Meseguer calls this a crip practice advocating for crip spaces: pockets of nonnormative time and space, where body minds are welcome to rest. She argues that rest enables a different mode of being and of experiencing the world: a more poetic orientation that shouldn't be hidden away but that can be a proud and central parts of our cityscapes. She finishes with a call for the design of beautiful and exquisite resting spaces, that celebrate difference, and challenge cultural attitudes toward people who present differently in public.

References

Blunt, A., & Dowling, R. (2006). *Home*. Taylor & Francis.

Boys, J. (2014) *Doing disability differently. An alternative handbook on architecture, disability and designing for everyday life*. Routledge.

Boys, J. (2021). *Shifting the Ground: An essay on re-thinking access and inclusion Molonglo*. Accessed 25 June 2022. Available from https://molonglo.com/theory/series.

Cachia, A. (2017). Along disabled lines: Claiming spatial agency through installation art. In J. Boys (Ed.), *Disability, space, architecture* (pp. 247–260). London: Routledge.

Costanza-Chock, S. (2020). *Design justice. Community led practices to build the worlds we need*. MIT Press.

Churchill, W. (1943). *Speech to the house of lords*. 28 October.

Delitz, H. (2017). Architectural modes of collective existence: Architectural sociology as a comparative social theory. *Cultural Sociology, 12*(1), 37–57. https://doi.org/10.1177/1749975517718435.

Dowling, R., & Power, E. (2012). Sizing home, doing family in Sydney, Australia. *Housing Studies, 27*(5), 605–619. https://doi.org/10.1080/02673037.2012.697552.

Duncan, N. (Ed.). (1996). *Body space. Destablising geographies of gender and sexuality* Routledge.

Duncan, J. S., & Lambert, D. (2004). Landscapes of home. In *A companion to cultural geography* (pp. 382–403). Oxford, UK: Blackwell Publishing.

Fox, W. (2012). *Ethics and the built environment*. Taylor & Francis.

Gadsby, H. (2022). *Ten steps to Nanette*. Penguin Random House.

Gleeson, B. (1964). *Geographies of disability*. Routledge.

Gorman-Murray, A. (2007). Contesting domestic ideals: Queering the Australian home. *Australian Geographer, 38*(2), 195–213. https://doi.org/10.1080/00049180701392766.

Hamraie, A. (2017). *Building access. Universal design and the politics of disability*. University of Minnesota Press.

Hendren, S. (2020). *What can a body do? How we meet the built world*. Riverhead Books.

Heylighen, A., Van Doren, C., & Vermeersch, P.-W. (2013). Enriching our understanding of architecture through disability experience. *Open House International, 38*(1), 7–19. https://doi.org/10.1108/OHI-01-2013-B0002.

Imrie, R. (1996). *Disability and the city*. Sage.

Jackson, M. A. (2018). Models of disability and human rights: Informing the improvement of built environment accessibility for people with disability at neighborhood scale? *Laws*, 7(1), 10. Retrieved from https://www.mdpi.com/2075-471X/7/1/10.

Kapp, S. K. (2019). *Autistic community and the neurodiversity movement: Stories from the frontline*. Springer Singapore.

Lautner, J. (2011). *The purpose of architecture*. Retrieved from http://www.johnlautner.org/wp/?p=710#:~:text=The%20purpose%20of%20Architecture%20is,flowering%20into%20a%20beautiful%20entity.

Limburg, J. (2021). *Letters to my weird sisters: On autism and feminism*. London: Atlantic Books.

Mainspring Arts. (2021). *Other words: Eight stories, eight unheard voices*. Unbound.

Marcus, S.J. (Ed.). (2002). *Neuroethics: Mapping the field*. Dana Press.

Matrix. (1984). *Making space: Women and the man-made environment*. Pluto Press (reprinted 2022 by Verso).

Mingus, M. (2011). Changing the framework: disability justice. In *Leaving Evidence*. 12 February. Retrieved from https://leavingevidence.wordpress.com/2011/02/12/changing-the-framework-disability-justice/.

Percival, J., & Hanson, J. (2009). Big brother or brave new world? Telecare and its implications for older people's independence and social inclusion. *Crit Soc Policy, 26*(4), 12.

Piepzna-Samarasinha, L. L. (2018). *Care work: Dreaming disability justice*. Arsenal Pulp Press.

Radden, J. (2019). Mental disorder (illness). In: Zalta, E.N. (Ed.), *The Stanford Encyclopedia of Philosophy* (Winter). https://plato.stanford.edu/archives/win2019/entries/mental-disorder/.

Rapoport, A. (1969). *House form and culture*. Prentice-Hall.

Rice, L. (2019). The nature and extent of healthy architecture: The current state of progress. *Archnet-IJAR: International Journal of Architectural Research, 13*(2), 244–259. https://doi.org/10.1108/ARCH-11-2018-0005.

Silberman, S. (2015). *NeuroTribes. The legacy of autism and the future of neurodiversity*. Avery.

Sins Invalid. (2019). *Skin, tooth and bone: The basis of movement is our people: A disability justice primer*.

Thom, J. (2020). *Acceptance and Identity: Seven Stages*. Touretteshero. Online (3rd September). Retrieved from https://www.touretteshero.com/2020/09/03/acceptance-and-identity-seven-stages/.

Titchkosky, T. (2003). *Disability self and society*. University of Toronto Press.

Watchorn, V., Larkin, H., Hitch, D., & Ang, S. (2014). Promoting participation through the universal design of built environments: Making it happen. *Journal of Social Inclusion, 5*(2), 65–88. https://doi.org/10.36251/josi.77.

Williamson, B. (2019). *Accessible America: A history of disability and design*. NYU Press.

Wong, A. (Ed.). (2020). *Disability visibility: First-person stories from the twenty-first century* Vintage.

Yergeau, M. R. (2017). *Authoring autism: On rhetorical and neurological queerness*. Duke University Press.

SECTION 1

Frameworks

CHAPTER ONE

The neuroethics of architecture

Judy Illes* and Camille Y. Huang
Neuroethics Canada, Division of Neurology, Department of Medicine, University of British Columbia, Vancouver, BC, Canada
*Corresponding author: e-mail address: jilles@mail.ubc.ca

Contents

1. Introduction

At a glance, the Corinne Dolan Alzheimer Center (CDAC) in Chardon, Ohio does not look like a typical assisted living facility. The tiny one-story building is triangular in shape, with two symmetrical wings of 12 beds each that enclose central public areas for living, dining, and cooking. Floor supports throughout the center are flexible, permitting the facile rearrangement of living space. The furnishings are minimalist and serene, and the open rooms are wrapped around a concentric path that minimizes both obstacles and long distances. With terracotta walls, skylights, and colorful quilts, the *New York Times* stated the building looked more like a "rustic corporate conference center" than a senior home (Lewin, 1990).

Why was the CDAC designed in this manner? The answer lies in the history of the site as one of the earliest model projects in the United States evaluating the connection between architecture and brain health—one that stood to benefit the 50 million people living with

Developments in Neuroethics and Bioethics, Volume 5
ISSN 2589-2959
https://doi.org/10.1016/bs.dnb.2022.06.001

Alzheimer's and other dementias worldwide. Here we explore how neuroethics—the study of the ethical implications of the neurological sciences—is aligned with a brain-based ethics for architecture that promotes effective, empathetic spaces, the widest range of engagement of thinkers and designers, and the integration of diverse perspectives throughout all forms of design definitions, process, and completion.

2. The CDAC as a case example

In the early 1990s, an interdisciplinary effort was underway to evaluate the impact of environmental interventions on neurological outcomes. Designed in collaboration with the architectural firm Taliesin Associated Architects founded by Frank Lloyd Wright, the CDAC site was conceived with interchangeable design elements to facilitate applied research. The main goal for the 9-person research staff was simple—to determine the environmental factors that supported independent and active living for early- and mid-stage patients with Alzheimer's disease.

From daily clinical observations examining the effect of design features on behavioral measures like disorientation, autonomy, and freedom, scientists pieced together general principles behind how design can be used to decrease neurological symptoms and maximize resident quality of life. Such principles included free access to the surrounding two-acre garden during daylight hours, an open circulation plan that facilitates ease of interaction and movement, and display cases filled with loved items and mementos. Together, the experiments in the CDAC from 1991-1992 are a clear statement about how the efficacy of healthcare is connected to the spaces in which it is delivered (Namazi et al., 1991; Namazi & Johnson, 1992). They make the CDAC, while small in stature, a foundational case study on how biophilic architecture can improve health outcomes while protecting patient dignity.

3. Neuroethics and the brain-environment connection

Adapting architecture to health and wellness is not an endeavor limited to patients with Alzheimer's disease or assisted-living centers. Architects and health professionals have long observed a deep connection between the built environment and well-being. In the past two decades, an increasing number of studies have demonstrated how altering simple features of the built environment—light, noise, and circulation among them—can have an acute impact on the efficacy of learning, healing, and stress reduction. For example, the mental health of psychiatric patients is positively correlated to environments that manage social interaction (Evans, 2003), focus lighting in third grade classrooms leads to improved reading fluency (Mott, Robinson, Walden, Burnette, & Rutherford, 2012), and patients whose rooms include a view of nature have shorter hospital stays (Huisman, Morales, van Hoof, & Kort, 2012; Ulrich, 1984). Similar to how a neurological condition does not concern only brain health but broader sociocultural implications, the architecture of wellness is a concept that implicates everyone. People spend 87% of their lives in buildings (Klepeis et al., 2001) where they set the stage for work, rest, and play. As Winston Churchill noted: "We shape our buildings; thereafter they shape us" (House of Commons Rebuilding, 1943).

People want to feel sheltered, safe, and inspired by the buildings that house them. They want their therapeutic and domestic spaces alike to convey respect, integrity, and solicitude for individual wellbeing. Arriving at this outcome relies on a two-step process: (1) an understanding of how the brain represents the built environment, and (2) using this understanding to support the populations the space serves. Crucial to mediating the connection between the brain and the outside world are the senses, which detect material cues such as shapes, materials, and symbols that are later crafted into an immaterial impression in the light of emotion and value. The brain-environment connection implies a deep integration between the material and immaterial—differing neurological profiles have varied ways of interpreting space and needs within it. It is invaluable for architects and other stakeholders to consider this variation when designing built spaces in the service of empathetic, effective architecture.

Neuroscience, psychology, and behavioral studies are coalescing today to give rise to a science-based framework on how architecture can be responsive to diverse neurological needs (Eberhard, 2009). Because of its potential influence on health and wellness, this is an investigation saturated with ethical implications—as such, the principles motivating it can be viewed through the field of neuroethics. Here, we outline the basics of how the brain interprets the built environment, and then propose a neuroethical framework based on the brain-environment connection. We conclude with a discussion of why neuroethics can assist architects as they consider neurological diversity in the design process.

4. The brain and the built environment

New developments are being incorporated into the evolving theory of the human brain at a rapid pace, generating new scientific and philosophical insights about the mind. Neuroscience incorporates all scales of the functional organization of the brain from neurons to networks. Individual neurons communicate to each other through electrical messages in the process of neurotransmission, the details of which are determined by the physical and chemical properties of neurons. These synaptic transmissions integrate across a subset of the 86 billion neurons in the brain, in patterns that lead to perception, learning, memory, and emotion. Neuroscience can be studied at many different levels of detail—a fine-grained investigation might look at the molecular basis of neurotransmission, whereas a broader one might focus on behavior through neuroimaging, cognitive science, and psychology. Moreno and Schulkin (2019) sum this up in *The Brain in Context* when they say "few fields of human inquiry… are more interdisciplinary than modern brain science" (p. 65).

A neuroscience of architecture focuses on the macro-scale perspective of how environments elicit different behavioral outcomes in the brain. It has to do with how neurons are organized and connected to one another in the nervous system—the brain, spinal cord, and peripheral nerves. Neurons are wired in a particular way to enable people to perceive the world through sight, sound, touch, taste and smell. The brain then takes the information, integrates it, and coordinates a response. In this way, the neuroscience of architecture connects to the most rudimentary function of the nervous system—to sense and react to the outside world.

The complexity of information processing in the brain makes understanding higher order behavioral phenomena, such as emotion and stress reduction, a difficult task. States of sadness, happiness, and relaxation might seem intuitive to appreciate. The subjective experience of them, however, incorporated into an individual's impression of a built space, are large-scale, integrated operations that enlist the entire nervous system. On a functional magnetic resonance imaging (fMRI) scan that correlates brain activation to oxygen flow, these states illuminate as a symphony of activation. Specific regions of the brain become active as if directed by a capable conductor in an orchestra that is the central nervous system.

To illustrate with a spatial example, imagine a person is on her way to work on a top-level office suite when, suddenly, the elevator that she is taking breaks down. Now she is trapped in a cramped, dark box. For most people, this will evoke a feeling of discomfort; for some, the experience will evoke feelings of claustrophobia. Why is this so?

First, the person perceives the confined edges, muffled breathing, and stuffy air. This sensory information is transmitted separately to specific parts of her cortex, after which they are processed and combined. Limbic centers of the brain including the amygdala and hippocampus assign emotional judgments and commit the event to memory respectively. Higher order centers of the brain themselves exert a top-down control of perception, based on context—running late for work, previous experience, and feelings such as anxiety. These sharpen the salience of the stimuli, which may make them seem much more threatening than they actually are. Moreover, psychological stimuli do not just affect the person's brain, but the entire body: when panic sets in, the autonomous nervous system sends a coordinated message to the body's organs that gears them up for fight or flight. The person is bathed in the stress hormone cortisol; the heart beats faster to provide tissues with more oxygen; the digestive activity slows to conserve energy.

How is the elevator example relevant to the present discussion?

The built environment is constructed from simple sensory elements. The brain chooses what is important based on both sensory cues and top-down control, processes it, and organizes it. Neuroscience then, connects the built environment to an individual's mental responses. This makes the discipline useful for the evidence-based practice of architecture in the interest of improving the effectiveness of built spaces. As Finnish architect Juhani Pallasmaa wrote in *The Eyes of the Skin* Pallasmaa (2012), people tend to privilege seeing things as the route to understanding them, which all but obscures the reality that people experience space through their bodies as an associated whole. The elevator example shows the impact of engaging the senses. All sensory information follows parallel streams of processing from the sensory organ to a mid-station known as the thalamus, which then relays this information to the cortex which combines and crafts it to a comprehensive impression. The only exception is olfactory information, which bypasses the thalamus to reach its cortical areas and limbic structures directly. This is why a particular smell—of a childhood bedroom, for instance—can elicit strong feelings like nostalgia and tranquility. By blending sensory information with personal memories, emotions, and values, it is possible for a built space to convey abstract moods like comfort, warmth, orientation, healing, and inspiration. It is possible for an impersonal structure to leave a personal impact, to impart a deep feeling of comfort and belonging—for a simple shelter to be called the home.

Neuroscience is about micro-scale research and macro-scale applications. Each individual experiment brings us closer to understanding the umbrella of brain and behavior, entailing ethical consequences, owing to the insight that neuroscience provides on age-old questions about how people think about themselves and relate to others. It is this ethical dimension of neuroscience that constitutes neuroethics.

5. Neuroethics within the architectural framework

Neuroethics considers the ethics of neuroscience and associated fields through its interdisciplinary lens on the ethical, social, and legal implications of monitoring, studying, and influencing the brain. The field grew out of the increasing capacity of scientists to predict and influence human behavior, from the scale of an individual neuron to that of a social network. In this chapter, we mainly focus on information from the middle of this spectrum: the perception of sensory elements and how they are organized in the brain. This organization and its implications for human health and wellness inextricably tie the built environment to ethics—the study of human values. We explore this position through two entwined strands: the *Neuro* strand and the *Ethics* strand. In the previous section, we sketched the relevance of the *Neuro* strand to architecture. Now, we will explore the relevance of the *Ethics* strand.

6. The ethics in neuroethics

A discussion of neuroethics would not be complete without reference to its ethical foundations. Ethics concerns the human values that govern morality. This can also be put in terms of the rightness or wrongness of an action. Generally, right actions are those that are conducive to human flourishing and wrong actions are those that hinder it. Ethics takes on applied form in neuroethics, which concerns itself with protecting and ensuring public good by aligning innovations in the brain sciences with human values. This means that the *Ethics* strand concerns not just abstract morals, but the legal and policy footprints that these morals leave in the societal landscape.

The question *What makes neuroethics different from bioethics?* is asked routinely. Some issues do overlap, particularly in a biomedical context. Examples include the ethics of genetic testing, stem cell research, and human subjects research. However, the brain also has a privileged status when compared to other organs in the body as it anchors the organism—it is responsible for thinking, feeling, and dreaming; it sustains consciousness and is the seat of the individual. Neuroscience can be said to be the biological window into what it means to have a lived experience. As such, central philosophical issues that are not inherently a part of bioethics are brought to the forefront of neuroethics, such as philosophy of mind, personhood, and free will.

Some conventional topics in neuroethics include the ethical, legal, and social issues raised by brain interventions such as psychopharmacology, neurological treatments, and novel neurotechnologies, all of which impact the understanding of the human mind and how it can be influenced. As brain imaging improves the ability to link patterns of brain activation to higher-order psychological states, unique concerns related to privacy and cognitive liberty arise. The study of disorders of consciousness, including patients in coma, vegetative, or minimally conscious states, stretch the current medical and scientific understanding of how conscious awareness might arise in the brain. Neuroethics concerns societal issues as well, including the rights and interests of various patient groups with neurological or mental health disorders, and cross-cultural perspectives on the brain and brain health. These issues are relevant to the wider population because ethics informs policy—the rules, regulations, and guidelines of society in which the brain sciences have a direct stake.

The brain is at the heart of what it means to be a sentient agent in a wider world, and neuroethics is about the broader philosophical issues associated with this notion. Ethics is fundamentally inseparable from the mind—humans are ethical agents because of the capacity for conscious experience and reason. To have a mind is to have access to moral codes. This foundational concept makes people deserving of the same rights, including the right to ethical environments that maximize their quality of life. For architecture then, neuroethics incorporates how neuroscience can illuminate an individual's experience of the built environment, and how it can inform design aligned with human health and well-being. This is what we call the *neuroethics of architecture*.

7. Neuroethics of architecture

A key premise underlying the neuroethics of architecture is that an individual is always embedded in their environment. People are contextualized within their broader surroundings—so much so that the mind may be mapped onto features in the outside environment. This idea in neuroethics is called the extended mind thesis, credited to philosophers Clark and Chalmers (1998).

When Clark was working at Washington University in St. Louis, his colleague Carolyn Baum, head of occupational therapy at the School of Medicine, was working with patients with Alzheimer's Disease who, by medical indications, would normally not be able to live alone. Yet, all of them led independent lives at home (Clark, 2003). How was this possible? The key was the detailed cognitive aids like notes, labels, and photographs, peppered across their home environments—with routines for catching the bus, buying groceries, and other daily tasks. In a way, Clark wrote, it was almost like the minds of these well-functioning Alzheimer's patients were extended throughout their home spaces. What was so different about recalling items from the mind versus from a note: only that one was internal and the other external?

In a functional sense, Clark thought this distinction was inconsequential. Since the cognitive aids allowed the patients to retain autonomy, orientation, and purpose, they became a carefully-calibrated scaffolding on which the biological brain could reside. They were innate to a natural system of memory and recall, rather than just a patchwork of compensatory props. Both Baum and Clark noticed that when these high-functioning individuals were put into care homes prematurely, they exhibited a decline in cognitive functioning (Clark, 2003). Forced re-allocation was clearly an ethical disservice. As Clark (2003) expresses, "The moral is: certain harms to the environment are simultaneously harms to the person" (p. 141). Overall, Clark's story generates two insights for brain-based architecture: cognition and the outside world are interlinked and inseparable; and the way that built environments are structured is ethically significant in that they can either help or hinder human flourishing.

Architecture has the same power as the notes in Clark's story. Of chief relevance is how material aspects in the physical environment are translated into immaterial experiences for its inhabitants. Similar to how simple cues such as post-it notes can provide necessary mental scaffolding to support functional independence for patients with Alzheimer's disease, architecture that maximizes the ability of people to live full lives is best supported by updated knowledge from neuroscience.

8. Adapting architecture to human diversity

Brain-based architectural design involves an understanding of how and why different people engage with space, and from it, deriving evidence-based guidelines for the design of buildings. For instance, sunnier hospital rooms can be used to supplement pharmacological treatment in patients with seasonal affective disorder, a type of depression tied to lack of daylight (Beauchemin & Hays, 1996). The same positive effects of daylight have been corroborated in non-seasonal depression for individuals with bipolar disorder (Benedetti, Colombo, Barbini, Campori, & Smeraldi, 2001). In autism, acoustics are the most influential architectural factor on behavior, followed by the spatial sequencing and layout of space (Mostafa, 2008). Certain neural mechanisms have been correlated to the benefits of environments designed for restoration and stress reduction (Martínez-Soto, Gonzales-Santos, Pasaye, & Barrios, 2013).

For patients with Alzheimer's disease, one essential factor in whether or not they can live independently is wayfinding (Passini, 1984). Wayfinding relies on many different mental processes, such as sensory integration and the ability to encode memories of familiar spaces. Both of these abilities become significantly impaired with Alzheimer's due to the degeneration of associated neural pathways (Rizzo, Anderson, Dawson, & Nawrot, 2000). The literature shows that wayfinding by patients with Alzheimer's disease can be facilitated by supportive environments (Allison, Fagan, Morris, & Head, 2016; Davis & Weisbeck, 2016) with colorful, clear, and personally significant cues placed at key points in the lived environment. The placing of personal items within them facilitates episodic memory and a sense of home within nursing facilities (van Hoof et al., 2016). Floor plans are crucial. Nursing homes with small units, short hallways, and clear circulation pathways are the most conducive to independent wayfinding (Marquardt, 2011). The CDAC, a small facility with a central circulation pattern and 12-bed wings, fits this description.

For people who live with special neurological needs and their families, a diagnosis is more than just pathology—it as much a social condition as a medical one. As such, interdisciplinary approaches can provide support at a more comprehensive level than singular ones in a way that reduces systemic barriers.

9. Conclusion

Explicitly integrating innate neurological differences into design is a commitment to improving human well-being and a recognition of the values conducive to human flourishing. Architecture provides the stage for daily life and is tasked with materializing this commitment. It is not just about the finished products, but also about the integration of diverse perspectives throughout the design process, completion, and beyond. In this way, the built environment is deeply connected to the ethical discussion, and reflects the underlying priorities of an equitable society that is committed to engage deeply and critically with assumptions about the brain and the purpose of the built environment.

What would a *neuroethics of architecture* look like explicitly? Answers to this question will be found in the chapters that follow, and may also draw inspiration from an equally burgeoning specialty of neuroethics called environmental neuroethics. Concerned with brain-environment connection in terms of its health, wellness, and justice, the lens of environmental neuroethics focuses on five factors (1) brain science—the emerging research linking the brain and its environment, (2) relational self—the interplay between mental factors and a person's experience of an environment, (3) cross-cultural factors—how these relationships are altered and compounded by cultural differences, (4) social policy—the role of policymakers, designers, and public health officials in addressing challenges in the environment through brain and population-centric approaches, and (5) public discourse—facilitating conversations between professional disciplines and the public to raise awareness and consensus (Cabrera, Tesluk, Chakraborti, Matthews, & Illes, 2016). These factors inject human considerations into the empirical objective of translating neuroscience research to both the natural and built environment, and function as a guide for implementation. As such, ethics is deeply interwoven throughout, foregrounding human rights, diversity, health, and quality of life into every step of the design and translational process.

All told, brain-based architecture can both illuminate the needs of the population for which a space is designed, as well the ethical and societal components of how a population will experience design choices. The relationship is bidirectional: architecture has always served to provide spaces that are not simply beautiful but functional. It is a foundational goal that intersects with emerging perspectives and research that entwine the brain, ethics, and environment. Just as neuroscience and ethics may be of use to architects, neuroscientists and ethicists may look to the long history of architecture to better understand how people dwell and relax, work and play, suffer and heal, and learn and live in the service of a deep understanding of the brain, diversity, and wellness.

References

Allison, S. L., Fagan, A. M., Morris, J. C., & Head, D. (2016). Spatial navigation in preclinical Alzheimer's disease. *Journal of Alzheimer's Disease*, *52*(1), 77–90. https://doi.org/10.3233/JAD-150855.

Beauchemin, K. M., & Hays, P. (1996). Sunny hospital rooms expedite recovery from severe and refractory depressions. *Journal of Affective Disorders*, *40*(1–2), 49–51. https://doi.org/10.1016/0165-0327(96)00040-7.

Benedetti, F., Colombo, C., Barbini, B., Campori, E., & Smeraldi, E. (2001). Morning sunlight reduces length of hospitalization in bipolar depression. *Journal of Affective Disorders*, *62*(3), 221–223. https://doi.org/10.1016/S0165-0327(00)00149-X.

Cabrera, L. Y., Tesluk, J., Chakraborti, M., Matthews, R., & Illes, J. (2016). Brain matters: From environmental ethics to environmental neuroethics. *Environmental Health*, *15*(1), 20. https://doi.org/10.1186/s12940-016-0114-3.

Clark, A. (2003). *Natural-born cyborgs: Minds, technologies, and the future of human intelligence*. Oxford University Press.

Clark, A., & Chalmers, D. (1998). The extended mind. *Analysis*, *58*(1), 7–19. https://doi.org/10.1093/analys/58.1.7.

Davis, R., & Weisbeck, C. (2016). Creating a supportive environment using cues for wayfinding in dementia. *Journal of Gerontological Nursing*, *42*(3), 36–44. https://doi.org/10.3928/00989134-20160212-07.

Eberhard, J. P. (2009). Applying neuroscience to architecture. *Neuron*, *62*(6), 753–756. https://doi.org/10.1016/j.neuron.2009.06.001.

Evans, G. W. (2003). The built environment and mental health. *Journal of Urban Health: Bulletin of the New York Academy of Medicine*, *80*(4), 536–555. https://doi.org/10.1093/jurban/jtg063.

House of Commons Rebuilding. (1943). https://api.parliament.uk/historic-hansard/commons/1943/oct/28/house-of-commons-rebuilding.

Huisman, E. R. C. M., Morales, E., van Hoof, J., & Kort, H. S. M. (2012). Healing environment: A review of the impact of physical environmental factors on users. *Building and Environment*, *58*, 70–80. https://doi.org/10.1016/j.buildenv.2012.06.016.

Klepeis, N. E., Nelson, W. C., Ott, W. R., Robinson, J. P., Tsang, A. M., Switzer, P., et al. (2001). The national human activity pattern survey (NHAPS): A resource for assessing exposure to environmental pollutants. *Journal of Exposure Science & Environmental Epidemiology*, *11*(3), 231–252. https://doi.org/10.1038/sj.jea.7500165.

Lewin, T. (1990). *Alzheimer's and architecture: A search for order*. The New York Times. https://www.nytimes.com/1990/05/02/us/alzheimer-s-and-architecture-a-search-for-order.html.

Marquardt, G. (2011). Wayfinding for people with dementia: A review of the role of architectural design. *HERD: Health Environments Research & Design Journal*, *4*(2), 75–90. https://doi.org/10.1177/193758671100400207.

Martínez-Soto, J., Gonzales-Santos, L., Pasaye, E., & Barrios, F. A. (2013). Exploration of neural correlates of restorative environment exposure through functional magnetic resonance. *Intelligent Buildings International*, *5*(Suppl 1), 10–28. https://doi.org/10.1080/17508975.2013.807765.

Moreno, J. D., & Schulkin, J. (2019). *The brain in context: A pragmatic guide to neuroscience.* Columbia University Press.

Mostafa, M. (2008). An architecture for autism: Concepts of design intervention for the autistic user. *International Journal of Architectural Research, 2*(1), 189–211.

Mott, M. S., Robinson, D. H., Walden, A., Burnette, J., & Rutherford, A. S. (2012). Illuminating the effects of dynamic lighting on student learning. *SAGE Open, 2*(2). https://doi.org/10.1177/2158244012445585. 2158244012445585.

Namazi, K. H., & Johnson, B. D. (1992). Pertinent autonomy for residents with dementias: Modification of the physical environment to enhance independence. *American Journal of Alzheimer's Care and Related Disorders & Research*, 7(1), 16–21. https://doi.org/10.1177/153331759200700105.

Namazi, K. H., Whitehouse, P. J., Rechlin, L. R., Calkins, M. P., Johnson, B. D., Brabender, B. A., et al. (1991). Environmental modifications in a specially designed unit for the care of patients with Alzheimer's disease: An overview and introduction: Article 1 in a series of 9. *American Journal of Alzheimer's Care and Related Disorders & Research, 6*(6), 3–9. https://doi.org/10.1177/153331759100600602.

Pallasmaa, J. (2012). *The eyes of the skin: Architecture and the senses* (3rd ed.). Wiley.

Passini, R. (1984). *Wayfinding in architecture.* Van Nostrand Reinhold.

Rizzo, M., Anderson, S. W., Dawson, J., & Nawrot, M. (2000). Vision and cognition in Alzheimer's disease. *Neuropsychologia, 38*(8), 1157–1169. https://doi.org/10.1016/S0028-3932(00)00023-3.

Ulrich, R. S. (1984). View through a window may influence recovery from surgery. *Science, 224*(4647), 420–421. https://doi.org/10.1126/science.6143402.

van Hoof, J., Verbeek, H., Janssen, B. M., Eijkelenboom, A., Molony, S. L., Felix, E., et al. (2016). A three perspective study of the sense of home of nursing home residents: The views of residents, care professionals and relatives. *BMC Geriatrics, 16*(1), 169. https://doi.org/10.1186/s12877-016-0344-9.

CHAPTER TWO

Science and Technology Studies (STS) and the neuroethics of architecture

John Gardner*
School of Social Sciences, Monash University, Clayton, VIC, Australia
*Corresponding author: e-mail address: John.gardner@monash.edu

Contents

1. Introduction

The well-known US architecture scholar Thomas Fisher remarked: "buildings arise out of our brains, and our brains—and our bodies—spend on average 87% of their existence in buildings" (Fisher, 2016). Fisher's statement, which is by no means out of the ordinary in its foregrounding of "the brain," inevitably raises a number of important questions about our relationship to the built environment. At one level, for example, we might ask: how exactly do our brains respond to the built environment? What elements of the built environment have relevance for our cognitive and emotive capacities, and how does this relate to our sense of well-being? We might also therefore ask: what design elements will actually improve our well-being? Or more specifically, what design elements will foster desirable attributes, such as creativity, efficiency, or a sense of calm and tranquility, among inhabitants and users? And how might such considerations be balanced with the many other factors (e.g., economic, aesthetic) that must be taken into account when designing and constructing our built environment? It is questions such as these that have underpinned a flourishing of

Developments in Neuroethics and Bioethics, Volume 5
ISSN 2589-2959
https://doi.org/10.1016/bs.dnb.2022.06.002

scholarship on the neuroscience of architecture over the last two decades. Aided by a rapid growth in brain imaging methodologies and new scientific insights into cognition and emotion, neuroscientists, architects, designers and others have formed a common interest in the nature of human responses to the built environment, and in particular the neural and cognitive bases of human responses to the build environment (Papale, Chiesi, Rampinini, Pietrini, & Ricciardi, 2016). Some have referred to this still nascent but very active area of scholarship and design as "neuroarchitecture" (see, for example, Ritchie, 2020), and a number of specialized neuroarchitecture collaborations and initiatives have been established, most notably the Academy for Neuroscience for Architecture based in San Diego.

Fisher's statement, that buildings arise from brains and our brains reside in buildings (to paraphrase) also raises another set of questions. We might ask, for example, from whose brains are such buildings actually emerging? Who, in other words, is responsible for creating the built environments that many of us share in our day-to-day lives, and what impact might this have on people whose brains differ in some significant way? What sort of differences in brain functioning are indeed relevant? To what extent should neuroscientific understandings of "typical" and "atypical" brains inform design and architecture processes, and how might such understandings be balanced with other, potentially valuable knowledge, such as the lived experiences of those who are "atypical"? These are, of course, questions about the relationship between architecture and *neurodiversity*. Indeed, neurodiversity is also the explicit purview of many scholars and practitioners within the neuroscience and architecture movement, and there is a strong commitment within some strands of the movement to create environments that recognize and accommodate *neurodivergent* individuals (for example see de Paiva & Galvao, 2021). The premise of this chapter, however, is that the neuroarchitecture movement—or at least significant elements within the movement—are not necessarily well-suited to such an ethical endeavor. Ultimately, this is a shortcoming that relates to the way in which neuroscience is framed and utilized, not just within elements of the neuroarchitecture movement itself, but in much scientific discourse and in the public realm.

Specifically, there are two aspects of the framing and utilization of neuroscience that are potentially problematic, both of which have ethical implications. The first is the reductive tendency of some neuroscientific representations of human action, or what has elsewhere been referred to as neuro-reductionism or neuro-essentialism (Gold, 2019; Reiner, 2011).

In such framings, human behavior is reduced to brain-based, stimuli-sensor-response mechanisms (Schultz, 2018). The risk is that when human behavior and action is understood solely or predominantly through such a lens, we are disinclined to engage with other understandings and knowledges (Porter, 2020), such as the neurodivergent lived-experiences, when creating new built environments. The risk, in other words, is that in neuroscientific engagements with architecture, "the brain" becomes a stand-in for neurodivergent individuals and communities. The second and related ethically problematic aspect is the tendency of neuroscientific and particularly biomedical discourse on brains, bodies and people to perpetuate binaries of "function/dysfunction," "normal/abnormal" and "pathological/healthy" (Rose & Abi-Rached, 2013; Williams, Katz, & Martin, 2011). The risk here is that atypical neural or cognitive mechanisms may be deemed abnormal and/or dysfunctional with little consideration of the individual and their context. This is especially problematic, given the efforts of contemporary advocates to reframe various neurodivergent modes of existence and expression as atypical but reflective of normal variation in humans. Overall, then, the risk is that neuroscientific knowledge can be deployed and wielded in a way that undermines efforts to create built environments that can genuinely support the well-being of a diversity of people.

To be clear, this is not an inevitable consequence of engaging with the neurosciences. There is much being done within the neuroscience and architecture movement to explore how neuroscientific knowledge can be combined with the perspectives and lived experiences of neurodivergent communities, and of course, the perspectives of design and architecture practitioners (e.g., Spanjar & Suurenbroek, 2020). The risks I outline here relate to the potential of neuroscience and understandings of the brain to be framed and deployed *in particular ways* when engaging with architecture and the built environment. In light of this, and in order to buttress those aspects of neuroarchitecture which are indeed aiming to support and celebrate neurodivergence, here I will propose a conceptual framework that provides an alternative framing of neuroscience, and brain-architecture interactions.

The framework I articulate in this chapter is derived primarily from Science and Technology Studies (STS). STS is a field of scholarship that examines the processes that constitute scientific endeavor, and the ways in which science and technology are represented, used and adapted in wider society (Sismondo, 2011). The general finding of the field is that techno-scientific enterprise is an intrinsically value-laden social endeavor, rather than, say, an objective pursuit of discovery and fact-finding. The ethical motivation of STS is to expose the values and norms that shape scientific endeavor, and in doing so, to open a space for scientists, communities and other stakeholders to explicitly discuss what values and norms *should be* enacted within and through scientific and technological endeavors (Sismondo, 2011). This ethical focus on science and technology within STS has generated important insights on the nature of reality, agency, and personhood (see Law, 2008; Mol, 2002); insights that reflect a general *assemblage theory* conceptual "framework" for understanding the world. In this chapter, I will argue that such an understanding can help us to see the potential for neuroscientific engagements with architecture to genuinely create more caring and enabling built environments for neurodivergent people. However, the realization of this potential, according to the general *assemblage theory* framework I outline here, necessitates meaningful engagement with neurodivergent advocates and communities, and indeed other stakeholders, at various levels of the design and creative process. To summarize my endeavor here, I outline a particular neuroethical approach to architecture and the built environment that draws on a general STS understanding of the world and its corresponding ethical orientation.

To do this comprehensively requires that I bring together several threads of STS scholarship. First, I will begin by providing a more comprehensive description of the STS "assemblage theory" conceptual approach for understanding the world. I will then illustrate how it can be used to make sense of neuroscience as an enterprise that has both potentially reductive (or constraining), *and* enriching (or enabling) socio-ethical implications. Second, the chapter will illustrate how we can use the general STS conceptual approach to make sense of the brain itself, or more specifically, the neurological system. The importance of such an understanding, I will argue, is that it enables us to recognize the possibility of neurological difference and neurodivergence, and the value of neuroscientific knowledge, without having to adhere to some sort of neuro-reductionism or neuro-essentialism. When viewed through an STS lens, we are also compelled to make sense of the neurological system in terms of its remarkable dynamic sensitivity to various phenomena, including those relating to the built environment. More generally, it encourages us to ask: what can neurological systems do, rather than, what is the essence of neurological system?

The third section of this chapter focuses more specifically on the neuroethical relevance of architecture and the built environment. I draw on arguments put forward by STS scholars to illustrate the significance of the built environment in shaping our sense-of-self and personhood, and the role of the built environment in perpetuating idealized understandings about who we are and how society should be arranged. The built environment, I argue, has neuroethical relevance because it is dynamically entwinned with our neurological system, and thus our sense of self and our capacity for agency. It reflects social norms and values, including those that privilege neurotypical modes of existence, and it therefore participates in enabling and supporting people who approximate "typical" individuals while disabling and constraining those who do not. When viewed in this way, architecture and the built environment are clearly important targets for neuroethical intervention. The final section thus provides a general description of the types of neuroethical interventions that can bring about built environments that accommodate neurodiversity and support and celebrate neurodivergent modes of existence and expression. This includes interventions that aim to raise awareness about the role of the neurotypical biases inherent in our built environment; incorporating the experiences of neurodivergent people when designing built environments; and it includes activities, such as immersive art and theater, that aim to explore, and experiment with, what a *neurological system can actually do*, when presented with unique sensory and experiential encounters.

The STS approach I describe in this chapter asserts a relational understanding of neurodivergence. In this regard it aligns closely with the theorisations of personhood and disability that have emerged from a number of social sciences, including disability studies, sociology and anthropology, and feminist scholarship (e.g., Garland-Thomson, 2011; Roseneil & Ketokivi, 2015; Tolhurst, Weicht, & Kingston, 2017). Relational conceptualisations that foreground the interactional generation of personhood and disability in everyday life have often appeared at odds with neuroscientific or biological understandings (due in large part to the often-reductive language that characterizes neuroscientific understandings). An important contribution of this chapter, therefore, is its articulation of how neuroscientific knowledge can add value *within* a relational understanding of neurodivergence, without causing some sort of epistemological or ontological discord. Neuroscientific knowledge of "human responses to the built environment" has an important role to play in informing the design of built environments that are more equitable and enabling. It does not, however have the epistemic grounding to supplant other knowledge, especially knowledge which arises from the diverse lived experiences of neurodivergence. This chapter, then, can be seen as adding another voice to the calls for meaningful co-design processes (Benton, Vasalou, Khaled, Johnson, & Gooch, 2014; Hendriks, Slegers, & Duysburgh, 2015; Trotman & McGinley, 2018).

2. The world—and neuroscience—according to STS

STS as a scholarly field has its origins in the confluence of several academic disciplines—sociology, anthropology, feminist theory, and others—brought about by a shared interest in the intersection of science, technology, and society (Felt, Fouche, Miller, & Smith-Doerr, 2016). It is a diverse area of scholarship with multiple foci and schools of thought. In broad terms, however, STS scholars have challenged the common assertion that science produces a representation of the world "as it really is," while at the same time distancing themselves from the proclamations of some social scientists that science is the imposition of meanings on an otherwise unknowable world (Latour, 2005; Sismondo, 2011). Instead, STS scholars envisage science as being something akin to a highly specialized craft: scientists and their instruments produce a particular kind of knowledge by intervening in the world (e.g., via experiments, trials) in a way that incites detectable responses between artificially-isolated entities (Barad, 2007; Latour & Woolgar, 2013). This enables these entities to be categorized as having known qualities—as specific subatomic particles, genes, or neural mechanisms, for example. In general terms, scientific enterprise is understood as the crafting of *assemblages* of various elements including scientific apparatuses, embodied scientists themselves, and the entities under investigation.

The consequence of these crafted scientific assemblages is a reality "snapshot," and for STS scholars, such snapshots can enrich the world. Scientific knowledge can provide the basis for new action and intervention (e.g., research, technological development), social organization, and even self-identification. Scientific knowledge can become problematic, however, when we ignore that such snapshots are the product of a carefully organized set of constraints—scientific apparatuses—that function by foregrounding specific phenomena and eliding many others (Latour & Woolgar, 2013). When the conditions of knowledge production are ignored, these reality snapshots are equated with constituting some sort of "God's eye" view of the world as it really is', that must, therefore, supplant other understandings (Haraway, 1988). When this happens, the opportunities for action, intervention and self-perception are reduced rather than multiplied. Much of STS scholarship, then, can be seen as a critique of the way in which science is commonly represented and championed in the public realm.

We can use a very general example neuroscience and neurodivergence to provide some detail to this description of an STS orientation. Contemporary neuroscience involves a vast array of tools and techniques (e.g., CT and PET imaging, fMRI, neuropsychological testing, questionnaires and rating scales). These generate neuroscientific knowledge by intervening in the body, inciting bodily phenomena—a neural network or cognitive mechanism, for example—so that it becomes detectable and classifiable. This typically includes a comparison to pre-established norms and, consequently, the identification of neural or cognitive divergence. From the perspective of STS scholars, neuroscientific knowledge of divergent mechanisms is interactionally generated by a neuroscience-body assemblage: it is the product of an engagement between the neuroscience enterprise with its tools and techniques, aspirations, funding priorities and so on, *and* the fleshy body (or brain) of individuals (for example, see Hollin, 2017). Neuroscientific knowledge tells us something significant about the brain's capacity to interact and become intelligible when *incited to do so in specific conditions*. The resulting knowledge of neural or cognitive aberrations can be enriching. The emergence of the neurodivergence advocacy is of course an example of this: scientific understandings of neural and cognitive divergence have become the basis for new forms of self-understanding, social organization, creativity, and political activity (Ortega, 2009). In such cases, neuroscientific understandings of atypical neural and cognitive mechanisms have captured something of the timbre of diverse peoples' every-day experiences. The result of this is new assemblages that bring together neuroscientific understandings, elements of personal and social life, and which constitute novel modes of experience, expression and action.

But as I stated earlier, neuroscience can also have reductive—or neuroreductive—consequences (Porter, 2020). We can easily imagine, for example, scientific (or popular scientific) accounts in which brain mechanisms are seen to determine human behaviors and difference, and which ignore the interplay of social, cultural or other biological factors. In such cases, neuroscientific knowledge and diagnostic categories are presented as a representation of "the brain" or "the individual" as they really are, independent of context. From the perspective of STS, this is an epistemologically and ethically problematic. It is epistemologically problematic because it elides the many potential ways in which individuals and bodies can be incited to respond and to become knowable (or make themselves knowable) by other assemblages, and it is ethically problematic because it cuts-off the modes of experience and action, expression and knowledge-production enabled by these other assemblages.

This general STS critique of the reductive consequences of scientism is underpinned by a commitment to what others have referred to as a *relational understanding of epistemic agency* (Catala, Faucher, & Poirier, 2021). An individual's capacity to act, and their capacity to produce, adapt and share knowledge, is the product of their dynamic engagement with an assemblage of enabling and constraining and transmuting entities which commonly include other bodies and individuals, language, technologies and tools (Mol, 2008). Indeed the very sense of being an "individual"—including the sense of being "able-bodied" or "disabled"– is interactionally generated within assemblages (that include, as I will argue further on, elements of the built environment) (Latour, 2005). Accordingly, STS as a field of scholarship has broadly sought to identify and support those knowledge-producing practices (scientific and mundane) that enrich the world by provide new opportunities for self-understanding and expression.

In this general vein, STS and related scholarship on neurosciences has taken several approaches. These include examinations of neuroscience as a field, including the reasons for, and consequences of, its remarkable ascendency as an area of strategic focus and huge investment in the US, Europe and elsewhere. Rose and Abi-Rached, for example, have traced neuro-reductionist tendencies within the field, while also drawing attention to the emergence neuroscientific ideas that could, potentially, bring about productive collaborations between neuroscientists and social scientists (Rose & Abi-Rached, 2013): the notion of the "social brain" for example, posits a brain with structure and function dynamically entwinned with collective social life. Scholars have also looked closely at specific neuroscientific practices, and neuroscientists themselves (e.g., Brosnan & Michael, 2014; de Boer, te Molder, & Verbeek, 2020; Fitzgerald, 2017). This work foregrounds the tensions and ambiguities that characterize the field, and the hopes and emotions that animate it, as neuroscientists grapple with the immense complexity of human behavior, the brain and its functioning. Other scholars have focused more specifically on the interface between neuroscience and publics (see edited volume by Pickersgill & Van Keulen, 2012). This includes explorations of how and why neuroscientific discourse is deployed in social policy (Broer & Pickersgill, 2015), how experts such as addiction practitioners strategically deploy neuroscientific frames of reference when working with clients (Barnett, Pickersgill, Dilkes-Frayne, & Carter, 2020), explorations of lay-peoples' perspectives on the relationship between themselves, their brain and their behavior (e.g., Pickersgill, Cunningham-Burley, & Martin, 2011; Singh, 2013) and the perspectives and experiences of the families of children with diagnoses of ADHD, autism or similar (Rapp, 2012). Other work has focused more specifically on the impact of peoples' engagements with emerging neurotechnological developments, such brain-training (Pickersgill, Broer, Cunningham-Burley, & Deary, 2017) or brain stimulation technologies (Gardner, 2017; Gisquet, 2008). Collectively, this scholarship highlights the persuasive power and allure of neuroscientific explanations within clinical contexts and in the public realm. It also illustrates, however, the pragmatic ways in which people adapt neuroscientific concepts and terms to make sense of themselves and others. This ubiquity of neuroscientific ideas demonstrates that there is much at stake in terms of defining and enacting and intervening in social life, but the ascendence of neuroscience can certainly not be characterized as an inevitable march toward neuro-reductionism. The cultural impact of neuroscience can perhaps be better understood as a "plenitude of emerging socialities" (Gardner, Warren, & Mason, 2018). It has the potential, in other words, to become entwined in various enriching and diverse assemblages.

The intersection of neuroscience and architecture has yet to receive much in the way of critical attention from STS scholars. There has been a proliferation of neuroscientific engagements with the architecture, as architects, neuroscientists and others have expressed great enthusiasm for an endeavor which, they hope, will shed light on the neural and cognitive mechanisms that characterize human responses to the built environment. These neuroscience-architecture engagements are varied, and include, for example, approaches from affective neuroscience, experimental phenomenology, and sensorimotor psychophysics, thus reflecting the diversity of neuroscience itself (Gepshtein & Snider, 2019). This has involved the elicitation of stimuli-sensory relationships that could no-doubt lend themselves to overly reductive, mechanistic understandings of brains, humans their environment: but we might also understand the endeavor as contributing to an understanding of the neurological system which foregrounds its remarkable dynamic sensitivity. My task here is not to demonstrate this by examining specific instances of this neuroscience-architecture engagement. Rather, my argument is simply that reframing neuroscience as I have done here encourages us to see that neuroscience can have enriching—as well as reductive—consequences. This, I will argue further on, also then compels us to recognize the neuroethical relevance of architecture and the built environment for people and communities who identify as neurodivergent.

3. Rethinking the brain: What can a neurological system do?

Thus far we have explored how neuroscience as an enterprise can be understood from the perspective of an STS conceptual framework. But how, then, might the brain itself, and thus neurodiversity and neurodivergence, be understood within this perspective? What, in other words, can we say about the nature of the brain and its diversity *a priori* to neuroscientific representations of it? Here, I will argue that we can understand the brain—or more precisely, the neurological system—as having innumerable potential inclinations, some of which are elicited within assemblages. These include potential inclinations that are shared among all humans, many that are shared by large groups of humans and not others, and some that may be shared by small numbers of individuals. A very small number of these inclinations are elicited and made-known by neuroscience, and some are, of course, enacted and experienced via the day-to-day activities of individuals themselves. Conceptualizing neurological difference as I do in this section helps to assert the relevance of architecture and the built environment to the neuroethics enterprise, and it also illustrates that neurodivergence—however it might be experienced by individuals—need not be reduced to, or equated with, neuroscientific representations of it.

While there has been a significant amount of STS scholarship on the ascendance of neuroscience within contemporary society, and on neuroscientific practices themselves, there has been a general reluctance among STS scholarship to conceptualize the brain in-itself. This is no doubt a reflection of the commitment in STS methodologies to "follow the actors"; that is, to examine how people and objects and technologies and other entities interact and assemble to make aspects of the world knowable (Latour, 2005; Mol, 2002). Assemblages constitute the world as we know it and experience it, and hence, it makes little sense from a conventional STS orientation to postulate the nature of things, *a priori* to their articulation in assemblages. As Marres notes, this has meant that within conventional STS scholarship "ontological questions cannot be settled by theoretical means... such questions require detailed empirical investigation of social and political practices"

(Marres, 2013). A consequence of this approach is that it gives the impression that neurological differences and divergencies can be reduced to neuroscientific or other representations of them.

Nevertheless, the general STS approach—the assemblage theory conceptual framework—does make assumptions about the nature of things, and we can use this to make sense of neurological systems and neurological difference. The premise of assemblage theory is that the world is composed of entities with innumerable *capacities for interaction*, (or what I will call potential inclinations). These can be thought of as inclinations to interact with other entities in certain ways, and it is the expression of these inclinations within assemblages that enables the entity to become discernible and knowable (DeLanda, 2016). A single entity can, then, become discernible and knowable in multiple ways, as different assemblages will prompt the expression of different capacities or inclinations (Mol, 2002). (The convention in western metaphysics is to label these expressed capacities as distinct and defining qualities or characteristics of the entity itself, rather than recognizing them as being interactionally generated). This general understanding has been used to make sense of the human body within STS and in related schools of thought across sociology, philosophy and feminist scholarship. Scholars such as Nick Fox (2012), for example, have drawn on Deleuze's' notion of the Body-without-Organs (BwO): a concept that refers to whatever the body is prior to its articulation via engagement with other entities. Such engagements might occur within medical assemblages (which typically deem the body intelligible in terms of organs—hence a body *with* organs), or in various assemblages of everyday life, such as staircases and doors (which generate a locomoting bipedal body) (Fox, 2012). Fox argues that the BwO is not some sort of blank canvas: It is, rather an innumerable set of potentials (inclinations and obduracies) that reflect its complex physical-materiality—a physical materiality that can only ever be partially known via the elicitation of specific inclinations within assemblages (Fox, 2012). The significance of this conceptualisation is that it encourages to ask *what can a body do*, rather than *what is* the body? The ethical implication of this is that it encourages us to create environments—or more specifically, assemblages—that enable individuals to express and realize the various potentials of their body (Fox, 2012).

The neurological system can also be understood in this way. I suggest it can be understood as a set of innumerable potential inclinations to engage with other entities (stimuli). A very small number of these inclinations have been expressed and delineated in neuroscientific studies (or assemblages), such that the neurological system becomes intelligible as having particular qualities or characteristics. Many of these same potentials are indeed expressed, and thus experienced subjectively, by individuals as they go about their day-to-day lives. And of course, many of the potentials that are expressed and experienced in everyday life have yet to be expressed and delineated in neuroscientific settings. Innumerable inclinations may indeed never be experienced or expressed at all in any settings. This simple reframing of the neurological system encourages us to recognize the value of neuroscience as a means of elucidating new neurological inclinations and obduracies, without reductively equating the neurological system, and individuals themselves, to scientific knowledge of that system. This provides a space for a neuroscientific knowledge that can supplement and augment, rather than supplant, the knowledge derived from diverse peoples' lived experiences.

The potential inclinations of the neurological system can be seen as a reflection of its complex physical-material form. Our tendency here is to make sense of such form in terms of known morphological features, such as brain structures, neural pathways and neurones of various types, and in terms of neurotransmitters and receptors, but we must also keep in mind that the intelligibility of these features emerges from our conventions for categorizing specific neurological inclinations (i.e., qualities)—inclinations that are commonly elicited and delineated within neurological and clinical settings (e.g., within magnetic resonance imaging). Regardless, much of the physical-material form of the neurological system will be shared by all humans, reflecting its entwinement with other qualities that are, by convention, human-defining, such as the approximately 25 thousand genes that constitute the human genome. Some of the neurological system's form, however, will differ between groups of people and individuals, reflecting's some genetic differences, age related degeneration and disease, and also the system's sensitivity to the very complex sets of stimuli the characterize socialization and interpersonal relations, education, trauma, and other features of every-day life. Because of this, some potential neurological inclinations and obduracies are shared among humans, whereas other reflect unique and shared life-histories and are particular to individuals or groups. This is of course the basis of neurodiversity.

Some neurological inclinations will be expressed and experienced by people as perceptual habits, in which kinaesthetic or auditory or visual stimuli prompt specific embodied responses and affective states that, in turn, prompt specific thinking patterns and behaviors. One way to understand neurodivergence is that it is characterized by atypical perceptual habits and corresponding behaviors that are in some way significant to individuals themselves and their families. Neuroscientific knowledge of these perceptual habits and behavior responses may be useful to atypical individuals and their families, but it need not be essential for designating or defining neurodivergence.

We can see here, then, how an STS-informed conceptual approach can enable us to recognize the possibility of neurological diversity and neurodivergence, and the value of neuroscientific knowledge. The potential inclinations of the neurological system are innumerable and will always immensely exceed those which are elicited and delineated within neuroscientific settings. There are several neuroethical implications of such a reframing. The first, as noted above, is that it encourages us to recognize that neuroscientific knowledge of the brain is not a stand-in for the knowledge derived from the lived experiences of individuals. This of course emphasizes the importance of meaningfully engaging with neurodivergent people and those closest to them when designing policies and environments that will impact them. The second relates more directly to architecture and the built environment as a target for ethical reflection and intervention. Both neuroscientific knowledge and the knowledge derived from lived experience illustrates the remarkable sensitivity of the neurological system to elements of the built environment. In the next section I will argue that the built environment is, in effect, a means by which aspects of our neurological systems are dynamically exposed to social norms and values. For most of us, these norms and values become encoded in common perceptual habits and behavioral responses. In some people, however, these social norms "misfit" with their ways of being in the world, likely reflecting atypical neurological inclinations of complex and perhaps unknowable etiology, such that neurotypical and normative contexts can be experienced as perplexing and traumatic. In the final section of this chapter, I will therefore emphasize the importance of creating atypical spaces which care for, and facilitate, neurodivergent modes of existence and expression. This of course requires the active involvement of neurodivergent people. The third neuroethical implication of conceptualizing the neurological system as a set of innumerable potential inclinations, is that it encourages us to ask: *what can a neurological system do*, rather than: *what is* the neurological system? In the final section of this chapter, I will argue that this encourages us to create challenging and experimental built environments.

4. Why architecture matters: Architectural—Neurological entanglements

A useful way to conceptualize the dynamic interconnectivity between social norms and values, the built environment, and neurological systems is to begin with an argument presented by STS theorist Bruno Latour. In his well-known paper "How to Talk About the Body" (2004), Latour argues our capacity to know, think and act is ultimately dependant on our body's capacity to sensorially register contrasts. We can also understand this as referring to the neurological system's capacity to register contrasts, including those relating to the built environment.

Latour begins his argument with a very useful description of the training undertaken by perfume experts to enhance their sense of smell. The general process involves an odor kit containing a gradient of fragrances. Trainees begin by being exposed to strongly contrasting "pure" fragrances, and once they become attuned to these, they are exposed to a succession of more-subtly contrasting fragrances, until eventually, they can discern the composition of sophisticated perfumes (Latour, 2004). Latour describes this assemblage of the odor kit-nose as a process that composes a world: The trained individual acquires the capacity to register and experience and respond to elements (aromas) that most of us would fail to discern. In this respect, they have acquired a world that is richer than, and different from, those of us who are untrained. We can reframe this slightly as: via the odor-kit nose assemblage, their neurological system has acquired a new set of inclinations to engage with and respond to otherwise unnoticed external stimuli. Importantly, Latour points out, at the same time, the nose as an olfactory apparatus has also become more discernible and prescient to the trainee: their perception and experience and expression of their body, then, is also different for them, compared to those who are untrained (Latour, 2004). For Latour, it is in this way that the body and the world become intelligible for all of us. Assemblages such as the odor kit-nose prompt an awareness of specific bodily aspects, and the sense of having a "whole" body emerges from the innumerable assemblages that constitute daily life—assemblages that can include combinations of fragrances, objects and technologies, other bodies or body-parts, and indeed words and phrases. It is, in other words, the body's capacity (or more precisely, the neurological systems' innumerable capacities) to assemble and to learn to be affected—to register and respond—that produces the worlds that we each exhibit (Latour, 2004).

Latour's articulation of assemblage theory proposes that it is through our embodied engagement with other entities that those very entities, and the body itself, become intelligible and utilizable. An obvious feature of human life is that much of our embodied engagement involves the built environment. Indeed as Latour has argued elsewhere, it is the built environment, along with other non-human entities such as tools and technologies that collectively provide contemporary society with a degree of structure (Latour, 2005). An odor kit might be used to train dozens of perfumers, all of whom consequently share some common neurological inclinations, embodied perceptions and responses, but many millions of us share embodied engagements with particular architectural forms, from the very mundane (walls, steps and stairs), to the complex (shopping centres, airports). Many of our shared perceptual habits and the neurological inclinations that underpin them are partly the result of these common embodied engagements with built environments. Architecture matters because it is dynamically entwinned with our neurological system, sense of self and capacity for agency, *and* because the ubiquity of some architectural forms results in common patterns of human understanding and behavior—some of which we might deem neurotypical. This is what Foucault has referred to as the disciplinary qualities of architecture (Foucault, 1977/1991). Many bodies, subject to similar architectural forms, are disciplined into registering and responding to specific contrasts. Populations of individuals emerge with common neurological inclinations, embodied dispositions and associated understandings.

Architectural forms are thus a significant means by which social norms and conventions are perpetuated. Obvious examples include the two- or three-bedroom domestic house that reflects a dominance of a particular family structure, or the ubiquitous presence of steps and staircases that reflect prevalent beliefs about how a human body should move. In effect, the built environment is the materialization of general views about how humans should perceive and act and how societies should be arranged. This, then, raises the question: whose views are being expressed in the built environments that we inhabit? This is not to imply that the built environment is the straightforward imposition of a particular elite viewpoint. The built forms that characterize our every-day life are better seen as the coming together of multiple views, operationalized in, for example, building regulations, aesthetic trends, and the professional habits and responsibilities of architects, designers and urban planners. Nevertheless, among such structuring forces there are common assumptions about how humans respond and how they should engage with those environments.

Jasanoff's and Kim's STS-informed concept of *Sociotechnical Imaginaries* is useful here (Jasanoff & Kim, 2009). The concept refers to the way in which dominant ideals (i.e., imaginaries) about how people should live, and how societies should be organized, become embedded and consolidated within sociotechnical systems, such as infrastructures and the built environment, and the human activities—social and professional—that are aligned with them (Jasanoff & Kim, 2015). The establishment of renewable energy infrastructures and associated professional roles, for example, is an example of a particular sociotechnical imaginary: a national aspiration to be a clean, green country. Once embedded in sociotechnical systems, such imaginaries and the norms and values that underpin them are much more likely to materialize, so that a particular future is more likely to eventuate (Jasanoff & Kim, 2009). Indeed the sociotechnical imaginaries concept reminds us that particular ideals, norms and values become powerful and durable precisely because they become embedded in the physical-material forms to which we orientate our lives. It also reminds us that our future trajectories are established via present-day decision-making about the form and function of built environments.

What has become increasingly clear is that much of our built environment reflects what could be understood as a dominant *neurotypical sociotechnical imaginary*. Much of our built environment assumes that people have very particular sets of neurological inclinations: general sets of sensitivities to certain environmental phenomena and insensitivities to others, particular perceptual habits, and embodied and cognitive responses, cognitive capacities, and associated behaviors (Mastafa, 2008). In accordance with Latour's argument, we can assume that many such neurological inclinations are to some extent entrained via our socialization to neurotypical environments. People who do indeed develop and possess these capacities will benefit from the affordances presented by the built environment. The resulting sense of congruence is generally taken for granted, but it is nevertheless essential to how neurotypical people develop a sense of themselves and the world around them. But people whose neurological inclinations differ, perhaps due to genetic differences, disease or old age, or unique socialization experiences (e.g., trauma), will experience a sense of discord with the built environment and with the many other people who appear to seamlessly engage with it. This can be experienced anywhere on a continuum from unsettling to deeply distressing.

From the general STS perspective I have outlined here, the built environment has neuroethical significance because it reflects and imposes neurotypical norms and ideals on the neurological inclinations of all of us. The built environment participates in the elicitation of neurological inclinations, some of which are perpetuated and expressed as neurotypicality or neurodivergence, and this has relevance for our sense of well-being and agency. This is of course an STS-informed relational understanding of neurodiversity, in which neurotypicality or neurodivergence are interactionally generated as people go about their everyday lives, engaging with architectures, objects, and each other. Architecture and the built environment are clearly, then, important targets for neuroethical intervention.

5. Neuroethical interventions for neurodiverse architectures

In this final section, I will provide some general comments on the forms of neuroethical intervention that could help us to collectively produce neurodiverse built environments. By this I mean shared built surroundings (e.g., public infrastructures) that can cater to both neurotypical and atypical modes of existence and expression, as well as spaces that are specifically designed for neurodivergent modes of existence and expression (e.g., particular educational settings). Much of what I advocate here is already being undertaken in various specific forms, although not as an explicitly neuroethical enterprise. In this section I will also highlight the importance for neuroscientific knowledge of human responses to the built environment to inform some interventions, provided it does not supplant other forms of knowledge such as lived experience. In effect, I argue that we should recognize the importance of engaging with neuroscientific knowledge, *as well as other forms of knowledge*, when designing and implementing such interventions. Ultimately, what I am doing here is adding to the already considerable calls for meaningful co-design practices.

One important neuroethical task is to make explicit the role of the built environment in perpetuating neurotypical norms. Our built environment, as argued above, reflects dominant social-cultural ideals regarding human behavior and interaction. The consequence of this is that the majority of people—those who approximate "typical" humans—are enabled by the built environment with such mundane regulatory that it generally escapes notice. The built environment feels as if it's a backdrop within which we express autonomy, rather than a means by which autonomy is constituted. There is a need, therefore, to raise awareness about the enabling/disabling dimensions of our built environment in general, and more specifically, to publicize how the built environment assumes and supports sensitivities, cognitive competencies, and behavioral responses that are neurotypical. Raising such awareness is most obviously important for those groups of stakeholders involved in the design and creation of our shared built environment such as urban spaces and public infrastructures, and also our healthcare and educational settings. But it is also important to raise awareness among the public more generally, in the hope of encouraging support for greater investment into relevant infrastructure.

How might such awareness be achieved? Art clearly has an important role to play here. Across a number of countries there is now an emerging body of explicitly neurodivergent art—fictionalized or autobiographical narratives, visual art and sculpture, theater, dance and so on—through which neurodivergent modes of existence and experience are explored and celebrated. Such art can also help neurotypical people to gain at least some glimpse of what it might be like to experience our shared built environment in an atypical way. An example includes the *Neurodiverse City* stream of *The Big Anxiety* festival in 2017 in Sydney, Australia, organized by the University of New South Wales and the Black Dog Institute (for a summary see Sutherland, 2017). It included a number of significant pieces by the artist researcher Dawn-Joy Leong, including an audio-visual *Olfactory Map* of Sydney that, along with her written *Reflections on the Neurodiverse City* (Dawn-Joy, 2017), evoke and convey the experience of moving about the built environment with heightened sensitivities to smells, tastes, sounds and light. Immersive installations such Anthony Clarke's *Sound Field* (see chapter 4 of this volume) is also example of art that intentionally sets out to prompt a recognition of the neurotypicality of the built environment (by using a series of rotatable panels to create a confusing soundscape, similar to that which can be experienced by people with dementia). There is an emerging field of expertise in this area. In Melbourne Australia, for example, the *A_tistic* group has established itself to promote the inclusion of autistic lives, insights and experiences within theater (Aroni, 2018). This includes writing, directing and producing original works, working with other artists to include realistic neurodivergent characters and themes, and also importantly, running workshops with other organizations to advise on neurodivergence.

The second important neuroethical task is to create built environments that reflect and cater for neurodiversity. This, I argue, should include architectures that *care* for both neurotypical and neurodivergent modes of existence. Architectures that *care* refers to built environments that have been designed to support the well-being of users. Annemarie Mol (2008) has described care as a complex process involving a network of people and technologies and objects (including those that constitute the built environment) which recognize, and are tailored to, the specific needs and sensitivities of recipients. In a general sense, we can understand all built environments *caring* for recipients in this way, in that they serve basic needs (e.g., shelter, warmth, security). Architectures that care for neurodivergent modes of existence are those that are tailored to atypical sensitivities, embodied and cognitive responses and associated behaviors. Already there is a significant body of expertise in how to produce such spaces (as illustrated by contributions in this volume), particularly spaces (educational, domestic, therapeutic, play and recreation) for people with autism, but also for people with attention deficit disorders, and for people with cognitive impairments such as dementia. Caring architectures, for example, is achieved by reducing or eradicating stimuli that may otherwise be experienced as distracting overwhelming (e.g., noise reverberation, glaring light), or by accentuating and foregrounding other stimuli which can provide comfort or anchoring (e.g., tactile surfaces), or—in the case of dementia—by arranging rooms and passage ways and visual cues in a manner that can reduce disorientation (see Tanya Petrovich's account of the Virtual Dementia Experience in chapter "The virtual forest"). Much of this expertise on how to achieve such spaces has been codified in the form of design guidelines and recommendations: there are now a number of resources for stakeholders wishing to create spaces for people with autism spectrum disorder for example (e.g., Bowes & Dawson, 2019; Gaines, Bourne, Pearson, & Kleibrink, 2016; Mastafa, 2008; Owen, 2016).

Neuroscientific knowledge of human responses to the built environment has played, and can continue to play, an important role in highlighting new ways in which the caring capacity of architecture might be enhanced. In the language I have used above, neuroscientific studies, particularly those that explore neural stimuli-response mechanisms, can be understood as telling us how neurological systems respond in vary specific settings. It tells us, in other words, something relational about what the *neurological system can do*. However, it does not tell us what the neurological system *actually is*, and for this reason, it should not be deployed in a deterministic fashion when designing and implementing built environments. If, for example, new neuroscientific knowledge were to suggest a new design response, that design response should be presented as a possibility that is then subject to ongoing refinement via engagement with the neurodivergent people who will occupy the resulting spaces, and depending on the nature of the purpose of built environment, other stakeholders and advocates (e.g., in the case of educational settings, educators). This need for engagement reflects the remarkable dynamic sensitivity of our neurological systems and the fact that any lived experiences of the built environment is the product of a complex elicitation of numerous neurological inclinations, well beyond those that can be delineated in a neuroscientific context.

One other neuroethical task is indeed to explore what the *neurological system can do*, via various exploratory mechanisms. This includes via neuroscientific endeavors of various kinds, but also through other means. Again, this is where art can play an important role. Immersive art, theater, cinema, and new media such as virtual reality have the capacity to temporarily subject participants (neurotypical and neurodivergent) to unique and experimental sensory (and multi-sensory) experiences. The intention here is to prompt and explore the resulting cognitive and emotive responses. Experimentation can sensitize participants to elements—including elements of the built environment –of which they may otherwise have been unaware, which in turn can prompt participants to engage in new avenues of self-exploration and expression. One example of this was *Sensory Experiments,* a 2019 Melbourne Design Week collaboration between Justin James, Hattie Malloy and the Sydney Opera House. The immersive experience combined sense, sound and performance; a "fleshy live experience" aimed at fostering "sensory intelligence" (Clift, 2019). Such art has not necessarily engaged directly with architecture or neurodivergence, but it can be seen as a means for creatively probing the intersection of our sensory and cognitive habits, our personhood, and our (built)-environment. Ultimately, such endeavors help us to imagine what a future that genuinely supports and celebrates neurodivergence might look like.

It is through the construction of our built environments that particular futures become much more likely to materialize. Drawing on an STS conceptual approach, I have argued in this chapter that architecture and the built environment is a means by which neurotypical norms and ideals become consolidated in a durable form. We risk a future in which our built environment enables and supports and cares for some people, while it disables and constrains many others. The built environment, then, is a critical target for neuroethical intervention. By undertaking the types of neuroethical interventions I have briefly described here (e.g., raising awareness of the intrinsic atypical norms, creating architectures that genuinely care, and the use of experimental, exploratory spaces), we open the possibility of a future that recognizes the full range of neurodiversity, and which is much more supportive and celebratory of various neurodivergent modes of existence and expression. The complexity of our built environments and the various socio-political dynamics that accompany them mean that this is of course an extremely difficult feat. Much of our shared built environments have multiple stakeholders with often differing, if not diverging, aims and interests: financiers and investors, governments and public servants, professionals, users or inhabitants, and others. Proposals for a more caring, neurodiverse architecture will therefore inevitably need to be coordinated with a myriad of concerns and aims. I end here then, with the suggestion that we require more exploration, not only into what neurodivergent built environments might look like, but also into how such built environments can actually be implemented and achieved through the interaction of various stakeholders. Such knowledge will be extremely valuable to designers, architects and others as they work with neurodivergent communities to create a neurodiverse future.

References

Aroni, H. (2018). *A_tistic: Theatre, Workshops, Consultation*. Retrieved 29th April, 2022, from https://www.a-tistic.com.au/.

Barad, K. (2007). *Meeting the universe Halfway: Quantum physics and the entanglement of matter and meaning*. Duke University Press.

Barnett, A. I., Pickersgill, M., Dilkes-Frayne, E., & Carter, A. (2020). Neural imaginaries at work: Exploring Australian addiction treatment providers' selective representations of the brain in clinical practice. *Social Science & Medicine, 255*, 112977.

Benton, L., Vasalou, A., Khaled, R., Johnson, H., & Gooch, D. (2014). Diversity for design: A framework for involving neurodiverse children in the technology design process. In *Proceedings of the SIGCHI Conference on Human Factors in Computing Systems* (pp. 3747–3756). Toronto, Ontario, Canada: Association for Computing Machinery.

Bowes, A., & Dawson, A. (2019). *Designing environments for people with dementia: A systematic literature review*. Emerald Publishing Limited.

Broer, T., & Pickersgill, M. (2015). Targeting brains, producing responsibilities: The use of neuroscience within British social policy. *Social Science & Medicine, 132*, 54–61.

Brosnan, C., & Michael, M. (2014). Enacting the 'neuro' in practice: Translational research, adhesion and the promise of porosity. *Social Studies of Science, 44*(5), 680–700.

Catala, A., Faucher, L., & Poirier, P. (2021). Autism, epistemic injustice, and epistemic disablement: A relational account of epistemic agency. *Synthese, 199*(3), 9013–9039.

Clift, T. (2019). *Sensory Experiments*. Retrieved 29th April, 2022, from https://concreteplayground.com/melbourne/event/sensory-experiments.

Dawn-Joy, L. (2017). Reflections on the neuodiverse city. *Artlink Magazine, 37*(3).

de Boer, B., te Molder, H., & Verbeek, P.-P. (2020). Constituting 'visual attention': On the mediating role of brain stimulation and brain imaging Technologies in Neuroscientific Practice. *Science as Culture, 29*(4), 503–523.

de Paiva, A., & Galvao, M. (2021). *Neurodiversity and accessibility: Insights from NeuroArchitecture*. NEUROAU. 18 March 2021.

DeLanda, M. (2016). *Assemblage theory*. Edinburgh University Press.

Felt, U., Fouche, R., Miller, C. A., & Smith-Doerr, L. (2016). *The handbook of science and technology studies* (4th edition). MIT Press.

Fisher, T. (2016). How neuroscience can influence architecture. *Architect: the Journal for the American Institute of Architects*. October 6.

Fitzgerald, D. (2017). *Tracing autism: Uncertainty, ambiguity, and the affective labor of neuroscience*. University of Washington Press.

Foucault, M. (1977/1991). *Discipline and punish: The birth of the prison*. Penguin Books.

Fox, N. J. (2012). *The body*. Cambridge UK: Polity Press.

Gaines, K., Bourne, A., Pearson, M., & Kleibrink, M. (2016). *Designing for autism Spectrum disorders*. Taylor & Francis.

Gardner, J. (2017). *Rethinking the clinical gaze: Patient-centred innovation in Paediatric neurology*. Basingstoke UK: Palgrave Macmillan.

Gardner, J., Warren, N., Mason, P. H., & Dominguez D, J. F. (2018). Neurosocialities: Anthropological engagements with the neurosciences. *Medical Anthropology*, 1–5.

Garland-Thomson, R. (2011). Misfits: A feminist materialist disability concept. *Hypatia*, *26*(3), 591–609.

Gepshtein, S., & Snider, J. (2019). Neuroscience for architecture: The evolving science of perceptual meaning. *Proceedings of the National Academy of Sciences*, *116*(29), 14404–14406.

Gisquet, E. (2008). Cerebral implants and Parkinson's disease: A unique form of biographical disruption? *Social Science & Medicine*, *67*(11), 1847–1851.

Gold, I. (2019). From mind to brain: The challenge of neuro-reductionism. In A. Raz, & R. T. Thibault (Eds.), *Casting light on the dark side of brain imaging* (pp. 85–87). Academic Press.

Haraway, D. (1988). Situated knowledges: The science question in feminism and the privilege of partial perspective. *Feminist Studies*, *14*(3), 575–599.

Hendriks, N., Slegers, K., & Duysburgh, P. (2015). Codesign with people living with cognitive or sensory impairments: A case for method stories and uniqueness. *CoDesign*, *11*(1), 70–82.

Hollin, G. (2017). Failing, hacking, passing: Autism, entanglement, and the ethics of transformation. *BioSocieties*, *12*(4), 611–633.

Jasanoff, S., & Kim, S.-H. (2009). Containing the atom: Sociotechnical imaginaries and nuclear power in the United States and South Korea. *Minerva*, *47*(2), 119–146.

Jasanoff, S., & Kim, S. H. (2015). *Dreamscapes of modernity: Sociotechnical imaginaries and the fabrication of power*. University of Chicago Press.

Latour, B. (2004). How to talk about the body? The normative dimension of science studies. *Body & Society*, *10*(2–3), 205–229.

Latour, B. (2005). *Reassembling the social: An introduction to actor network theory*. Oxford: Oxford University Press.

Latour, B., & Woolgar, S. (2013). *Laboratory life: The construction of scientific facts*. Princeton University Press.

Law, J. (2008). On sociology and STS. *The Sociological Review*, *56*.

Marres, N. (2013). Why political ontology must be experimentalized: On eco-show homes as devices of participation. *Social Studies of Science*, *43*(3), 417–443.

Mastafa, M. (2008). An architecture for autism: Concepts of design intervention for the autistic user. *International Journal of Architectural Research*, *2*(1), 189–211.

Mol, A. (2002). *The body multiple: Ontology in medical practice*. Duke University Press.

Mol, A. (2008). *The logic of care: Health and the problem of patient choice*. Routledge.

Ortega, F. (2009). The cerebral subject and the challenge of neurodiversity. *BioSocieties*, *4*(4), 425–445.

Owen, C. (2016). *Design Accross the Spectrum: Play spaces*. University of Tasmania.

Papale, P., Chiesi, L., Rampinini, A. C., Pietrini, P., & Ricciardi, E. (2016). When neuroscience 'touches' architecture: From Hapticity to a Supramodal functioning of the human brain. *Frontiers in Psychology*, 7.

Pickersgill, M., Broer, T., Cunningham-Burley, S., & Deary, I. (2017). Prudence, pleasure, and cognitive ageing: Configurations of the uses and users of brain training games within UK media, 2005–2015. *Social Science & Medicine*, *187*, 93–100.

Pickersgill, M., Cunningham-Burley, S., & Martin, P. (2011). Constituting neurologic subjects: Neuroscience, subjectivity and the mundane significance of the brain. *Subjectivity*, *4*(3), 346–365.

Pickersgill, M., & Van Keulen, I. (2012). *Sociological reflections on the neurosciences*. Emerald Group Publishing Limited.

Porter, D. L. (2020). Neuroessentialism and the rhetoric of neuroscience. *Philosophy, Psychiatry, & Psychology*, *27*(3), 239–241.

Rapp, R. (2012). A child surrounds this brain: The future of neurological difference according to scientists, parents and diagnosed young adult. In M. Pickersgill, & I. van Kuelen (Eds.), *Sociological reflections on the neurosciences* (pp. 3–26). Bingley UK, Emerald: Press.

Reiner, P. (2011). The rise of Neuroessentialism. In J. Illes, & B. Sahakian (Eds.), *Oxford Handbook of Neuroethics*. Oxford UK: Oxford University Press.

Ritchie, I. (2020). *Neuroarchitecture: Designing with the mind in mind*. Wiley.

Rose, N., & Abi-Rached, J. M. (2013). *Neuro: The new brain sciences and the Management of the Mind*. Princeton University Press.

Roseneil, S., & Ketokivi, K. (2015). Relational persons and relational processes: Developing the notion of relationality for the sociology of personal life. *Sociology*, *50*(1), 143–159.

Schultz, W. (2018). Neuroessentialism: Theoretical and clinical considerations. *Journal of Humanistic Psychology*, *58*(6), 607–639.

Singh, I. (2013). Brain talk: Power and negotiation in children's discourse about self, brain and behaviour. *Sociology of Health & Illness*, *35*(6), 813–827.

Sismondo, S. (2011). *An introduction to science and technology studies*. Wiley.

Spanjar, G., & Suurenbroek, F. (2020). Using neuro-architecture to reinforce Plarticipatory planning and design. In *Concious Cities Anthology 2020: To Shape and Be Shaped*. https://doi.org/10.33797/CCA20.01.05. 7 October 2020.

Sutherland, K. (2017). Autism and the arts: Making a space for different minds. *The Conversation*.

Tolhurst, E., Weicht, B., & Kingston, P. (2017). Narrative collisions, sociocultural pressures and dementia: The relational basis of personhood reconsidered. *Sociology of Health & Illness*, *39*(2), 212–226.

Trotman, N., & McGinley, C. (2018). Design and the mind engaging and collaborative workshops for the Neurodiverse. *Studies in Health Technology and Informatics*, *256*, 223–235.

Williams, S. J., Katz, S., & Martin, P. (2011). Neuroscience and medicalisation: Sociological reflections on memory, medicine and the brain. In M. Pickersgill, & I. Van Keulen (Eds.), *Vol. 13. Sociological reflections on the neurosciences* (pp. 231–254). Emerald Group Publishing Limited.

CHAPTER THREE

Disability studies, neurodivergence and architecture

Jos Boys*

Learning Environments Equality Diversity and Inclusion Centre (LEEDIC), The Bartlett Faculty of the Built Environment, University College London, London, United Kingdom

The DisOrdinary Architecture Project, London, United Kingdom

*Corresponding author: e-mail address: j.boys@ucl.ac.uk

Contents

1. Introduction

Since the 1970s, disability studies has become a strongly emerging discipline—growing out of the activism of diverse disabled people, a commitment to challenging everyday negative attitudes to disability, and a focus on "the social norms that define particular attributes as impairments, as well as the social conditions that concentrate stigmatized attributes in particular

Developments in Neuroethics and Bioethics, Volume 5
ISSN 2589-2959
https://doi.org/10.1016/bs.dnb.2022.08.012

populations" (Schalk, 2017). As someone who is non-disabled, neurotypical and architecture trained, I have learned a huge amount from such work. I am interested in amplifying the voices, experiences and analyses of disabled scholars, campaigners and creatives; and in exploring how non-disabled people can better understand the ways we are implicated in the continued marginalization of disabled people, and what we can do about it. In this chapter, I will first outline some key aspects of disability studies to consider how this field is intersecting with many varieties of neurodivergence and neuroethics. I then explore connections with architecture—both as a disciplinary practice, and as actual built spaces. Of course—like other disciplines—disability studies is multilayered, amorphous and contested, with various perspectives on disability as a concept and disabled people as a constituency. It has many associated subfields and offshoots, including mad studies (Beresford & Russo, 2022; Ingram, 2016) and neurodiversity studies (Bertilsdotter-Rosqvist, Anna Stenning, & Chown, 2020). Many alliances exist across work related to the environmental humanities (Ray & Sibara, 2017), such as around non-human and animal rights (Chen, 2012; Taylor, 2017), environmentalism (Clare, 2017; Kafer, 2013) and in discard studies (Gissen, 2009; Liboiron & Lepawsky, 2022). In this context, advocacy and scholarship about neurodivergence and neurodivergent people—what Walker (2013) calls neurominorities—is similarly complex. Partly this has to do with the history of disability activism and scholarship itself. In the United Kingdom and the United States, the disability rights movement initially grew in the 1960s and 70s, from physically disabled people who had been put in institutions and campaigned to live independently. In the United Kingdom, the Union of the Physically Impaired Against Segregation (UPIAS), together with the Disability Alliance, produced a pamphlet called "The Fundamental Principles of Disability" (1976) that emphasized how it was society rather than individual impairments that disabled them, both through attitudinal and environmental barriers. During the 1980s, this became known as the Social Model of Disability. It is often seen as an important alternative to the Medical Model of Disability, where disabled people are seen only as impaired individuals, in need of a cure to make them more "normal" (see also chapter "Introduction" to this volume).

The Social Model of Disability was meant to be a starting point in disability campaigning that challenged stereotypes of, and assumptions about, disabled people. It is best seen as a tool that enables disabled people and their allies to campaign together to improve lives, rather than as an overall explanation for disabled people's diverse experiences and perspectives.

From the start, there were worries about making a simplistic and oppositional divide between disability and impairment, and which tended to focus only on societal issues. Many disabled feminist scholars argued that this divide was problematic because it downplayed the difficult realities of living with impairments, compared to the impact of external factors and undervalued the significance of these to disability activism (Morris, 1991, 1996; Crow, 2013; Thomas, 1999). There was also a lack of exploration of the different playing-out of disabled people's lives as inherently intersected across a diverse range of impairments, as well as race, class, culture, gender, sexuality and other factors.

Then, as disability activism and scholarship expanded, the social model was revisited and extended beyond its initial focus on physically disabled people, to those with invisible disabilities—such as environmental sensitivities—and increasingly to self-advocating neurodivergent people, including autistics and people with learning disabilities. This, in turn, has led to new insights about affinities and differences, in relation to the variety of non-normative ways of being; and to how various impairments are framed differently by a "normal" world. This has also been richly informed—and challenged—by important work that decentres white and global north assumptions (Erevelles, 2011; Meekosha & Soldatic, 2011).

However, I suggest some shared principles that underpin these various conceptual frameworks, perspectives and agendas. These are first, a challenge to the deficit model of disability and neurodivergence, which assumes being "normal" and neurotypical is inherently a naturally preferred and superior way of being. Second, it is about aiming toward a future where bodily and neurodiversity is valued, and seen as a straightforward part of everyday life. And third, "it is about the work that's taking things to the next level—[that] aims higher still, toward a future in which we engage with neurodivergence in ways that unleash previously undertapped creative potentials of individuals, communities, and humanity as a whole" (Walker, 2021, p. 8).

2. Thinking through a disability studies lens

What are some of the tools, then, that disability studies gives us to explore both the normative processes through which neurodivergence in its myriad variations is framed societally; and to rethink neurological variation as a potentially creative generator and as a powerful critique of social, spatial, material and design injustices? Here, I will briefly outline three core themes from disability scholarship and activism (with particular relevance for access to, and inclusion in, the built environment) before going on to explore these in more detail in their inter-relationships with, and interrogation through, neurodivergence scholarship.

2.1 Beyond binaries: Fitting and misfitting

The first principle is about developing alternative terminologies beyond stereotypical oppositions between disability and ability. Challenging mainstream common sense that disability is an individual tragedy requires carefully unraveling of the very concepts we use to categorize people and normality; and the exploration of alternative frameworks (Cryle & Stevens, 2017). This in turn, opens up to question the processes through which people come to be treated differentially because of bio- and neuro-differences. Many scholars across race, class, gender and queer scholarship, as well as disability studies, have shown how everyday language is often framed around simplistic binary oppositions that historically have implied superior versus inferior bodies and minds—white/black, men/women, straight/gay, normal/abnormal, able/disabled. In fact, the history of disability is one of endless attempts at definitive categorization of mental and other "abnormalities" and a separation out of those deemed inferior, both attitudinally and in terms of equality of access to resources, through the institutionalization and societal removal of the mad, bad and sad. At the same time, normative common sense treats a particular type of body (stereotypically white, male, young, able) as unmarked, unremarked-on and thus non-problematic. This type, though, is also labeled with preferred characteristics—mobile, active, unencumbered, independent, autonomous, rational, competent and self-regulating—which in turn has effects on what counts as acceptable behavior more widely and on who gets valued and enabled by everyday social, spatial and material practices.

In addition, disability studies challenges simplistic divisions between mind and body. For instance, by using the term bodymind (Dychtwald, 1978; Price, 2014), our minds are understood as always an embodied phenomenon:

> *The mind is encoded in the brain as ever-changing webs of neural connectivity. The brain is part of the body, interconnected with the rest of the body by a vast network of nerves. The activity of the mind and body creates changes in the brain; changes in the brain affect both mind and embodiment. Mind, brain, and embodiment are intricately entwined in a single complex system.*
>
> **(Walker, 2021, p. 6)**

By unraveling the complex, partial, situated (and inherently impossible) attempts at demarcating and fixing categories of difference, disability studies is deliberately "disruptive of any boundaries that claim to police distinctions between disabled/non-disabled subject positions" (Erevelles, 2014). Rather, it centers on relationality. Who counts as disabled, and how disability is interpreted societally, *depends* on the specificity of encounters and settings. Rosemarie Garland-Thomson proposes the concept of fitting and misfitting to capture something of this fluidity. Misfitting, as she explains it, is a concept that directs attention to "the co-constituting relationship between flesh and environment" (Garland-Thomson, 2011, 594). As Margaret Price summarizes:

> *Beginning with the straightforward image of a square peg attempting to fit into a round hole, Garland-Thomson then complicates that image to show that misfitting always involves a shifting context: it does not describe a static experience, but rather a contingent relationship perpetually in motion. [...] Fitting, then, is a function of relations of power, for a person's fit or misfit is "sustained" or not by changing contexts and attitudes. (pp. 4–5)*

2.2 Nothing about us without us

The second principle is the centering of disability-led and disability-oriented participatory processes. This is best summed up by the campaign slogan "Nothing about us without us" (Charlton, 2000). This slogan, also used for ensuring a collaborative approach to research with indigenous and other marginalized people, aims to communicate the idea that no actions about a group should be decided without the full and direct participation of members of the group(s) affected by that action. In the United Kingdom, doing disability research through a deliberately emancipatory and disability-led approach was being actively explored and campaigned for from at least the 1990s (Barnes & Mercer, 1997; Barton & Oliver, 1997). Research and medical experimentation on disabled people without their permission, was all too normal in this period (Dolmage, 2017). Together with the deeply problematic history of eugenics, forced sterilizations, the Nazi's T4 euthanasia program (Caplan, 1992; Silberman, 2017) and the locking up of various "categories" of disabled people through history, it is not surprising that disabled people continue to be suspicious of medicalization of, and control over, their bodies and minds. Equally crucially, disabled people continue to face unconscious bias from researchers and designers, such that ableist assumptions about 'inherently' unsatisfactory life experiences can take precedence over actually paying attention to disabled people's own narratives:

> *The misrecognition of disabled people is often tied to the fact that non-disabled people are unfamiliar with disabled life or assess it inaccurately. As Barnes (2016) notes, "a vast body of evidence ... suggests that non-disabled people are extraordinarily bad at predicting the effects of disability on* perceived *well-being" (p. 71). Yet few non-disabled people are aware of their deficits in understanding what disabled life can be, given prevailing norms, media, and the like. Relying on standard tropes of disability, they try to imagine themselves in an impaired body and make negative judgments based on what they expect their quality of life would be.*
>
> ***(Goering and Klein, 2020, p. 8)***

2.3 Access is love: On care and interdependence

A third theme from disability studies is emerging ideas around valuing interdependence and care over individualism, self-actualization and competition. This aims to articulate disability, access and inclusion as an ongoing, never finished and always partial collective act of care—well captured by another slogan "access is love" (Mingus, 2018). Such a reframing is potentially radical for society as a whole. Yet in many ways, it captures something of the way disabled people already support each other, as well as the interlocking connections of care between disabled people, their households and friends (Piepzna-Samarasinha, 2018). Again, the aim is to go beyond binary divisions, this time of active caring versus passively being cared for, which locates disabled people as a burden to others and to society. By raising awareness about accessibility as a shared and truly intersectional activity, the aim is to "encourage people to incorporate access into their everyday practices and lives" (Disability Intersectionality Summit, 2018). In the process, access and inclusion are also reframed—not as a requirement to create simple functional or impairment category-based design solutions to assumed problems, but instead through developing appropriate and collective *practices* in support of the concept of disability justice (Mingus, 2017; Sins Invalid, 2019). And rather than individual normativity and agency, ethical judgments become about human flourishing, redefining what it is that constitutes a good life (Johnson, Walmsley, & Wolfe, 2010).

3. Disability studies, neurodivergence and ethics

How then have these disability studies themes been interrogated and inflected by neurodivergent scholars and activists, and their allies? Here, I will outline just two ethical considerations that seem central to neurodiversity studies—first, definitions of what it is to be human; and second, neurodivergence and difference as a creative generator. Savarese and Savarese (2010)—focusing specifically on autism—suggest that:

> *the field of disability studies has been slow to take up cognitive disability, and it has done so with some discomfort—in part because notions of social construction, while important, seem inadequate to the task of assessing physiological differences in the very organ of perception, and in part because these differences seem such a threat to what most makes us human" (p. 23).*

Margaret Price also critically engages with the tendency toward social construction (as embedded in the social model of disability) because it tends to focus on social and spatial interactions, rather than affective relationships and responses. Exploring cognitive impairments she proposes that Garland-Thompson's fitting/misfitting model only really speaks to the physical lived realities of impairment and difference (Price, 2011, 2014). She outlines an example of someone having a meltdown episode (revealed later in the piece as herself)— someone who wants to do themselves harm "for no reason other than being compelled to do so" (2011, 6). For Price, in this situation, "to be 'considered equally valuable' when I am in the midst of a violent break means to be treated as someone who is having a meaningful experience, even if my actions are not always safe (and thus sometimes need to be curtailed)" (Price, 2011).

This focus on the painful and complicated aspects of intellectual disability leads Price to the vital importance of interdependence, non-judgemental human valuing and disability access as care; and to the importance of a more considered engagement with the possibilities of cure than can sometimes occur in disability studies literature, with its often focus on disability positivity. As Price explains "the problem is not so much turning down medical advances that could alleviate or eliminate particular impairments for the people who desire it, but that there remains an assumption in medical science that 'curing pain not only desirable but also compulsory' (Patsavas, 2014, 208)" (2011, 9).

Crucially this is about recognizing complexity and nuance in relation to personal positioning—being born with a disability, for example, compared to developing one in later life; social context, that is how particular impairments are framed by society; and the considerable heterogeneity in experiences across and between different non-normative neurologies. As Savarese and Savarese also write:

> *We need to make room for difference, even, or especially, when that difference seems incapable of producing prized forms of competence. We need to be hospitable, however tired or frustrated caregivers and service providers might be*

> *(or focused on the bottom line politicians might be). There are myriad ways to be present, connected, and alive; myriad ways to have relationships (2010, 10).*

3.1 Who counts as human

As all these authors make clear, neurodivergence in all its multiple forms intervenes very immediately, intimately and disruptively into normative assumptions about what it is to be human. Since the early western philosophers, the very definition of being human—and thus superior to animals—has been through the concept of reason. This can lead to hierarchies of capability as morally significant psychological properties, and the positive valuing of individual rationality, intellectual competence, personal independence, ability to participate, verbal dexterity and social communication. In this context, mental capacity and neurological functioning become a central basis not only of assumed rationality but as a marker of who deserves to be considered human, or to have personhood or agency. When neurodivergence is assumed a pathology (Walker & Raymaker, 2021) this raises important ethical implications, not only because of the context of the historical dehumanization, institutionalization and extermination of neurodivergent people (alongside other disabled people) but also because of contemporary arguments that continue to insist that some neurodivergent people do not even come in the category of human. McMahan (2003, 1996), for example, has argued for a two-tier morality—one for persons and one for non-persons, the latter being "the profoundly cognitively impaired [who] are incapable ... of deep personal and social relations, creativity and achievement, the attainment of the highest forms of knowledge, esthetic pleasures, and so on" (1996, 8). For him, as for moral philosopher Peter Singer (1979), the appropriate comparison group is non-human animals; in fact, the moral status of animals is to be itself elevated above "non-persons." In response Eva Feder Kittay writes:

> *[a]s a philosopher who is the mother of a wonderful woman with severe intellectual disabilities, I finally determined that I should subject philosophical treatments of cognitive disability to a critique based on my knowledge and personal experience as a mother. [...] In fact, much of philosophy depends on being able to make such claims about distinctive capacities, and many claims of political ideals of justice, autonomy, equality are grounded on a set of competences or potentials, many my daughter most likely does not possess. (Kittay (2009): 607)*

For Kittay (deliberately making a philosophical argument) the fact that she has a true and loving mother and child relationship itself requires moral recognition, which in turn demonstrates her daughter's moral personhood, as for all cognitively disabled people, however severe their impairment. For many neurodivergent scholars and campaigners, this willingness to brand non-normative minds as less than human urgently needs challenging:

> *The idea that there is one 'normal' or 'healthy' type of brain or mind, or one 'right' style of neurocognitive functioning, is a culturally constructed fiction, no more valid ... than the idea that there is one 'normal' or 'right' ethnicity, gender, or culture."*
>
> **(Walker, 2014, online)**

In a seminal 1962 paper "On the Politics and Sociology of Stupidity in our Society" Lewis Anthony Dexter argues that it is precisely seemingly obvious problems such as of mental deficit—taken as self-evident facts—which actually need to be critically analyzed and "unlearned" (p. 221). He uses the thought experiment of a society that is not bothered by differences in intelligence, but rather focused on clumsiness, labeling "gawkies" as the deficit group:

> *Let us assume [...] such a society stresses grace and style in movement as we stress intellectual skill. Let us assume that people are taught to abhor clumsiness as many people in our society are taught to abhor stupidity. [...]*
>
> *Naturally, under such circumstances, the marginally clumsy, permitted to remain at large in the community, would always be conscious of having two strikes already called against them. They would be liable to be institutionalized if they did anything unusual. Naturally, too, clumsy children would become social rejects and isolates, and instead of the moron jokes, beloved in this country, there would probably develop pantomime jokes, directed against the gawky.*

Normative mental competency, then, is not inherently a marker of humanity, rather it is integral to the power dynamics of social inequalities.

3.2 Neurodivergence as an alternative way of seeing

For Dokumaci (2019) quoting Wahlberg (2009), "disability or neurodivergence are not 'normality' gone wrong, but 'kinds of living' in and of themselves with their own 'vital norms' and their own 'quality of life'" (p. 185). Such non-normative ways of being in the world, then, can be a creative challenge to current values as well as a means to "construct the category of the human in the most capacious manner possible" (Savarese & Savarese, 2010, p. 19). As already outlined, normative mental competency tends to be associated with individual embodied acts such as independence, mobility, self-regulation, social capabilities, verbal dexterity and other communication skills. However, there is already a richly expanding literature by, and in collaboration with, diverse neurodivergent people that aims to both explain and validate other forms of knowledge and competency.

This is across academic publications (Johnson et al., 2010; Kittay, 2019; Manning, 2013; Savarese & Savarese, 2010; Yergeau, 2018); memoirs and collections (Bascom, 2012; Gadsby, 2022; Higashida, 2021; Kapp, 2020; Pang, 2021) artists' blogs and twitter feeds (Adams, 2016, 2017; Boue, n. d; Brown, 2021; Thom, n.d.; Trotman, 2020) as well as many campaigning websites. Here, I will outline just a few threads, to signpost how taking notice of our diversity of ways of being in the world can be a potential creative generator for thinking about disability, neurodivergence and architecture.

From the perspective of contemporary autistic scholarship, such as by Remi Yergeau, autistic people find themselves defined precisely in and against the "compulsory sociality" of neurotypical people; as being marked inferior, and even non-human, precisely by appearing to have traits "that not only defy social order, but fail to acknowledge social order's very existence" (2018:27). Yet, as authors in this volume demonstrate, there are many ways of being social, that may not conform to normative assumptions—for example, concerning eye contact, body language or verbal interactions—but nevertheless both involve social bonding and offer different kinds of processes for collaborating together (see for example, chapters "Cultural commentary: Being autistic together" by Sinclair, "Not all surfaces catch the light at the same time" by Boue, and "Equalities design: Toward post-normative equity" by Trotman, in this volume). As Dind (2021) notes in her review of Alsya Hillary's work, the problem lies in assuming a norm and deviance or deficit from that norm;

> *[Hillary, 2020] argues that while principles of cross-cultural communication and mutual understanding could be applied to cross-neurotype communication, there is instead a tendency to blame miscommunication between neurotypical and neurodivergent people "on neurodivergent people's perceived 'social deficits,' including in 'Theory of Mind'" (p. 93). That is, whereas instances of miscommunication across cultural differences are conceived as an interpersonal problem, miscommunication across neurotypes is regarded as an individual problem: "that of the neurodivergent individual" (online).*

Similarly, Prince (2010) critiques the way that neurotypicals have narrowly promoted one kind of language "and in the process denigrated nonspeaking people with autism, animals, and the natural world we all inhabit" (Savarese & Savarese, 2010, p. 19). Many autistic and other neurodivergent commentators and advocates have explored these issues, as well as outlining some positive values to being neurologically different. As autistic Elizabeth (Ibby) Grace suggests:

> *It is a surprise that in our society it is not more disabling to live your whole life unaware that folding the symbolic kissing cherry blossoms together just so creates a regular pentagonal dodecahedron. But I am glad people can live fulfilling lives that way, because does anybody wish ill to people who are without that inner visual ability to properly spin things?*
>
> ***(quoted in Yergeau, 2018: p. 25)***

As Amanda (Mel) Baggs, outlines in their YouTube video *In My Language* (Baggs, 2007), for a non-normative mind, language and communication do not necessarily need to use conventional frameworks of expression and representation but are rather a way of thinking in their own right—"it is being in conversation with every aspect of my environment, reacting physically to all parts of my surroundings." As they also note "this is not a look-at-the-autie gawking freakshow as much as it is a statement about what gets considered thought, intelligence, personhood, language, and communication, and what does not" (Baggs, 2007).

Of course, factors assumed to indicate deviations from normative rationality vary across impairment categories. People with learning disabilities, for example, have been historically defined as having an IQ of under 70, what Altermark notes is an arbitrary and non-evidenced based measure originally aimed at differentiating between educatable and non-educable "idiots"' (Altermark, 2014, p. 1463). Yet during the 20th century, mean IQ tests for people with Down Syndrome increased by 30 points caused, he suggests, by the improving opportunities to thrive and lead a good life rather than being isolated and institutionalized. For Altermark this suggests the importance of neuroscience of intellectual disability exploring brain plasticity rather than centering of an assumed fixed pathology in need of a cure or other solution.

Other disabilities, such as senility, are indicated by an increasingly loss of ability to perform in normative ways. As Shannon Mattern writes about her mother (2021, 9) most clinical treatments do not pay enough attention to embodied experiences of dementia. "While the person with late-stage dementia may be engaging in behaviors and expressions that make sense for their reality, our conventionally rational, logocentric logics might lead us to see them as vanished" (Mattern, 2021). When people with dementia fail to recognize others, their social worth is also seen as disappearing. As Lynn Castell Harper writes about her mother "[o]nly if she is able to recognize others is she a candidate for greater social recognition." (Harper, 2021 p. 164 quoted in Mattern, 2021, 9). Here, the authors are arguing for recognizing non-normative minds as a way of being that can open up possibilities rather than close them down.

4. Neurodivergence and architecture

In the previous sections, I have explored some key threads in how disability and neurodivergent scholarship and activism challenges common sense understandings of our diverse bodies and minds. Next, I will first outline some of the ways our built surroundings and their occupation are framed conceptually and methodologically within the discipline of architecture, before considering how neurodivergence and non-normative bodyminds have been increasingly brought into play in these debates. I will suggest that as well as architectural design projects understood as meeting the "special needs" of neurodivergent people, there are others that instead start from difference as a creative generator for building design, and still others that that aim toward a future where starting from neurodivergence enables alternative ways of thinking and doing architecture. This in turn, enables some final thoughts on how neuroscience, architecture and the intersections between then, can learn from disability studies, activism and creativity.

Across the disciplines related to architecture (in the widest sense) there are a multiplicity of theories and approaches that conceptualize how built space itself works and how it is inhabited. Behaviorism, environmental psychology and the neurosciences, for example, look to map how our neurons, physiological states and/or behaviors are wired, to better understand how people perceive the world through our senses and emotions (see chapter "The neuroethics of architecture" by Illes and Huang, this volume). In this model, our relationship to the sensory and spatial world is centrally mediated through sophisticated stimuli and response processes. Analyzing how this process operates is aimed at enhancing the evaluation of the performance of the existing built environment, providing evidence-based design decisions, and informing both intended design and behavior changes and enhancements for "improving the quality of life of humans and the society" (Karakas & Yildez, 2020). However, in their systematic review of a neuroscience approach to the built environment, these authors also note that there remains a gap between the wide range of concepts and theories informing the field, and the lack of verification through experimental studies, with ongoing challenges in measuring and associating the data (pp. 244–5).

In the fields of architecture and design the concept of affordances—developed from Gibson's, 1966 seminal book *The Senses Considered as Perceptual Systems*—is often used to extend our understandings of behavioral and cognitive engagements with the world, by also incorporating what the environment offers "back to" the individual. Here, material space is framed as inherently relational, as humans interact with their surroundings both functionally and socially. As Rietveld and Rietveld (2018) write:

> *Many interpretations of this theory of affordances are tied to motor behavior, such as the fact that something—like a cup or a book—can be grasped because of its dimensions, shape, texture, etc., or that the relatively horizontal, structurally supported, elevated surface of what we call chairs allows one to sit on them. However, affordances do not only depend on the way our environment is designed, but also on people's abilities and thus on the patterns of activity that have been cultivated by sociocultural practices. The affordance of sitting on a chair created by a carpenter, for instance, exists only against a wider background of sociocultural practices of sitting on chairs, rather than sitting on the floor, on carpets, or perhaps living in a non-sedentary way altogether.*

This expansion beyond more limited stimuli-response behavior models opens up analysis to the impacts of societal attitudes, and social, material and spatial encounters; and thus engages with disability studies concepts of "fitting" and "misfitting" in both physical and affective forms. To go back to the chair as an example. Dokumaci (2019) offers other interconnections with different bodyminds—where "the surface itself would be 'bump-into-able' in relation to a blind person, 'climb-on-able' in relation to a toddler, and 'get-underneath-able' in relation to a worm" (2019, p. 397). To this I would add, as a sitting place that does not support the way you sit; an object that can be perceptually manipulated for sensory pleasure; as "something more alive than the people who sit on it" (Adams, 2016); as something you might feel compelled to bang your head against; or as an object whose meaning and use is slippery to you. As Dokumaci also notes, if existing built environments are not seen to offer "a field of possibility" as the affordances literature often assumes, but rather a field "of difficulties to negotiate" (Leder, 1990, 81) this is not because neurodivergence or disability is the problem, but that architects tend to assume a "normal" use; and then only add on the "abnormal"—such as a space for a wheelchair user—requirements at the end of the design process, as merely a technical and legalistic issue (Boys, 2014).

In fact, an approach that recognizes and equally values the capacious range of our human bio- and neuro-diversity in negotiating built spaces not only complicates such conventional architectural design processes but also disrupts some of the assumptions built into Universal and Inclusive Design frameworks. There is already a vital developing literature that understands the language of building accessibility and inclusion, not as technical, functional or even common sense principles but as historically and geographically specific negotiations between and across non-disabled and disabled people (Gissen, 2022; Hamraie, 2017; Titchkosky, 2011). This is because most disability studies and neurodivergence scholarship starts from social, spatial and material practices, rather than from perceptual or behavioral responses. Whether through Science and Technologies Studies (STS, see chapter "Science and Technology Studies (STS) and the neuroethics of architecture" by Gardner, this volume), sociology of the built environment, anthropology and ethnography, or autoethnography (see chapter "Autoethnographic reflections on architectural design for neurodivergence" by Clarke, this volume), such accounts also aim to challenge reductivist uses of behaviorism and neuroscience (both perceived and real) in either evaluating or proposing improvements to built settings. At the same time, as also noted previously, the complexities, tensions and contradictions across the needs, preferences and experiences of different forms of neurodivergence and disability necessarily complicate any over-simplified social model of disability in relation to the material world.

For the rest of this chapter, I will outline how some of these different approaches have affected both built environment practices and designed environments. This will begin with a consideration of the subset of buildings designed specifically for those with "special needs," for example, schools or care homes and other institutional settings. I will then provide an overview of scholars, artists and designers who are critically and creatively exploring intersections of neurodivergence with the built environment to offer alternative approaches and adaptations, together with some associated examples. Finally, I will engage with scholarship that starts from neurodivergence and disability to critique current built environment practices by "neuroqueering" architecture itself (Walker, 2021).

4.1 Designing for special needs

There are already many good examples of design guidance related to neurodivergence, particularly autism (Mostafa, 2018, 2021), as well as architectural projects designed for specific categories of neurodivergence, such as the Corinne Dolan Alzheimer Center (CDAC) in Chardon, Ohio (see chapter "The neuroethics of architecture" by Illes and Huang, this volume); the Mossbrook School classroom of the future for autistic children by Sarah Wigglesworth Associates (see chapter "Designing with neurodiverse children and adults: learning a different lesson with every engagement" by Wigglesworth and Bond, this volume); and the Alzheimer's Respite Center in Dublin, designed by Niall McLaughlin, together with a related website *Losing Myself* and installation for the Venice Architecture Biennale (see chapter "Dialogic Drawing" by Manolopoulou, this volume). All of these—by dint of being category specific—are able to aim at integrating non-normative bodyminds from the outset, often both as participants in the design and development process and as a set of "needs" to be meet.

How then, have such examples articulated the design of built space as a means to frame and support particular categories of neurodivergence? De Hogeweyk, in the Netherlands, is "the world's first dementia village" (Adams & Chivers, 2021). While designed to appear as a small residential settlement with a shopping street, it is in fact a secure care facility. The aim is to create experiences of home and community that produce a sense of belonging in its inhabitants. For Adams and Chivers this exemplifies an increasing trend in healthcare and aging design away from institutional settings, that is, where the realities of being in a locked space are hidden, as is the necessary medical support—here a "'[B]ackstage'—[where] the administrative and medical infrastructure of the nursing home, the charts and the records, the medications and the stretchers—is kept out of sight" (Mattern, 2021, 8). In designing such environments, the focus is often necessarily upon surveillance cameras, movement sensors, and other security measures to prevent wandering, and/or, as in the case of De Hogeweyk, to enable limited safe wandering. These are not merely appropriate functional solutions, they are value judgments about who matters and how space and resources are allocated; made in the context of wider attitudes about what counts as being human and within larger frameworks of the social and political funding and management of care. Harper (2021) challenges how dementia is articulated as a form of disappearing, both in terms of rational communication and of forgetting. She argues that there can be enriching relationships that do not rely on being remembered, and that "the closer I came to people with dementia, the more my assumptions vanished, the more they became individuals rather than a blur of disability" (p. 13). This is an example of aiming to center mutual supportiveness, acceptance and the insights that come from valuing non-normative bodyminds, as well as supporting their flourishing. For Mattern, reflecting on her own mother's developing dementia and its changing relationships to particular domestic and local outdoor spaces—disorientation and the desire to wander, as well as the need to "retreat to small spaces of containment" (2021, p. 5)—it is crucial to understand these as negotiated and complex responses by people to their own experiences of neurological changes and changed realities. Thus, she follows Harper's view "that there is much to gain by attuning ourselves to dementia's own affect and reason" (2021, 9).

Particular built examples, such as the ones given above, of course have to operate within existing societal practices (as well as the economic context of getting projects built and being financially viable). For Mattern, considering the actual options available for her mother, this wider context is relevant as it reveals differential access to resources and spaces by people in need, and exposes wider assumptions about divisions between normative independence and competence, vs interdependence with its assumed burden of care. Even the De Hogeweyk dementia village, then, is about the containment of vulnerable bodyminds, because the world beyond its walls is hostile to the appearance or expression of madness, dementia or other non-normative neurodivergent behaviors. Instead, Mattern wishes for "an external universe [that is] less hostile to and more inclusive of different abilities" (p. 10).

5. Neurodivergence as a creative generator

As well as exposing existing normative attitudes and spaces, much disability scholarship and activism has articulated diverse impairments as having their own strengths. While, as already mentioned, it is important to take care not to have a simplistic focus on disability positivity, I will next briefly outline some built space processes and products that start from neurodivergence as positive contribution to society and as a creative generator. Whether through disability-led arts practice for people with learning disabilities, such as the Creative Growth Center in the United States (Katz & Ludins-Katz, 1990) or the Rocket Artists in the United Kingdom (Fox & Macpherson, 2015); through examples of good practice developed by neurodivergent people themselves, (see for example, chapters "Relax and resist: Reflections on the Touretteshero Relaxed Venue methodology" by Renel and Thom, and "Not all surfaces catch the light at the same time" by Boue, this volume); or spaces that start from diverse ways of being rather than from conventional users; the aim is toward conceptual and material spaces that recognize, value and support non-normative bodyminds. Here, I will outline just two examples—the spaces of Autreat annual conferences developed in the United States by Autism Network International (ANI) and the Heart N Soul Hub at the Wellcome Collection in London United Kingdom, led by learning disabled artists.

5.1 Autreat

For Jim Sinclair, a renowned autistic activist and scholar, neurotypical assumptions about the kinds of spaces diverse autistic people need can be challenged by the shared spaces created by autistic people themselves. In their seminal piece "Cultural Commentary: Being Autistic Together" (Sinclair, 2010, see also chapter "Cultural commentary: Being autistic together" by Sinclair, this volume) Sinclair outlines the development of Autism Network International (ANI) as probably the first autistic-run organization and describes the creation of physical gatherings run by and for autistic people, particularly Autreat, a retreat-style conference. Rather than having to deal with the challenges of occupying neurotypical conference spaces, with bland individual hotel rooms and crowded meeting and socializing spaces, participants camped on an empty upper floor of a large empty office building nearby. Because this was a building being renovated, all the interior walls on that floor had been removed. Attendees bought their own equipment and sleeping bags, as well as "some empty refrigerator boxes for anyone who wanted to sleep in one or needed to retreat alone to a dark enclosed space for a while" (2010 online). This was not, however, a straightforward solution for as Sinclair writes, rather than no interpersonal or environmental difficulties, it simply has some different difficulties. This is not just about adjusting away from the experiences of coping with NT space, it is also because autistic people are themselves very diverse. As Sinclair goes on to say:

> *...autistic people have in common (by definition, as part of the diagnostic criteria for autism) abnormalities or "impairments" in communication and social interaction. But we're not all abnormal in the same ways. Our ways of interacting are different from NT norms—and also different from each other's!—and the* amount *of interaction we want also varies across an even wider range than it varies among NTs. We range from not wanting any social contact at all (which is the common stereotype, but rarely the reality), to wanting some social contact but needing more down time than most NTs (even introverted ones) need, to wanting an NT-normal amount of contact, to wanting or needing* constant *contact, even more than "normal" extraverted NTs can tolerate.*

The flexibility of large open spaces, together with a variety of much more enclosed retreat zones aimed to take into account this variation. Other neurodivergent authors have also explored the problematic space of conferences and suggested methods for rethinking and adapting such spaces, both in terms of design and services (Price, 2011; Yergeau et al., 2013).

5.2 Heart n Soul at the Hub

The *Inclusive Future* project was the final part of a 2 year residency by Heart n Soul, a creative arts company. Funded by the Wellcome Trust in London and created in collaboration with the Helen Hamlyn Center for Design and commissioned artists, it aimed to offer a space for imagining alternative versions of the future, by asking people with learning disabilities and autistic people what they would most like to change about the world (Heart n Soul, 2021; Renel et al., 2020). This was undertaken via a series of workshops and an exhibition. Through a co-creation workshop, four learning disabled artists worked with artist Ben Connors, to imagine objects, environments and services that they would like to change. Each co-designer then selected one idea to present to the public. Through an exhibition, the public were invited to consider the artworks and respond by describing or illustrating what they would like to change. Ideas templates were provided, encouraging people to write or draw their ideas and add them to the exhibition (with 245 contributions made.). To develop these activities, two futuring workshops were then held which engaged a larger group, resulting in 33 participants with learning disabilities and/or autistic people. A visual story was created for each event, providing a clear understanding of what to expect.

> *Inclusive Futures is about disabled people designing things that work for them and hopefully for other people so that we can all be included in the world' (participant Robyn).[...] In a world that is increasingly designed around specific ways of being and doing that gives priority to particular bodies and minds, it feels more important than ever to imagine genuine alternatives that consider divergent experiences.*
>
> ***(Renel et al., 2020, p. 142)***

The results of this project are explored, and communicated in accessible form, through a website.

6. Architecture as an emergent process of collective care

An important development that disability studies and activism has been exploring for many years—outlined above as "access is love"—is an understanding of built space as a dynamic and contested process, rather than a merely a physical output. Access and inclusion become framed as a multiplicity of diverse needs and preferences, that occur differentially in relation to specific and situated encounters, and that can both be in alignment and contradiction across various requirements. Rather than architects' often held belief in, or hope for, a functional, technical one-size-fits-all solution, this centers on layered data about access needs, for example, through the crowd-sourcing and other approaches of the Critical Design Lab (Hamraie, 2018). It also suggests a generosity in enabling multimodal approaches that offer different types of engagement with our built surroundings in the same space; a concern with longer-term adaptations to changing and better understood circumstances; and a willingness to incorporate services committed to buildings enabling collective care and equality of resources long-term. This is always a partial work in progress, not a "perfect" answer or singular design solution.

Seeing access as a multiplicity of requirements needs to be underpinned by what Yergeau et al. (2013) call an ethics of accessibility. This requires paying responsible and respectful attention to the differences people bring to a situation "that allow[s] the broadest possible range of people to make meaning in ways that work best for them" (online). Yergeau et al. also explore how multimodality needs to be more than just the additive combination of different elements (particularly when this retains a retrofitting mentality). We need always to be exploring how those access layers can be commensurable, without a differential quality of experience, whether these are the same or different. Such an approach does demand a spatial and resource generosity—not just one accessible toilet but several with different facilities; corridors spacious enough for comfortable passing; doors wide enough to get through easily in a mobility scooter, or with children, or when carrying packages; and support for people who need rest or escape. Thus, committing to multimodality is a deliberate re-distribution of resources toward those who currently face barriers across objects, spaces, encounters and services.

Once buildings and spaces have been constructed, they can be understood as co-evolving through ongoing processes of fitting and misfitting. Occupants, dependent on their degrees of agency and control, can alter spaces to try out novel material arrangements that disrupt "built-in behavior patterns and other forced expectations" (Kullman, 2019, 9). Commissioning and managing organizations need to find ways to respond to these adaptations, what are often called disability lifehacks (Williamson, 2012, 2019) as indicators of possible positive changes. This is precisely because access and inclusion are an always emergent, uneven and unfinished collective responsibility, that don't stop with a building being designed, but require ongoing monitoring and adaptation, care and repair.

6.1 Ed Roberts campus, Berkeley California

The Ed Roberts Campus at Berkeley in California is an interesting example of what this approach can mean in practice. Designed by Leddy Maytum Stacy Architects, it opened in 2011 and remains an iconic example of an accessible building, led and operated by disabled people. The scheme is named in honor of Ed Roberts (1939–1995), a pioneer of the disability rights movement. After his death, a disabled group came together to develop an inclusive campus for organizations that provide services to disabled people. More recently however, its many and diverse users are widening and shifting ideas about inclusive design. As Kullman (2019) shows, a building that was designed out of a particular historical moment of disability activism has had to shift and supportively adapt to, for example, the experiences of autistic people and of people with environmental sensitivities. For these newer generations, disability is not so obviously framed around identity categories such as wheelchair users, blind or deaf people, or focused on physical barriers that can be designed out. Kullman suggests that the people with environmental sensitivities he talked to were much more concerned to explore the intersections between their vulnerabilities, and the requirements of other disabled people in the building; recognizing tensions and multiple needs as a means to develop improvements, rather than by first defining the needs of different impairment groups and then somehow adding these together to produce a universal design. For example, a series of moveable screens enabled autistic occupants to control their own workspaces when needed.

Kullman's research also highlights the vital role of ongoing service support in managing these tensions productively and illustrates how building maintenance and caretakers—mostly learning disabled people at the Ed Roberts Center in a supported employment scheme—can respond to complex differences positively and creatively. This is ongoing detailed work that is often framed as problematic—as a "wasted" resource in conventional building services and management—because it takes time and connectedness to support individual adaptation of, for example, particular spaces or building-wide ventilation, heating and lighting systems. This project's continuing commitment to meeting diverse needs is a great illustration of access as collective care.

7. Neuro-queering architecture

In this final section, I will outline, through an example, how starting from neurodivergence can be a critique of normative ways of making buildings and spaces. The term neuroqueer was coined by Nick Walker and Athena Lynn Michaels-Dillon, and further developed by M. Remi Yergeau in their groundbreaking 2018 book *Authoring Autism: On Rhetoric and Neurological Queerness*; with the aim of disrupting and reshaping normative assumptions about disabled people. For Walker, neuroqueering involves "[e]ngaging in practices intended to 'undo' one's cultural conditioning toward conformity and compliance with dominant norms, with the aim of reclaiming one's capacity to give more full expression to one's neurodivergence and/or one's uniquely weird potential and inclinations" and as also a way to "[w]orking to transform social and cultural environments" (2013, online).

7.1 Madlove Asylum

An example of what this means in practice is the Madlove Asylum project, started in 2014 by James Leadbitter aka the vacuum cleaner, and exhibited as part of an exhibition in London entitled *Bedlam: The Asylum and Beyond* (Harris, 2017). With Madlove Asylum, Leadbitter aims to blend research, design, building and exhibitions as a means of reimagining mental health support and the environments this support happens in, such as art spaces, hospitals and community spaces.

> *It ain't no bad thing to need a safe place to go mad. The problem is that a lot of psychiatric hospitals are more punishment than love... they need some Madlove. Is it possible to go mad in a positive way? How would you create a safe place in which to do so? If you designed your own asylum, what would it be like?*
>
> ***(Leadbitter, 2014, online)***

The intention, then, is to rethink institutional and medical settings—often experienced as oppressive and unsafe—with more positive spaces for those who experience mental distress. To do this, Leadbitter brought together people with and without mental health experiences, mental health professionals and academics, artists and designers—"and everyone else on the spectrum." The aim was to build

> *"the most crazy, bonkers, mental asylum we dare dream of: a desirable and playful space to 'go mad', countering the popular myth that mental illness is dangerous and scary. Together we are attempting to create a unique space where mutual care blossoms, stigma and discrimination are actively challenged, divisions understood, and madness can be experienced in a less painful way. This temporary structure will be a reflexive and responsive space for exploring and redesigning madness".*
>
> ***(Leadbitter, 2014)***

As Leadbitter goes onto note, this is "not the lunatics taking over" existing asylum spaces, it is about thinking through how they can be in charge, to design, build and manage asylums. As with previous examples discussed, this is also a project about mutual care, both in how people work together and in how building spaces are intended to be occupied (Fig. 1).

Fig. 1 "Madlove: A Designer Asylum" is a prototype of what a psychiatric ward designed by patients might look like. Design by Benjamin Koslowski and James Christian. *Image courtesy of The Vacuum Cleaner.*

8. Conclusion: Toward architecture for a good life

As some of the authors in this collection suggest (see Illes and Huang, and Gardner) neuroscience offers opportunities to interrogate how our minds interpret, and respond to, environmental stimuli so as to inform architectural design. What disability studies shows is that for researchers and practitioners—whether in neuroscience, or in built environment disciplines—this involves several important ethical considerations.

First, it needs careful delineations between descriptive and normative uses of data. When particular kinds of bodyminds are assumed both the norm and of greater value—more human—than bodyminds framed as impaired, then experimental methods and findings can reproduce the same deficit model bias. This can lead to a focus on cure (in medicine) and on design solutions (in architecture)—both aiming to restore normative function, that is, aimed at making non-normative people better fit society. Acting ethically in fact requires a critiquing of this binary logic of normal/abnormal, human/less-than-human. It asks for a decentering of reason, mental competence and individual agency as the preferred norm, as well as the development of ways of improving lives that center a more capacious understanding of what it is to be human, on the importance of interdependence and care, on enabling human flourishing in all its variations, and on rethinking what counts as a good life.

Second, it is also about not assuming that cure and/or normalization is necessary and compulsory, outside of disabled people's own involvement and control. This is a vital point. Where personhood and agency is assumed only in relationship to specific forms of mental rationality and self-regulation, then personal autonomy can be—and has been—overridden. As Kittay notes, "[p]hilosophical bioethics can itself be unethical in its methodology and import" (2009, p. 608.) And she goes on to say, this is not merely an abstract problem:

> *"In matters dealing with cognitive disability, philosophical positions formed in the absence of any representation either by people with disabilities or by their families find their way, via bioethics, into healthcare and policy decisions. Such positions have posed the possibility of truly deleterious consequences" (2009, p. 612).*

In 'Disability Ethics: from theory to practice' (2017) Rosmarie Garland-Thomson outlines how disability studies has grown since the 1960s, in parallel and intersectionally with gender and race civil and human rights campaigns and research; and thus has enabled disability as a concept, and disabled people as a constituency, to critique and reframe a previous "narrow focus in in medical science and health studies" (p. 324). However, she also argues that the "insights and knowledge of disability studies still need to be *applied*—to be brought into—medical science as a knowledge base and to its practitioners" (p. 325) and proposes bioethics as the appropriate arena for rethinking deficit and negative models of disability:

> *What is needed, then, is to implement a robust disability bioethics that would translate the concept of conserving disability into a capacious yet specific set of principles, practices, policies, and competencies that can affect biomedical decision-making and life decision-making by people with disabilities (2017: 331).*

Garland-Thomson calls this disability cultural competence. Central elements would be developing "language and ways of talking about disability that are accurate yet nonprejudicial" (p. 333); supporting learning "how to be disabled—how to achieve well-being and good life quality as disabled people" (Garland-Thomson, 2011); and providing tools in healthcare training environments, including "curricula, training, presentations, expert patients, certification, speakers, exhibits, media products" (p. 334).

The third and related point from disability studies is that exploring neuroscience and architecture also requires challenging existing modes of practice *within* these disciplines, where these reproduce normative assumptions. Manolopoulou writing in this volume, argues for the importance of dialogic practices—engaging directly with neurodivergent individuals and groups, paying attention to, and valuing diverse perceptions and experiences of making sense of, and surviving in, the world. The DisOrdinary Architecture Project, of which I am co-director, brings together disabled, Deaf and neurodivergent artists, designers and architects with normate and neurotypical architectural students, educators and practitioners to collaborate on ways of addressing disability more creatively across the built environment disciplines (Boys, 2018). Here, the focus is also on rethinking modes of practice, including ways of thinking, design methods, subject curricula and professional forms of organization. Ultimately, this is also about imagining alternative types of built space, beyond functional access and inclusion and toward what some disability activists call social, spatial and material justice (Nussbaum, 2006; Sins Invalid, 2019). To do this, we need to start asking different kinds of questions. As Walker writes:

> *In terms of discourse, research, and policy, the pathology paradigm asks, "What do we do about the problem of these people not being normal," whereas the*

> *neurodiversity paradigm asks, "What do we do about the problem of these people being oppressed, marginalized, and and/ or poorly served and poorly accommodated by the prevailing culture?"*
>
> **(2021, p. 7)**

Whether working in neuroscience, architecture, or across these disciplines, those of us who—like me—are non-disabled, neurotypical people have a responsibility for being part of this shift; by rethinking who gets valued in our society, by questioning how normality and mental competence are defined, by properly listening to the diverse experiences of neurodivergent and disabled people, and by working with those who are most marginalized in access to power and resources. Ultimately this requires moving beyond a focus on access and inclusion, or normalization and cure, and toward social, spatial and material justice.

References

Adams, J. (2016). *Jon Adams on autism, diagnosis and creativity*. National Autistic Society. 18th January https://www.autism.org.uk/advice-and-guidance/professional-practice/creativity.

Adams, A., & Chivers, S. (2021). *Deception and design: The rise of the dementia village. e-flux Architecture*. https://www.e-flux.com/architecture/treatment/410336/deception-and-design-the-rise-of-the-dementia-village/.

Adams, J. (2017). *Flow observatorium*. https://flowobsad.wixsite.com/flow-observatorium. (Accessed 25 August 2022).

Altermark, N. (2014). The ideology of neuroscience and intellectual disability: Reconstituting the 'disordered' brain. *Disability & Society, 29*(9), 1460–1472.

Baggs, A. (2007). *My language*. https://www.youtube.com/watch?v=JnylM1hI2jc.

Barnes, C. (2016). *The minority body: A theory of disability*. Oxford University Press.

Barnes, C., & Mercer, G. (Eds.). (1997). *Doing disability research* The Disability Press. https://disability-studies.leeds.ac.uk/publications/doing-disability-research/.

Barton, L., & Oliver, M. (Eds.). (1997). *Disability studies: Past present future* The Disability Press. https://disability-studies.leeds.ac.uk/publications/disability-studies-past-present-and-future/.

Bascom, J. (Ed.). (2012). *Loud hands: Autistic people, speaking* Autistic Self-Advocacy Network.

Beresford, P., & Russo, J. (Eds.). (2022). *The Routledge international handbook of mad studies* Routledge.

Bertilsdotter-Rosqvist, H., Anna Stenning, A., & Chown, N. (2020). *Neurodiversity studies: A critical paradigm*. Routledge.

Boue, S. (n.d.) https://www.soniaboue.co.uk/.

Boys, J. (2014). *Doing disability differently: An alternative handbook on architecture, dis/ability and designing for everyday life*. Routledge.

Boys, J. (2018). The DisOrdinary architecture project: A Handy guide for doing disability differently in architecture and Urban Design. In *The funambulist 19: The space of ableism*. August 31st https://thefunambulist.net/magazine/space-of-ableism/disordinary-architecture-project-handy-guide-disability-differently-architecture-urban-design-jos-boys.

Brown, L. X. Y. (2021). *Autistic hoya*. https://www.autistichoya.com/.
Caplan, A. L. (Ed.). (1992). *When medicine went mad: Bioethics and the holocaust* Humana Press.
Charlton, J. (2000). *Nothing about us without us: Disability oppression and empowerment*. University of California Press.
Chen, M. Y. (2012). *Animacies. Biopolitics, racial mattering and queer effect*. Duke University Press.
Clare, E. (2017). *Brilliant imperfection: Grappling with cure*. Duke University Press.
Crow, L. (2013). *Lying down anyhow: Disability and the rebel body (Online)*. http://www.roaring-girl.com/work/lying-anyhow-disability-rebel-body/. (Accessed 25 August 2022).
Cryle, P., & Stevens, E. (2017). *Normality: A critical genealogy*. University of Chicago Press.
Dexter, L. A. (1962). On the politics and sociology of stupidity in our society. In H. S. Becker (Ed.), *The other side: Perspectives on deviance* (pp. 37–49). New York: Free Press.
Dind, J. (2021). Review of: Bertilsdotter Rosqvist, Hanna, et al., editors *neurodiversity studies: A new critical paradigm*. Routledge, 2020. *Disabilty Studies Quarterly*, *41*(1 Winter). https://dsq-sds.org/article/view/7900/5902.
Disability Intersectionality Summit. (2018). *Access is love*. (online isabilityintersectionalitysummit.com/access-is-love/#:~:text=Access%20Is%20Love%20aims%20to,their%20every-day%20practices%20and%20lives).
Dokumaci, A. (2019). People as affordances: Building disability worlds through care intimacy. *Current Anthropology*, *61*, S000. https://doi.org/10.1086/705783.
Dolmage, J. T. (2017). *Academic ableism: Disability and higher education*. University of Michigan Press (Open Access at:https://www.fulcrum.org/epubs/nz8060437?locale=en#/6/2[Cover]!/4/2[cvi]/2/4/2[Dolmage_front-desc]/2/1:0).
Dychtwald, K. (1978). *Bodymind*. New York: Jove Publications.
Erevelles, N. (2011). *Disability and difference in global contexts: Enabling a transformative body politic*. Palgrave Macmillan.
Erevelles, N. (2014). Thinking with disability studies. *Disability Studies Quarterly*, *34* (2 online).
Fox, A., & Macpherson, H. (2015). *Inclusive arts practice and research. A critical manifesto*. Routledge.
Gadsby, H. (2022). *Ten steps to Nanette: A memoir situation*. Allen and Unwin.
Garland-Thomson, R. (2011). Misfits: A feminist materialist disability concept. *Hypathia*, *26*(3), 591–609.
Gibson, J. J. (1966). *The senses considered as perceptual systems*. George Allen and Unwin Ltd.
Gissen, D. (2009). *Subnature: Architecture's other environments*. Princeton Architectural Press.
Gissen, D. (2022). *The architecture of disability. Buildings, cities and landscape beyond access*. University of Minnesota Press.
Goering, S., & Klein, E. (2020). Neurotechnologies and Justice by, with, and for disabled People. In A. Cureton, & D. T. Wasserman (Eds.), *The Oxford handbook of philosophy and disability* Routledge.
Hamraie, A. (2017). *Building access. Universal design and the politics of disability*. University of Minnesota Press.
Hamraie, A. (2018). Mapping access: Digital humanities, disability justice, and socio-spatial practice. *American Quarterly*, *70*(3), 455–482.
Harper, C. L. (2021). *On vanishing: Mortality, dementia, and what it means to disappear*. Catapult Books.
Harris, B. (2017). Bedlam: The asylum and beyond. *Bulletin of the History of Medicine*, *91*, 434–435.
Heart n Soul. (2021). *A different kind of future with heart n soul*. YouTube video Wellcome Collection 5th March https://wellcomecollection.org/events/YCLmSxIAACUAY7Cv.
Higashida, N. (2021). *The reason I jump: One boy's voice from the silence of autism*. Sceptre.

Hillary, A. (2020). Neurodiversity and cross-cultural communication. In H. Bertilsdotter Rosqvist, N. Chown, & A. Stenning (Eds.), *Neurodiversity studies: A new paradigm* Routledge.

Ingram, R. A. (2016). Doing mad studies: Making (non)sense together. *Intersectionalities: A Global Journal of Social Work Analysis, Research, Polity, and Practice*, *5*(3), 11–17 (ISSN 1925-1270. Available at: <https://journals.library.mun.ca/ojs/index.php/IJ/article/view/1680>. Date accessed: 09 Jul. 2022).

Johnson, K., Walmsley, J., & Wolfe, M. (2010). *People with intellectual disabilities: Towards a good life?*. Policy Press.

Kafer, A. (2013). *Feminist, queer crip*. Indiana University Press.

Kapp, S. K. (Ed.). (2020). *Autistic community and the neurodiversity movement. Stories from the frontline* Palgrave Macmillan.

Karakas, T., & Yildez, D. (2020). Exploring the influence of the built environment on human experience through a neuroscience approach: A systematic review. *Frontiers of Architectural Research*, *9* (236e247).

Katz, E., & Ludins-Katz, F. (1990). *Art and disabilities. Establishing the creative arts center for people with disabilities*. Downloadable pdf from https://creativegrowth.org/symposium-resource-guide.

Kittay, E. F. (2009). The personal is philosophical is political: A philosopher and mother of a cognitively disabled person sends notes from the battlefield. *Metaphilosophy*, *40*(3–4), 606–627. Special Issue: cognitive disability and its challenges to moral philosophy.

Kittay, E. F. (2019). *Learning from my daughter: The value and care of disabled minds*. USA: Open University Press.

Kullman, K. (2019). Politics of dissensus in geographies of architecture: Testing equality at ed Roberts campus, Berkeley. *Transactions. Institute of British Geographers*. https://doi.org/10.1111/tran.12276.

Leadbitter, J. (2014). *Madlove: A designer asylum*. http://www.thevacuumcleaner.co.uk/madlove-a-designer-asylum/.

Leder, D. (1990). *The absent body*. The University of Chicago Press.

Liboiron, M., & Lepawsky, J. (2022). *Discard studies. Wasting, systems and power*. The MIT Press.

Manning, E. (2013). *Always more than one: Individuation's dance*. Duke University Press.

Mattern, S. (2021). Concealment and compassion. *Places Journal*. https://placesjournal.org/article/architectural-and-urban-design-for-dementia-care/#0.

McMahan, J. (1996). Cognitive disability, misfortune, and justice. *Philosophy & Public Affairs*, *25*, 3–35.

McMahan, J. (2003). *Ethics of killing: Problems at the margins of life*. Oxford University Press.

Meekosha, H., & Soldatic, K. (2011). Human rights and the global south: The case of disability. *Third World Quarterly*, *32*(8), 1383–1397.

Mingus, M. (2017). *Access intimacy, interdependence and disability justice*. (*Leaving Evidence* blog April 12th https://leavingevidence.wordpress.com/2017/04/12/access-intimacy-interdependence-and-disability-justice/).

Mingus, M. (2018). *Disability justice is simply another term for love*. (*Leaving Evidence* blog November 3rd https://leavingevidence.wordpress.com/2018/11/03/disability-justice-is-simply-another-term-for-love/).

Morris, J. (1991). *Pride against prejudice: Transforming attitudes to disability*. London: Women's Press.

Morris, J. (Ed.). (1996). *Encounters with strangers: Feminism and disability*. London: Women's Press.

Mostafa, M. (2018). Designing for autism: An aspectss™ post-occupancy evaluation of learning environments. *International Journal of Architectural Research*, *12*(3), 308.

Mostafa, M. (2021). *Autism design and architecture for all: Architecture for a differently abled world*. IQD.

Nussbaum, M. (2006). *Frontiers of justice: Disability, nationality, species membership*. Harvard University Press.

Pang, C. (2021). *Explaining humans: What science can teach us about life, love and relationships*. Penguin.

Patsavas, A. (2014). Recovering a cripistemology of pain: Leaky bodies, connective tissue, and feeling discourse. *Journal of Literary and Cultural Disability Studies*, *8*(2), 203–218. https://doi.org/10.3828/jlcds.2014.16.

Piepzna-Samarasinha, L. L. (2018). *Care work: Dreaming disability justice*. Arsenal Pulp Press.

Price, M. (2011). *Mad at school: Rhetorics of mental disability and academic life*. University of Michigan Press.

Price, M. (2014). The Bodymind problem and the possibilities of pain. *Hypathia*, *30*(1 Winter), 268–284 (Cambridge University Press).

Prince, D. (2010). The silence between: An autoethnographic examination of the language prejudice and its impact on the assessment of autistic and animal intelligence. *Disability Studies Quarterly*, *30*(1). https://dsq-sds.org/article/view/1055/1242.

Ray, S. J., & Sibara, J. (Eds.). (2017). *Disability studies and the environmental humanities: Toward an eco-crip theory*. Lincoln and London: University of Nebraska Press.

Renel, W., et al. (2020). The wall of change: Inclusive futures designed by people with learning disabilities and autistic *people*. In *D4H 2020 Proceedings*.

Rietveld, R., & Rietveld, E. (2018). *Affordances and architecture. e-Flux Architecture*. https://www.e-flux.com/architecture/superhumanity/179234/affordances-and-architecture/. (Accessed 25 August 2022).

Savarese, E. T., & Savarese, R. J. (2010). "The superior half of speaking": An introduction. *Disability Studies Quarterly*, *3*(1) (Autism and the Concept of Neurodiversity Accessed 17 April 2022).

Schalk, S. (2017). Critical disability studies as methodology. *Laterality*, *6*(1) (https://ia800708.us.archive.org/34/items/Lateral6-1/Sami%20Schalk%2C%20_Critical%20Disability%20Studies%20as%20Methodology_%20-%20Lateral.pdf).

Silberman, S. (2017). *Neurotribes: The legacy of autism and how to think smarter about people who think differently*. Allen and Unwin.

Sinclair, J. (2010). *Cultural commentary: Being autistic together. Disability Studies Quarterly*. https://dsq-sds.org/article/view/1075/1248. (Accessed 25 August 2022).

Singer, P. (1979). *Practical ethics*. Cambridge: Cambridge University Press (1980; second edition, 1993; third edition, 2011).

Sins Invalid. (2019). *Skin, tooth and bone: The basis of movement is our people: A disability justice primer*.

Taylor, S. (2017). *Beasts of burden. Animal and disability liberation*. The New Press.

Thom, J. (n.d.) Tourettes hero https://www.touretteshero.com/.

Thomas, C. (1999). *Female forms: Experiencing and understanding disability*. Open University Press.

Titchkosky, T. (2011). *The question of access: Disability, space, meaning*. University of Toronto Press.

Trotman, N. (2020). https://natashamtrotman.com/.

Walker, N. (2013). *Throw away the master's tools: Liberating ourselves from the pathology paradigm*. https://neuroqueer.com/throw-away-the-masters-tools/.

Walker, N. (2014). *Neurodiversity: Some basic terms & definitions*. https://neuroqueer.com/neurodiversity-terms-and-definitions/.

Walker, N. (2021). *Neuroqueer heresies: Notes on the neurodiversity paradigm, autistic empowerment, and postnormal possibilities*. Autonomous Press.

Walker, N., & Raymaker, D. M. (2021). Toward a Neuroqueer future: An interview with Nick Walker. *Autism in Adulthood, 3,* 1. https://www.liebertpub.com/doi/10.1089/aut.2020.29014.njw.

Williamson, B. (2012). Electric moms and quad drivers: People with disabilities buying, making and using Technology in Post-war America. *American Studies, 52,* 15–30 *(Excerpted in Boys, J, (2017) ed. Disability, Space, Architecture: A Reader. Routledge).*

Williamson, B. (2019). *Accessible America: A history of disability and design.* NYC Press.

Yergeau, M. (2018). *Authoring autism: On rhetoric and neurological queerness.* Duke University Press.

Yergeau, R., Elizabeth Brewer, M., Kerschbaum, S., Oswal, S., Price, M., Howes, F., et al. (2013). Multimodality in action: Disability and Kairotic spaces Kairos. *A Journal of Rhetoric, Technology, and Pedagogy* (18.1. Fall https://kairos.technorhetoric.net/18.1/coverweb/yergeau-et-al/).

CHAPTER FOUR

Autoethnographic reflections on architectural design for neurodivergence

Anthony Clarke*
BLOXAS, Melbourne, VIC, Australia
*Corresponding author: e-mail address: anthony@bloxas.com

Contents

1. Introduction

"Making sense of the world by way of design starts with the designer's self."[a]

Over the past several years, through my work as a registered architect, PhD candidate, and university educator, I have been consciously operating using auto-ethnography as a form of data collection, design interpretation, communication, and dissemination (also see Harder, this volume). It has been, and continues to be both incredibly challenging, and immensely rewarding, fundamentally re-articulating the way in which I practice, teach, and contribute to the field of design.

The real and hypothetical projects I have undertaken involve working with, and alongside individuals and groups with cognitive impairments and/or atypical sensory perceptions. These include, but are not limited to, ASD (Autism Spectrum Disorder), Dementia, ME/CFS (Myalgic Enchephalomyelitis/Chronic Fatigue Syndrome), MS (Multiple Sclerosis), and varying degrees

[a] Louise Schouwenberg, Michael Kaethler, "The Auto-ethnographic turn in design', Valiz, Amsterdam, 2021. P27.

Developments in Neuroethics and Bioethics, Volume 5
ISSN 2589-2959
https://doi.org/10.1016/bs.dnb.2022.05.001

of physical and mental trauma. I do not live with these conditions myself, however, and I attempt to only ask explorative and relevant questions, remain open minded and curious, and continue to challenge my own bias and privilege through a critical understanding of the power architecture and design has on people's lives. I do not claim that my architectural approach solves problems, only that such projects may offer another mode of communication, another method through which to experiment with built spaces in relation to neurodivergence. I do my best to stay open minded about both the opportunities and difficulties of spatial outcomes, and to take ownership of these as a participant in the built environment's evolution.

I will commence this chapter by introducing myself, both as a person, and as an architect. I will then introduce and briefly discuss two selected, built projects. The first of these is a completed residential project for a family with a young son on the Autism spectrum. The second is an experimental installation project looking into the effects of sound on those living with Dementia. Finally, I will conclude with a brief synopsis of my current academic research. The importance of including this research component, and in thinking of it like a "project" within itself, is that it offers time and space for in-depth analysis and reflection. While I believe architecture in its built form is capable of illustrating an "auto-ethnography" process, it can often be difficult to express the rapid re-direction of thought into design realization. The built result is just that, one element at the end of a long and complicated set of activities and thoughts. Therefore, understanding this through an academic research project can be incredibly revealing, and important to include in this chapter.

My architectural practice, BLOXAS, was established in 2010. Its philosophy is to value personal connections. However, this was initially attuned more towards landscape and place than humans. I revered local (Australian) designers including Richard Leplastrier,[b] Brit Andresen[c] and Esmond Dorney[d] who all foster(ed) a strong appreciation for the elements and derive(d) deep reference from culture.

In 2012, BLOXAS undertook a residential project whereby the attributes of a young boy with ASD occupied an integral part of the brief (Fig. 1). At the time, I was too inexperienced and naïve to understand the full intricacies beyond the visible and straightforward, however the

[b] Richard Leplastrier, "Spirit of Nature Wood Architecture Award", Wood in Culture Association, University of Michigan, 2004.

[c] Andresen O'Gorman, "Texts and Commentaries", UME, Queensland, 2011.

[d] Paddy Dorney, "Cusp: James Henry Edmond Dorney, Architect", Esmond Dorney Trust, 2022.

Fig. 1 A BLOXAS residential project. A strong formal gesture through a singular material composition allows this residential project to be easily recognized and understood by both its inhabitants and the suburban context. *Image credit: Peter Bennetts.*

unique nature of the project began to change the way I thought about the relevance of architecture, and my position within it. As the practice has continued to develop a specialized portfolio, and my personal interests and passions developed, I have valued the opportunity of formally exploring the ethical and emotional tensions architects such as myself may face, particularly when designing and delivering projects alongside "people of difference.[e]"

The process of reflection offers an inevitable connection to the past. Sensitivities that continue to develop through my role as an architect are inextricably linked to my upbringing, something the projects discussed here, and other specialized projects I am involved with increasingly make me cognisant of. I grew up in Tasmania. My parents divorced when I was six, and their calendar possession of me was common for the time. I lived with my mother during the week, and my father most weekends. I saw and experienced Tasmania differently on weekends: on the back of a motorcycle. My father would collect me from the end of a suburban driveway on a

[e] Kathleen Connellan, "Parresia. Between Principles and Practices in Design", Design Principles and Practices: An International Journal – Annual Review, Common Ground Publishing, Illinois, USA, 2013. p22.

Fig. 2 This image represents my childhood. I spent a lot of time with my father, on the back of a motorcycle camping around Tasmania.

bike loaded with camping gear. Summer, winter, rain, or shine I would squeeze between my dad and all the necessary paraphernalia for several days of travel (Fig. 2). I loved it. I experienced the elements in all their unrelenting detail, developing a thirst for what then represented freedom. We would meet and talk with strangers and have endless time in each other's silent presence. My father was difficult, sick. He had had a very bad road accident when he was a teenager and had been on copious medications ever since. He had epilepsy, depression, heart disease, recurring cancer, and multiple strokes. These culminated through a very volatile mood disorder. I grew up quickly, particularly on weekends. At twelve, I moved in with my dad full-time. This lasted several years before it became too difficult, and I was moved back with my mother. Those weekends, and years with my father gave me an appreciation for sensory driven experiences, and the value of submersing yourself in uncertainty. In retrospect, I believe it also gave me a strong degree of empathy, and the ability to engage with challenging expectations, particularly when carrying a sense of being grossly underqualified (imposter syndrome). In addition, these experiences provided important insights into the difficulties of both physical and mental health conditions, exposing the importance of the physical environment to individual needs, desires, and expectations.

I finished school at the end of year ten. I was fifteen. I was passionate about drawing; however following years of incessant bullying, I became unrelenting in my opposition to continue secondary education. My mother encouraged me to apply for a role as a drafting assistant at a local Hobart architecture practice—Bush, Parkes, Shugg and Moon (BPSM). I reveled in it. The butter paper sketches bulldog-clipped to walls of stainless-steel mesh, mounds of dust covered magazines, and a constant buzz of creative conversation that I was too young to appreciate. I worked during the day and studied to become a draftsman through evening Polytechnic courses. This refuge fostered and grew my passion for the visceral expression of drawing. It shaped and encouraged this interest while balancing it with a rapid appreciation for the power of built space, and its ability to either enhance or disable individual interactions.

At twenty-one I moved to Brisbane to commence studying architecture at the University of Queensland. After completing my undergraduate studies, I drove around the country, stopping in Western Australia to complete my bachelor's degree at the University of Western Australia. Upon completion, in 2005 I drove to Sydney and commenced my first full time position as a graduate of architecture. By 2008 I was living in Paris, and in 2009 Japan. The following year I returned to Australia and established BLOXAS.

These life experiences have inevitably fostered traits that have increased in relevance throughout the course of my professional life. These, I believe, are attributes and consequences of my particular upbringing that are now progressively central to the project examples outlined here. These characteristics have helped me understand the critical importance of the built environment, and its ability to act as both an enabler and/or disabler. Now, having the opportunity to work alongside those who experience the built environment in such an intense and astonishing way has offered me the opportunity to challenge the design tensions posed by architecture, and begin to find innovative ways of developing a more valuable, and empathic approach to design inclusivity. The projects briefly discussed below are just two such examples.

2. Hungry hands—Project 01

Quite often, individuals contacting my practice are amazed to discover that architects could be interested in listening and learning about their unique ways of experiencing the built environment, let alone designing for them. There is a regular embarrassment and restraint in the disclosure of neurodivergent information on the basis of judgments being made, or their difference being seen as problematic or incongruent to an architectural process or outcome. Frequently, designers are not fully informed of their client's unique requirements through fear of receiving a compromised, or medicalised outcome, or potentially being seen to impinge on a designer's recognizable style or brand. While this is incredibly unfortunate, its reality highlights the importance, and dire need, to question traditional client and architect relationships, and to begin to establish new methods of redress these imbalances.

This project represents an interesting example on the cusp of this dilemma. At the time of this project's inception, the clients were at the very beginning of understanding their journey with ASD. They required a more functional and suitable family home but were unsure what degree their son's recent diagnosis would, or should, impact their architectural brief. They were in the early process of collating articles discussing the positive aspects of color, nature, and play in design, however they found their relevance difficult to prioritize. Together, through open, honest, and at times incredibly vulnerable conversations, we were able to establish a curiosity focussed approach, continually sharing, and discussing relevant articles and experimental approaches to design.

One of the most valuable pieces of work carried out on this project was by an independent anthropologist. Prior to any design taking place, Carmen Cummings, an anthropologist with experience working with autistic children, was commissioned by BLOXAS to write a paper specifically for this project. The paper illuminated the importance of considering independence, self-esteem, safety, and well-being through the experiential themes of "a deeply sensory world," and "a mentally challenging world," focussing primarily on building for deeper sensory integration. It also worked to bring both the client and myself as the architect together, solidifying a tone of voice and language that could be shared and understood by all parties, equally and without hierarchy. Resulting from this article a series of broad questions could be asked which in turn began to establish project brief priorities.

How does architecture provide the need for someone to experience daily life through physical sensation? Sensory-seeking is an environmental feedback response often present within individuals on the Autism Spectrum whereby physical feelings are constantly being sought in order to stimulate a starved nervous system. The necessity for these active responses became a key component of this projects brief, and a key requirement for its clients.

Rigorous spatial planning and sequence of movement mapping was vital for success. The textural and sensory properties of building materials and a sense of hierarchy among building elements was essential. Consisting of eleven solid brick blades, this project guides its occupants through a careful sequence of spaces both internally and externally. Each brick blade is orientated to take full advantage of its context and associated environmental parameters. Each brick blade is detailed and expressed to be touched, climbed, and remembered over time. Where you walk, where you pause, where you sit, stand and work are all considered critical operations in the realization of how this project is read and more importantly, physically understood, and experienced (Fig. 3).

Although there are many key components of this project making it highly bespoke for a young boy with Autism and his family, there are three primary characteristics I would like to discuss here. These are materials, planning, and longevity.

2.1 Materials

27,000 Recycled bricks were used in this project, visually manifesting the necessity for self-regulation through offering a multitude of physical and sensory engagements. They were sorted into meticulous piles: those which were relatively smooth, those with moderate texture, significant texture, and those with extreme "crag" imperfections (Fig. 4). The bricklayer, Simon, was an old-school craftsman. He took pride in his work and understood

Fig. 3 An early project model cast of ink dyed concrete. This model was an early project representation focused on sensory exploration.

Fig. 4 A close-up detail of the higher textural qualities of the over 20,000 recycled bricks used throughout this project.

the critical importance of his material as not just a cladding for protection against the elements, but as a critical component in an occupant's personal education, safety, independence, self-esteem, sense memory, and overall well-being. The placement of these bricks, both in plan and elevation, allows a growing boy to utilize these walls as a way of alleviating internal pressure and anxiety, a constant source of relief that would continue to be there, reliably, predictably for many years to come (Figs. 5 and 6).

In addition to these bricks acting as two-dimensional (floor plan) regulators, it was also important for the project to work in the third dimension. Being able to climb the building in a safe and monitored way allowed the project to act as an apparatus, an additional object for play and activity. This was done through larger inclusions such as using stepping bricks and adding ladder rungs (for roof garden access), however it was also achieved through smaller, hidden details (Figs. 7 and 8).

Between each brick are mortar joints. These are both horizontal and vertical and can be seen in every brick or blockwork wall. There are various styles of traditional mortar detailing and finishing (concave, struck, weathered, flush, raked) to name a few. Generally, this component of brickwork construction is not given a great deal of consideration, since mortar simply acts as the connector, the spaces between the main material. In this project

Fig. 5 Rough bricks were strategically placed, offering opportunity for explicit levels of physical self-regulation. *Image credit: Trevor Mein.*

Fig. 6 Front project façade. This project is full of sensory focused materials, both internally and externally. *Image credit: Trevor Mein.*

Fig. 7 A detail of the higher textural qualities of recycled bricks. These bricks offer opportunities for the project facade to be scaled and climbed upon.

Fig. 8 This image illustrates ladder access onto a roof top vegetable garden. Vegetables can be winched down via a fabricated pulley system. *Image credit: Trevor Mein.*

however, they were seen as an opportunity to allow a more intimate possibility for climbing and interaction. Selected wall zones throughout the project were given deeper than normal (20 mm as opposed to 5–10 mm) mortar recesses. These deeper setbacks allowed small fingers to be embedded into the wall face just far enough to gain climbing purchase and allow these walls to be scaled until the young boy's weight and size prevented its continuance. Quite often these details were not disclosed but added as a form of discovery, continuing this project's narrative of being alive and ever changing.

Although these bricks establish the aesthetic precedent for the project, they work in collaboration with a myriad of additional sensory heightened materials. Pressed tin ceilings in the entry undercroft and library offer a refracted view of the ground plane below. Their fish scaled texture picks up every surrounding detail, glistening it back like an active mirror ball (Fig. 9). Glass blocks blur and pixelate the green landscape beyond, offering an

Fig. 9 Internally, the project has a combination of pressed tin, and timber ceilings. The pressed tin allows the external landscape to be gently reflected internally. *Image credit: Trevor Mein.*

Fig. 10 The material palette is simple but incredibly robust. Featuring concrete, bricks, and hoop-pine plywood. *Image credit: Trevor Mein.*

engaging and distorted view of the outside gardens from within the lower-level washroom (Fig. 10). Exposed plumbing pipework allows temperature to be felt, and things which are normally hidden to be exposed and understood (Fig. 11). The ground plane is covered with sensory provoking shrubs and plants, including Grevilleas, Banksia, pea shrubs, ferns, jasmine, aloe, and mushroom logs, to name a few. Grasscrete, provides a textured and layered experience under foot, while also establishing the importance of an auditory presence (Fig. 12). On the upper-level cork flooring adds warmth and softness. Timber windows and joinery offer a homely and familiar touch while also offering a strong sense of smell.

The material composition of this project has thus been carefully crafted to offer an intense, yet holistically calming environment. Individually, each material offers some form of sensory relief, together however these materials offer an environment full of possibility and change, ensuring the building, and its landscape is clearly understood, honest, and most importantly designed to help and not hinder (Figs. 13 and 14).

2.1.1 Planning

Independence, connection, privacy, choice, control: these were all critical considerations in the spatial configuration and design strategy for this project. For example, the kitchen is quite often described as the heart of the home. It's the place for haphazard discussion, and for coming together around food production and consumption. The kitchen can also act as an office, a classroom, a library, and a play space. The kitchen can offer unique

Fig. 11 The lower-level washroom uses exposed copper pipework presenting complete transparency in how the project operates and has been constructed. *Image credit: Trevor Mein.*

Fig. 12 Pressed tin ceilings continue through to the external spaces. Native sensory focused gardens were also a critical component to this project's compositional layout. *Image credit: Trevor Mein.*

Fig. 13 Building and landscape are of equal importance. With no formal front door, occupants are directed to the projects center, the kitchen and living spaces. *Image credit: Trevor Mein.*

Fig. 14 A combination of textural external ground coverings offers a heightened visual and oral awareness. *Image credit: Trevor Mein.*

connections and vantage points between itself and other surrounding activities, both internally and externally. The kitchen can become the linchpin around which everything revolves. It can signify time through smell, sound, and light. In this particular project the kitchen is not a destination, but very much a connector. It can be approached, and engaged with from varying sides and angles, visually penetrating to the far corners of the site. Its material composition is highly textured and robust dragging the external conditions into the internal, disrupting any simplistic divisions into functional zones or categories (Fig. 15).

There is no front door, and there are no hallways. There is no hierarchy between indoor and outdoor spaces, and there is no traditional or pre-determined procedure for how this home is activated. Each space has been designed to blur into the next, allowing occupation and family connection to be of primary importance. Availability is always offered. In contrast to this however, moments of individual repose, quiet, and respite are carefully incorporated. Specifically located window seats are positioned to offer

Fig. 15 The kitchen space offers warmth and muted tones. *Image credit: Trevor Mein.*

different and changing perspectives on the nearby and surrounding occupation, taking full advantage of the morning, afternoon, and evening light. A sunken lounge provides physical separation from thoroughfare activities while still offering auditory and visual connection. Steps, and varying height bench space provide areas for instantaneous play and activity. A separate workshop to the North, and studio space to the South offer additional levels of full separation. This is a place to read or craft, or think, while never being too far from the home's hub, the kitchen (Figs. 16 and 17).

While the lower level is robust and open plan, with a strong connection to the external landscape, the upper level provides distinct private zones to retreat, day or night. The upper level is warm and calming with additional play and study areas attached to both the private and public zones. It offers

Fig. 16 Areas to sit, work, and play both individually and collectively are an important component in this project's overall design. *Image credit: Trevor Mein.*

Fig. 17 Natural materials frame moments of family connection and integration. *Image credit: Trevor Mein.*

panoramic views of the surrounding neighborhood, and a distinct appreciation for the landscape, and roof top vegetable garden, of which it forms a part. Importantly, this project is small in scale. It considers the garden, and built form, to be of equal value. No space is under utilized or ill considered. Opportunity for family connection is always the main priority (Fig. 18).

2.1.2 Longevity

During the Covid-19 pandemic lockdowns in Melbourne during 2020–21, this project's clients and I made regular contact. With home learning replacing in-person (i.e., at school) education, such a specific and tailored home like this became increasingly important. I was excited and humbled to learn that its intentions were being fully celebrated. The project offered a sanctuary, a place to feel safe, and an environment where sensory seeking,

Fig. 18 Materials on the upper level are kept simple and calming. Cork flooring offers warmth, texture, and softness whilst timber windows frame important views beyond. *Image credit: Trevor Mein.*

and all it entails, could be harnessed, and encouraged. The hope is for this project to aid in an individual's educational journey, one that can provide security in the fact that the built environment can show visual signs of caring. The project contributes to a building of self-esteem and trust.

3. Sound field—Project 02

Sound, as we know, is made up of vibrations and has the ability to propagate through a number of different mediums. Sound can be perceived and analyzed through a series of extremely complex states. How each of our brains receive sound (through pressure and time) offers diverse sensations and interpretations.

This project was designed to establish a new series of questions in relation to sound and dementia, not to simply answer existing ones, or re-define a past understanding. This project has been designed to challenge our phonic assumptions, looking at frequency, pitches, and the reaction of these on certain carefully selected materials. In so doing, it aims to help bridge the gap between civic and product design for dementia, creating human scale experimental and experiential sensory place-making. Given the sensory experience of living with dementia, this project aims to design spaces around sound, to see what we can learn and how these principles can form a broader topic for conversation.

In 2016 I flew to Essen, Germany on behalf of RMIT University to attend a global workshop called "Dementia Lab.[f]" Designers of all disciplines, from all over the world attended to discuss projects, research, current and future trends along with funding opportunities and experiential approaches to Dementia care. Upon returning to Australia, I was excited by the international knowledge and passionate connections I had gained and wanted to translate these into an architectural project. Unfortunately, at the time I was not met with the same level of eagerness from either the university or many of the industry corporations promoting their apparent leadership in experimental or curiosity focused research. So, with a small amount of seed funding (thank you to Prof. Richard Fleming from Dementia Training Australia), and an ambitious, and very well received crowd sourcing campaign (established by BLOXAS) we were able to instigate and deliver (alongside Dementia Australia and Arup acoustic engineers) an innovative installation project, discussing not only how those with Dementia perceive sound, but how this project could act as an empathy tool. This concept (Fig. 19) was intended to become a way for everyone to understand their actions within the built environment and to reflect on how these actions quite often have unseen, but critical and powerful reactions towards others around them.

Sound Field is comprised of twenty-four rotatable panels. Each of these panels are made up of different and specifically selected materials, some smooth and transparent, some semi-transparent and mildly textured, others completely opaque, and either sound absorbing or highly reflective. The reason for this spectrum of materials was to understand how their innate properties could be better understood when applied to questions around Dementia care and the built environment. It was also an important factor

[f] http://www.dementialabconference.com.

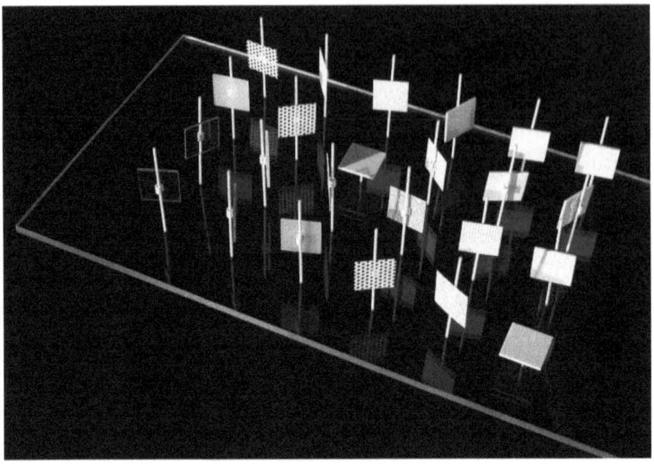

Fig. 19 Initial physical model for Sound Field using plastic and transparent Perspex sheeting.

for these materials to be shades of white or gray. This would enhance the focus of the project on sound, and not allow visual distraction to play a part (Figs. 20 and 21). In addition to these materials the installation relays a series of everyday sound recordings, from within the home and public spaces. These effects include children playing, trains arriving on a platform, sounds of a supermarket checkout, and a domestic kettle boiling. The individual "trees" work as a collective sound field to rebound and reflect these sounds. Visitors are encouraged to rotate the sound panels and experience how our daily aural environment, which goes unnoticed by most of us, can become distressing for people with dementia.

Arup's acousticians worked closely with BLOXAS to design the different panels within the sound field, and their proximity to one another. Each panel was placed 2.4 m apart. This allowed each sound to be projected, and reflected, with minimal transferred sound loss, while also ensuring the installation could be experienced by those using wheelchairs or mobility aids. Several of the panels included small hidden surface transducers (electronic devices that convert energy from one form to another) effectively turning those selected materials into loudspeakers—broadcasting selected sounds in all directions. Other panels incorporated integrated loudspeakers (Holisonics) that use a unique ultra-high frequency technology to create a very thin beam of sound. This can be "bounced" around from surfaces like a ball on a pool table (Figs. 22 and 23). The moveable panels allow visitors to direct sounds in different ways, creating ever-changing sound fields that

Fig. 20 Sound Field works as a forest of sounds. Each panel allows for 360-degree rotation. *Image credit: Coco and Maximilion.*

appear to come from multiple locations. This effect adds to the level of confusion and disorientation those living with Dementia quite often experience.

This project was originally conceived to be experienced by small groups, 3–5 individuals at a time. It was thought that by making the installation more controlled and pre-determined, people would be able to appreciate and understand it's intended effects. Interestingly however it was not until more than 15–20 individuals were within the space at any one time that it's intended value became clear. This was less about the installation itself, and much more about the interaction between the occupants, that was in part configured by the installation. Sound being invisible allowed people

Fig. 21 Sounds are dispersed and flown around between each panel. This offers a unique empathic opportunity and realization between users. *Image credit: Coco and Maximilion.*

to physically see the effect they were having on someone on the other side of the room. We were able to witness people apologizing to one another and rectifying their position of a panel in order to make someone else's experience calmer. In addition to the above realization, one of the other incredible outcomes of this project (besides it's process of design and construction) was how it was received within the community of dementia carers, healthcare professionals, friends and loved ones of those with Dementia. This project formed a critical component within the 17th Alzheimer's Australia Biennial National Dementia Conference "Be the change.[g]" Throughout this four-day conference we were extremely lucky to capture a series of user responses, including the following:

[g] 17th Alzheimer's Australia Biennial National Dementia Conference. 17–20 October 2017, Melbourne, Australia. https://www.dementia.org.au/conferences/alzheimers-australia-2017.

Fig. 22 Sound Field uses a series of vertical pivoting elements to test how sound interacts with varying hard and soft materials. *Image credit: Coco and Maximilion.*

Fig. 23 These sounds come from both "Holosonic" and "Transducer" speakers embedded within each frame. Their sound is manipulated by the turning and maneuvring of each panel. *Image credit: Coco and Maximilion.*

> *"The concept for sound as an important part of dementia friendly design has been overlooked for so long. Well done on creating this masterpiece—looking at both the positive and overwhelming spectrums of audiology. A beautiful concept integrating science, art, and health."*
> *"Fantastic, Insightful, Incredible. After a while I wanted to leave the room. How does one 'leave' behind dementia. Far more difficult to do."*
> *"This is what I experience day to day. It can be very uncomfortable at times."*
> *"Now I know how my mother feels, thank you."*

Following on from the Melbourne exhibition (2017), *Sound Field* was subsequently re-designed and re-realized at the 2019 Enabling Festival in Singapore. This second design iteration allowed the team to focus less on material properties and more on targeting the reflection and refraction of sound. The installation was also made monochrome highlighting the singular importance of the aural experience. Over 9900 individuals experienced this installation within a four-week period. It was incredibly well received once again and has continued to be an insightful research source for subsequent projects within my practice involving material properties, sound, and in-particular empathy (Figs. 24 and 25).

Fig. 24 Sound Field 2 was re-designed and re-built to form a critical component of Singapore's 2019 International Enabling Festival.

Fig. 25 Each "tree" was sculpted into a specific abstract form. These forms worked as a collective to illustrate how sound is reflected and refracted depending on their angle and location.

4. On "vulnerable" practices—Project 03/summary

"Auto-ethnographic design is both a writing and a material genre."[h]

First and foremost, I am a practicing architect. My practice continues to work on a diverse range of project scales from residential, to experimental installations, to larger scale community mental health projects. Each of these projects involves participating alongside individuals, and/or groups with varying levels of lived experience, primarily established through forms of trauma. Until relatively recently I considered my "designer" role quite separate from my personal self. I attempted to keep my own personal emotions, difficulties, and mental challenges separate to the processes I navigated in design. That was until I came across Auto-ethnography, a methodology derived from the social sciences, in-particular anthropology. As I began to consciously facilitate this approach, I realized separating these identities was impossible. At first, I found the concept of engaging with myself incredibly challenging and struggled to see its value within the more standard

[h] Louise Schouwenberg, Michael Kaethler, "The Auto-ethnographic turn in design', Valiz, Amsterdam, 2021. P76.

approach to architecture, the building of buildings. Being an architect, to me, had always been about the outcome. The architectural profession traditionally rewards the finished result, the stoicism, the professionalism, and to some degree the ego. How then can becoming vulnerable and expressing this through a design process be a worthwhile venture? Through my recent academic research and curiosity, I have been challenging the role of the architect—my role—to appreciate the benefits of reflecting on, and working with, the emotive dimensions of design and the design process. What is an architect's responsibility, both as a professional, and as a person? What makes an architect responsible? How do we understand our unique position within each commission we take on? Can the process of architecture be seen as valuable, or even more valuable, than any perceived outcome, and how can this be evaluated? How can architectural education be re-configured to accept this methodology, and what effects might this have on designing for neurodivergence?

The two "built" projects I discuss in this chapter were designed and constructed between 2016 and 2018. Since that time, I have been able to reflect on my previous processes, and contrast these with my current methodologies. While it will never be known if the outcomes would have remained similar, I can acknowledge that the questions I would have asked myself, and the way I would have approached these commissions would most certainly be different.

While these projects begin to highlight how direct personal experiences can be harnessed, and accepted, through thoughtful and empathic questioning of the built environment, they have only offered concepts through small, bespoke examples. So, how might larger, more public focused examples with multiple stakeholders benefit from this way of thinking and working? Is it possible to generalize and still create a responsible piece of architecture, or is it not about the architecture at all, and is much more about the mindset of the architect or designer that can have the most powerful impact?

Each year, I have the fortunate opportunity to conjure a new (hypothetical or real) design project for my university masters' students. Due to my own passion and curiosity these are exclusively focused towards designing for neurodivergence and trauma. Each semester I am both positively and negatively astounded. Positively, for the incredibly ambitious project outcomes the students produce, and their overall willingness and eagerness to engage with complex and challenging conversations. Negatively, for the fact that this willingness is most often driven by its perceived newness. That up until this point in their design education most students have no real understanding or appreciation for the people they are designing for, or the power they wield in being future shapers of the built environment. I would therefor conclude by suggesting that while architecture does indeed play a pivotal role in neurodivergence, a reflective ethics approach and designer vulnerability is equally as critical and deserving of re-evaluation.

Further reading

Connellan, K. (2013). Parresia. In *Between principles and practices in design, design principles and practices: An international journal – Annual review*. Illinois, USA: Common Ground Publishing.

Ellis, C. (2004). *The ethnographic I: A methodological novel about autoethnography*. United Kingdom: AltaMira Press.

Grandin, T., & Panek, R. (2014). *The autistic brain: Exploring the strength of a different kind of mind*. United Kingdom: Ebury Publishing.

James, O. (2008). *Contented dementia*. Ebury Publishing.

Pang, C. (2020). *Explaining humans: What science can teach us about life*. Penguin Random House: Love and Relationships. London.

Schouwenberg, L., & Kaethler, M. (Eds.). (2021). *The auto-ethnographic Tun in design*. Amsterdam: Valiz Publishing.

SECTION 2

Advocacy

CHAPTER FIVE

Cultural commentary: Being autistic together☆

Jim Sinclair*
Autism Network International (ANI), Syracuse, NY, United States
*Corresponding author: e-mail address: jisincla@syr.edu

Contents

1. Introduction

Autistic people are generally seen as lacking in ability to share common interests with others, disconnected from social participation and fellowship, and inaccessible to social transmission of behaviors and attitudes.

☆ This chapter presents an extract from a longer version that appeared in Sinclair, J. (2010). Cultural commentary: being autistic together. *Disability Studies Quarterly, 30*(1). https://dsq-sds.org/article/view/1075/1248. Printed with permission.

Developments in Neuroethics and Bioethics, Volume 5
ISSN 2589-2959
https://doi.org/10.1016/bs.dnb.2022.08.002

These are core aspects of what has been described as autistic "aloneness," "withdrawal," and "disconnectedness," autistic people "living in their own worlds," being "trapped" inside "shells" or behind "invisible walls," and many similar terms used by neurotypical (NT) people to describe their perception that autistic people are unable to be "together" with other people. (For examples see Claiborne Park, 1968; Frith, 1992; Kaufman, 1995; Maurice, 1994; Tustin, 1990).

For autistic people the situation is not so clear-cut. Some of us, at some times in our lives (generally early childhood), really aren't aware of other people or their activities or their attempts to connect with us. Others of us are aware of the people around us, but find them difficult or impossible to understand. Some of us have sensory sensitivities that make being around other people distressing to us, either because of the people themselves (unwanted touching, loud talking, perfumes and other overwhelming scents), or because of sensory features of the environments in which we encounter other people. Some of us aren't interested in the same things the people around us are interested in, and we don't have a natural impulse to take interest in things just because we observe that other people find them interesting. Some of us are keenly interested in other people and their interests and activities, but we don't know how to engage with other people around those shared interests.

By the time we reach adulthood, autistic people's experience of "togetherness" has likely consisted of some combination of: being intruded on by other people wanting us to engage with them, when we don't share that desire; being interested and curious about other people, but finding them confusing and overwhelming to be around; trying to engage with other people, and having frustrating and unsuccessful encounters; managing to engage "successfully" with other people, and finding ourselves drained and possibly even damaged as a result of what we had to do to "succeed."

After a lifetime of such distressing and discouraging experiences, it would be little wonder if autistic people did choose to withdraw from human contact. Yet during the past 20 years, facilitated in large part by the availability of the Internet, more and more autistic people have been reaching out and forming connections, creating community, and discovering our own styles of autistic togetherness.

As co-founder of Autism Network International (ANI, http://www.ani.ac) and as its coordinator since its inception in 1992, I have been blessed with the opportunity to witness much of this unfolding. To the best of my knowledge, ANI was the first autistic community to be created naturalistically by autistic people, and it remains the largest autistic-run organization to have regular physical gatherings of autistic people. I am primarily in a position to report about the culture that has developed within this particular autistic community. Other autistic communities, founded and populated by other "flavors" of autistics, may have very different cultures. I am more familiar with some of these than with others. Autistic connections have grown and expanded greatly since ANI was founded. There is now more autistic communicating and networking and socializing going on than any one person can keep track of.

2. Being autistic in one's own space

Autistic people who are fortunate enough to have their own private spaces, especially adults who have their own homes but to some extent also people living in shared homes who have their own rooms, have a refuge from many of the stresses of being in NT space. People in their own personal spaces have a great deal of control over what happens in those spaces. Within the limits of laws, zoning ordinances, and any rules set by landlords, people are free to arrange the environments the way they want them. I'm free to paint all my walls purple, replace eye-stabbing bright lights with low-wattage bulbs, play the same music over and over and over and over and over and over and over again (provided I don't play it so loudly that it disturbs my neighbors), and play with "age inappropriate" action figures, because it's my own home and nobody else gets to come in and dictate what I do in it. In fact with very few exceptions, nobody else gets to come in at all, unless I choose to let them in. Having one's own space means having control over the space itself, over what one does in the space, and over who has access to the space. There are minimal limits on what one may do in one's own space. One can stim, tic, perseverate, stay up all night, sleep all day, wear the same clothes for days at a time, or wear no clothes at all—one can do any number of things that most other people would find annoying or even disgusting, and if there's anyone else around to notice, it's because there's a mutual agreement and a shared comfort level in having that person in one's space.

But with that increased freedom in one's own space comes decreased external structure. Many autistic people have considerable difficulty managing routine self-care and maintenance functions without someone to remind and assist us. We can easily find our spaces and our lives devolving into chaos if we live completely on our own, with no one else coming into our space (Smith, n.d.-a; Zaks, 2006).

Being autistic in NT space and being autistic in one's own space both involve being "the only one"—the only autistic person in the environment. This is one major way in which being autistic in shared autistic space differs from being autistic anywhere else.

3. Autistic space vs. places for autistics

Some autistics have experienced situations in which autistic people are in the majority, but NT people are still in charge of creating structure and setting the agenda. Special education classrooms, sheltered workshops, day hab programs, residential facilities, specialized group therapy programs, and "special" leisure/recreational programs for autistic people are all examples of environments that NT people create for autistic people, according to NT people's perceptions of what autistic people need.

In the debate between proponents of inclusion and proponents of separate programs for autistic people, arguments in favor of special programs emphasize the benefits of having staff who are knowledgeable about the needs of autistic students/clients, and of having contact with autistic peers instead of being the lone autistic in a group of NTs (Broun, 1996; Sinclair, 1998). These can indeed be important benefits, regardless of who is directing and staffing the program. There are some NT educators and service providers who have good understanding of autistic people's needs and processing styles, and who provide excellent environments in which autistic people can learn, work, develop skills, and have a good time.

There are also some ignorant, misinformed, and misguided educators and service providers who create environments which can be very unpleasant and even harmful for autistic people (particularly for people who are "placed" in them involuntarily without being allowed a developmentally appropriate degree of choice) (Baggs, 2006a; Judd, 2009; Shapiro, 1991). Autistic processing is devalued and autistic functioning is undermined in these kinds of environments. At the same time people in such settings are struggling with those environmental impediments, they are surrounded by similarly stressed autistic peers who are suffering the effects of the same undermining and devaluation, and are reacting accordingly. People who experience such situations may develop negative associations about autism and about being in the company of other autistic people.

Whether good or bad, environments created and run by NT people are not autistic spaces, even if the majority of people within them are autistic.

A good NT teacher, therapist, job coach, life skills trainer, or other service provider can certainly create an environment in which autistic people have positive experiences and learn useful skills. But even such positive and helpful environments still have characteristics of "NT spaces," because NT people are in charge and NTs are making the rules. The very fact that NTs are creating and managing a program or a service, for the benefit of autistic participants, conveys the perception that autistic people are helpless and dependent on NTs to take care of us.

In a shared autistic space, autistic people are in charge. Autistic people determine what our needs are, and autistic people make the decisions about how to go about getting our needs met (Schwarz, 1999; Shapiro, 1991).

4. Being autistic together

Autistic spaces, whether small informal gatherings at people's homes, online discussion forums, or major autistic-run events like Autreat, are meant to provide autistic people with the benefits of contact and participation with other people, while accommodating autistic needs that can make NT spaces so difficult.

Being autistic in shared autistic space may be easier than being autistic in NT space or in one's own personal space—or it may be harder. People sometimes come to autistic space expecting a kind of utopia where there will be no interpersonal or environmental difficulties. In actuality it simply has some *different* difficulties.

One frequent difficulty that autistic people may have in their initial encounters with autistic space is that things are unfamiliar. Being in NT space or in one's own space may present difficulties, but by the time we're adults, we've gotten used to living with those difficulties. Most of us have developed strategies to help us function to some extent. People often arrive at a shared autistic space not knowing what to expect, or, worse, having expectations that don't match the reality. Our familiar coping strategies, which serve us well in NT space, may not work so well among other autistic people. They may even backfire and be received negatively.

5. Why autistic space is different from NT space

The reason autistic space is different from NT space is not merely that autistic people are different from NT people. Autistic people are also very different from *each other* (Spicer, 1998). In just about any area one can think of where humans vary, the range of variability among autistic people seems to be even *greater* than the range among NT people:

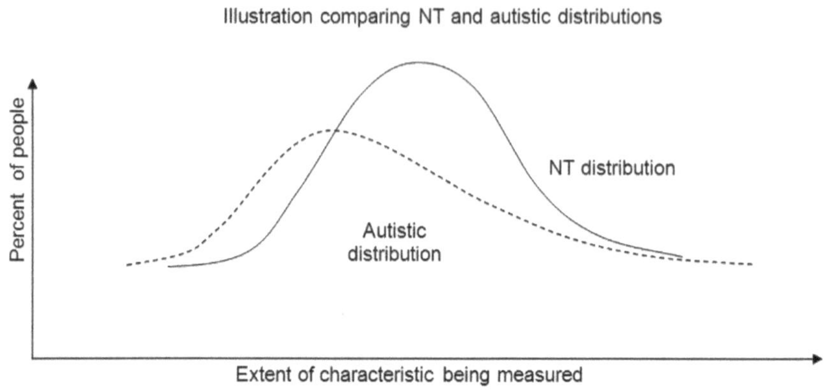

Sensory sensitivity and social interest may be the two best examples of this variability. The common stereotypes are that autistic people, across the board, are sensory defensive and are uninterested in social contact. If these stereotypes were accurate, it would be relatively simple to create comfortable autistic spaces. We would just have to develop customs and rules and environmental modifications so that there would be less sensory stimulation and less interpersonal contact than are the norms in NT environments.

The reality is that there are many sensory-defensive autistics who are easily overwhelmed by typical levels of stimulation in NT environments. But there are also low-registering autistic people who tolerate—even *need*—more intense stimulation than the most sensory-seeking NTs would find comfortable.

Moreover, "sensory defensiveness" is not uniform across sensory modalities. It is possible, for example, for someone to be easily overwhelmed by auditory stimuli but to seek out intense visual stimulation, or to be extremely tactile defensive but crave (and also create) a lot of loud sounds, or to avoid some types or ranges of visual/auditory/tactile/olfactory/gustatory stimuli while seeking out other types or ranges of stimuli, or any number of other combinations of sensory defensiveness and sensation-seeking within the same person.

This extreme autistic variability presents challenges in accommodating sensory needs in autistic space. The kinds of sensory stimuli that are hurtful to some autistic people may be necessary for others.

Similarly, autistic people have in common (by definition, as part of the diagnostic criteria for autism) abnormalities or "impairments" in communication and social interaction. But we're not all abnormal in the same ways.

Our *ways* of interacting are different from NT norms—and also different from each other's!—and the *amount* of interaction we want also varies across an even wider range than it varies among NTs. We range from not wanting any social contact at all (which is the common stereotype, but rarely the reality), to wanting some social contact but needing more down time than most NTs (even introverted ones) need, to wanting an NT-normal amount of contact, to wanting or needing *constant* contact, even more than "normal" extraverted NTs can tolerate.

Creating comfortable autistic social space is not a simple matter of removing NT social pressures and expecting that autistic people will naturally leave each other alone. Even with no NT influences at all, some autistic people will seek out social connections in ways that feel intrusive to other autistic people, and some autistic people will assert their need for personal space in ways that feel rejecting to other autistic people. And given that we also differ from each other, at least as much as we differ from NTs, in our verbal and nonverbal communication styles, we are not automatically able to guess each other's preferences and needs.

6. Virtual vs physical spaces

It is no accident that the concept of "autistic space" and the roots of autistic community originated in cyberspace. For those of us who are comfortable with written communication, online contact offers the opportunity to meet and interact with other people while remaining at home in our familiar environments. We don't have to go out into noisy crowded public spaces and risk uncontrollable sensory assaults. We can take in and respond to communications at our own pace, instead of having to keep up with realtime conversation. We can participate in online discussions with large groups of people, and since we're only reading one email at a time, we can keep track of who's saying what. We can catch everyone's input if we want to, or choose which ones we want to read and which we want to filter out. We don't miss out on what one person is saying while we're listening to someone else. It's fairly easy online to filter out input from people we find uninteresting or unpleasant to hear from. And many of the communication pitfalls we encounter in face-to-face communication with NTs—being expected to understand nonverbal signals, having people try to read meaning into our own appearance and nonverbal behavior—don't exist in text-only communication. (It still boggles my mind when NT people complain about this as a *drawback* of online communication!)

Gathering together with other people in physical space, even at events such as Autreat which are carefully planned to accommodate autistic people, cannot help but include some of the difficulties and challenges that online communication avoids. Leaving home to attend a physical gathering means experiencing the stress of travel, followed by the stress of being in an unfamiliar environment over which we have much less control than we have over our personal spaces at home. Being around other people inevitably means being exposed to the behaviors and sensory stimuli that other people produce. Being in a shared space means being required to modify our own behavior to accommodate other people's needs. Being around a lot of other people at once means it is impossible to attend to what everyone is saying without missing anything. Attending a "community" event such as Autreat (as opposed to one's own private event where one can hand-pick who is invited) means possibly encountering people one doesn't know, doesn't understand, and/or doesn't like. In terms of interpretations (and possible misinterpretations) of nonverbal behavior, autistic people vary as much in our nonverbal communication as we do in other regards. Misunderstandings based on appearance and nonverbal behavior can be minimized through awareness of our differences, but autistic people are not immune to misinterpreting each other.

Leaving our familiar personal spaces to come together in autistic physical space is not always easy or comfortable. Those of us who make the effort to come together physically (at least those who keep coming back) do it not because it's easy and problem-free, but because the benefits we get from it are worth the difficulties and the discomforts.

Many of the characteristics of autistic physical space can be experienced as either benefits or hardships, depending on one's point of view. Challenges can become opportunities if one is prepared to approach them as exercises in self-discovery.

7. Challenges and opportunities

7.1 Autistic similarities

Some autistic people have written moving, dramatic accounts of immediately feeling "at home" among other autistics, having a natural sense of "belonging," and recognizing other autistics as "their own kind" of people (Cohen, 2006; French, 1993; Williams, 1994).

My own words to describe the 1992 visit with Xenia and Donna during which ANI was founded were "feeling that, after a life spent among aliens,

I had met someone who came from the same planet as me." This "same planet" metaphor, along with metaphors about "speaking the same language" or "belonging to the same tribe," are very common descriptions used by autistic people who have had this experience of autistic space. One participant at the first Autreat in 1996 summed it up saying, "I feel as if I'm home, among my own people, for the first time. I never knew what this was until now."

For autistic people who have this kind of autistic-space experience, it is obviously a powerful and often life-changing experience. But it can also be profoundly disconcerting. Growing up autistic among NTs can lead to having a sense of identity and self-image formed around being "special"—the *only* person to have some particular autistic characteristic, the *only* person who has to struggle with some particular hardship, the *only* autistic person who has managed to achieve some particular accomplishment despite one's disability. NT society is inclined to treat disabled people with excessive pity for our difficulties and excessive admiration for our strengths, infusing our daily lives and routine activities with a melodramatic sense of tragedy over our problems and amazement over our successes (McNabb, 2001).

This feeling of "specialness" can be jarringly upset by encountering other autistic people who have a similar mix of autistic traits, have similar (or even worse) problems, and have attained similar (or even higher) achievements. In a genuine peer group, which most autistic people have never experienced, each person is *unique*, but no one is more "special" than anyone else.

The sense of "homecoming" and "belonging" among fellow autistics can also set people up to have idealized expectations of these newfound autistic compatriots. This is characteristic of the "immersion" stage of minority identity development (Helms, 1995), in which people embrace and idealize their own minority communities to the exclusion of other kinds of people. Inevitably this idealization is eventually followed by disappointment and disillusionment, when other members of one's own group turn out to have the same failings as the rest of the human race.

7.2 Autistic differences

The great diversity among autistic people means that while some people have these powerful autistic-space experiences of encountering people "like ourselves" for the first time ever, other autistic people do not. People who come to an autistic space expecting that everyone will like them and everyone will be "like them," everyone will understand them, and everyone will be easy for them to understand and get along with, are likely to be disappointed. If the autistic people in that particular group happen to be very different from them, they may end up feeling even *more* alienated and out of place than they feel in familiar NT environments. This can lead some people to question their own status ("If I'm so different from these people, maybe I'm not autistic after all"), or to doubt other people's status ("You're not like me, so you're not really autistic and you have no business calling yourselves autistic"), or to devalue some subgroups of autistic people relative to others ("Autistic people who are like me are better than autistic people like the ones in this group who are making me uncomfortable").

Some autistic groups try to minimize such discomforts by limiting participation to only certain "types" of autistic people. I've heard of incidents in which autistic people were excluded from groups for being "too low-functioning," not verbal enough, or otherwise not conforming to that group's idea of the sort of autistic people it wanted as members (Smith, n.d.-b; Baggs, 2006b).

ANI has taken a different approach. While our customs and our culture are heavily influenced by the shared characteristics that drew our early members to each other and enabled us to work together on building the infrastructure for community, we have always done our best to welcome and accommodate the widest possible range of autistic diversity (Autism Network International, n.d.-a; Autism Network International, 2009). (This does not mean that we'd never exclude any autistic person for any reason. People can be, and occasionally have been, excluded from participation in ANI functions—on the basis of behavior that victimizes other people, not merely for being too different or too severely disabled. See Autreat Planning Committee, n.d.)

The real "magic" of inclusive autistic spaces, such as we strive for in ANI, is not that every autistic person can automatically expect to find other people who are like him or her. The real "magic" is that *almost* every autistic person—everyone who is able to participate without violating other people's boundaries—can expect to be accepted for who he or she is. Some typical descriptions of this experience are:

> *"This was the first place I wasn't criticized for being different."*
> *"It showed me that being me was okay, and that my ways of doing things weren't 'wrong' or 'defective,' just different, and perfectly all right."*
> *"For the first time, I had my less than 'normal' attempts at communication recognized, and also accepted."*
> *"We loved the feeling of being accepted and liked for being ourselves!"*

Note that being accepted for who one is does *not* require being surrounded by people who are just like oneself. This is reflected in one participant's reflection on the first Autreat: "Here people who could paint and draw equally shared experiences with those who can't hold a pencil or a brush. People who are very articulate equally shared experiences and understood those who could only jump or clap their hands or point to letters on a letter board or picture board to respond to a question."

The associated challenge is that one must also be prepared to accept *other* people for who *they* are: It's okay for you to be weird, which is liberating, but it's also okay for other people to be weird. That can be startling and disturbing if their weirdnesses are very different from yours.

8. Spontaneous interaction

Autistic people may be less likely than NTs to spontaneously initiate interactions with anyone who happens to be nearby. Common complaints NTs have about autistics are that we seem unfriendly and standoffish and we don't express interest in other people (Aston, n.d.; Foden, 2008; Meyerding, 1998). Common complaints autistics have about NTs are that they seem pushy and demanding and they waste conversational energy on meaningless small talk (Adam, 2009; Muskie, 2002; Sinclair, 1995; WrongPlanet.net, 2008).

At autistic gatherings such as Autreat, there can be a tremendous amount of communication and interaction once people find a common point of interest. (Autreat presenters need to be prepared to limit questions and comments from the audience, or else be willing to accept the fact that they may get to cover only a fraction of their prepared material, because autistic participants get *extremely* participatory when our interest is engaged!) But there's not likely to be a lot of spontaneous initiation of social contact, particularly between people who don't already have an established social relationship.

This is an important factor for some autistic people's ability to be comfortable in autistic space. Many of us need undisturbed time to observe people and activities before we can decide whether or not we want to join in. Sometimes we'll decide to start participating after we've watched for a while. Sometimes we decide we don't want to participate, but we're still interested in watching what other people are doing. It's a great relief to be able to be among people, partaking of whatever aspects of the situation we're interested in, without pressure to do more socializing.

It can also be a great challenge for autistic people who are subject to getting stuck, and are used to relying on people around them to initiate and facilitate. If a person is hanging around in or near an NT social space without joining the group, someone will probably come over and invite the person to join in. Autistic people may come to rely on NTs to provide this kind of prompting, cuing, and initiation in social situations.

If a person is hanging around and not participating at Autreat or a similar autistic social space, other people will most likely just leave the person alone. Autistic people typically have a hard time attending to multiple people and things at the same time, so if we're attending to our own activity and to our interactions with other people who are participating in the activity with us, we may simply not notice someone sitting quietly off to the side. If we do notice, we can't assume the person wants to join in. We can't guess whether the person would welcome interaction or would feel intruded upon. Even if we suspect the person would appreciate some kind of interaction, we can't guess *what* kind of interaction would be welcome. Our own experience of unwelcome interactions has made most of us keenly aware of the pitfalls of trying to read other people's nonverbal signals!

For the person who does get stuck, and finds that nobody takes the initiative to come and help, there are many possible ways to process the experience as an opportunity. Some examples of how different people have processed this kind of experience at Autreat are:

> *I wasn't sure what to do or where to go or what was expected of me. I felt lost and very anxious. Then I realized that nobody was watching or judging me, and nobody cared whether or not I participated in the activities. I didn't 'have to' go anywhere or do anything unless I wanted to. I discovered that it's okay for me to take time to get my bearings, and it's okay to not know things, and it's okay to ask questions when I don't know what's happening.*

or

> *I wasn't sure what to do or where to go or what was expected. I felt lost and anxious and helpless. Nobody offered to help me. I felt abandoned and resentful. Then I realized that I could find out the information I needed by looking at my program packet, or at the signage in the building, or by asking other people where they were going. I discovered that I don't have to be anxious and helpless when I don't know what's going on. I don't have to wait for someone else to come along and rescue me. I can figure out what information I need, and empower myself to go out and get the information for myself.*

or

> *I wasn't sure what to do or where to go or what was expected, and [nobody offered to help me] [people did things that they said were 'help' but I did not find helpful].*

> *I realized that Autreat is not a comfortable experience for me. I am empowering myself by deciding [not to attend Autreat again] [to make my own arrangements to have a helper with me if I attend Autreat again].*

or

> *I wasn't sure what to do or what was expected, and nobody told me what I should be doing. I felt very lost and disconnected. Without external cues telling me what I should do and be, I started feeling as if I didn't exist at all. I realized that I've built so much of my life around trying to be what other people expect me to be that I don't actually know who I am, or even* if *I am, apart from other people's expectations.*

(The latter is an experience that I call "explosive decompression," and fortunately it doesn't happen very often. I've seen it happen to three or four people over the 20 years since I started seeking out other autistic people. It's very terrifying while it's happening. But it is survivable, and can be the first step toward discovering that one *does* exist independently of other people's perceptions and expectations.)

While it's more common for people to either appreciate or be bothered by a lack of spontaneous interaction in autistic space, some autistic people do initiate a lot of spontaneous interaction. Even when we initiate, we still don't tend to engage in NT-style small talk. We just start talking about whatever subject we happen to be interested in at the time.

When this happens in NT social contexts, instead of complaining about failure to initiate, NTs complain that autistics are approaching people inappropriately and are talking about odd or uninteresting things. In autistic social space, people are less likely to be put off by non-standard conversational topics. Personally, I would much rather have someone walk up to me and tell me some interesting fact I hadn't known before about grasshoppers, or helicopters, or forensic dentistry, than have someone approach me uninvited to tell me something I'm perfectly aware of, such as the fact that it's a sunny or a rainy day. If you're going to interrupt my train of thought and place demands on my cognitive processing to focus on you and comprehend what you're saying, at least say something that's intellectually engaging!

But for someone who is already struggling to remain oriented among unfamiliar people in an unfamiliar environment, any unexpected interaction at all, any disruption of a planned course through space and time, can be experienced as unwelcome and intrusive. And autistic people are not perfect; sometimes we do things that really *are* intrusive (i.e., that interfere with other people's ability to exercise their legitimate rights and freedoms), not merely unusual.

Given the great variability among the autistic population, both in what we want and in how we express ourselves, we can't expect to be able to interpret each other's cues or to guess each other's wishes. In autistic social space each person is responsible for explicitly communicating his or her wishes regarding interaction.

It's acceptable to choose not to interact. It's acceptable to choose to interact only with some people, or only at some times, and not with other people or at other times. It's acceptable to talk about things you're interested in. It's acceptable to tell someone else to stop talking to you. It's acceptable to approach people who differ from you in age, gender, race, ethnicity, education, income level, or other indices of social status.

It's acceptable, and expected, to say what you mean. It's not acceptable to expect people to figure out what you need, what you want, or what you mean, if you don't explicitly say it. And it is *very* unacceptable to try to tell other people what they need, want, feel, think, or intend. (NT readers, please take special note of this point of autistic etiquette: NTs seem to consider it an expression of "empathy" to label each other's wishes and feelings. Autistic people are not likely to appreciate being told how [you think] we feel. And if we don't take the initiative to label your feelings for you, that doesn't mean we don't care how you're feeling. It only means we don't presume to know how you're feeling, unless you tell us.)

In autistic space, communication is the responsibility of the individual, not the group.

9. Receiving and giving assistance

Many autistic people's social experiences have involved "all or nothing" dynamics: Either we are expected to function normally without needing or requesting any accommodations or assistance at all (Smith, 2005); or else we are led to expect that we are invariably less capable than everyone else around us, and that other people in general are more flexible than we are, available to assist us when we need help, and willing to adjust their behavior to accommodate our needs (Lane, 1992; Shapiro, 1991).

In autistic space, most of the people around us are themselves autistic. While it would be possible for a group of autistic people to hire support staff for an autistic event, and the staff would then work under the direction of autistic people, I have never seen this done in practice. In the shared autistic spaces that I am familiar with, there are neither designated official caregivers for autistic participants, nor is there a majority of nondisabled fellow participants.

In the absence of both official support staff and a surrounding nondisabled majority, autistic people in autistic space can have the opportunity to try doing things for ourselves. We may be awkward; we may be inefficient; we may fail initially and have to try again; we may eventually give up and ask for help. But we have a chance to *try*, without having uninvited "helpers" zoom in to tell us we're incapable and to relieve us of the opportunity to explore our capabilities. As a result, many autistic people discover in autistic space that we really are capable of doing things we'd never known we could do.

In autistic space it's also possible that someone who is accustomed to always being the one in need of help (or at least to always being perceived that way) ends up being able to provide help to a peer. One autistic person reported that this was such a powerful experience, she had to restrain herself from *thanking* another participant for having a problem and allowing her to feel good about being able to help.

For the person receiving the help, it can make a big difference whether the assistance comes from an autistic peer or from an NT. Being helped by an autistic peer allows one to see an autistic person in a position of strength and competence. Even if one is not feeling particularly strong or competent oneself at the moment, one is reminded that autistic people can be strong and competent. Peer assistance also presents the possibility of reciprocity and mutual support, rather than one category of people always being dependent and another category of people always being depended on.

The dynamic of peer assistance can change the attitudes and perceptions of NTs as well as of autistics. At one autistic event I observed an autistic person sitting on the floor in a hallway, sobbing. I did not immediately assume that she wanted to be interfered with, but following a brief assessment of the situation, I approached her and asked if I could help her. As we sat there conferring, some other autistic people saw what was happening, and sat on the floor with us and talked. One person walked by without saying a word and handed the crying person a visual stim toy. That gave me the idea of getting my MP3 player and playing a particular song for her. Eventually there were four or five people (as well as a stim toy and an MP3 player and a pair of speakers) sitting on the floor trying to help the person who was upset, and meanwhile blocking the hallway. An NT professional came by. As the professional picked her way carefully through all of us sitting on the floor, she remarked, "You're certainly giving the lie to everyone who says autistic people don't have empathy." Then she went on her way, leaving us to deal with our own distressed peer, clearly recognizing that we did not need her to come to the rescue.

While such instances of autistic peer support can be extremely powerful and meaningful both for the people receiving the help and the people providing it, autistic peers cannot be expected to have the same "helping" mindset that many of us have grown accustomed to in our dealings with NTs. Nondisabled people in general, and habitual caregivers in particular, often perceive any disabled person as being in need of their assistance. They swoop in to "help" regardless of whether or not the disabled person has asked for any help. (This can reinforce a disabled person's sense of helplessness and dependence, and interfere with learning to do things independently.)

In contrast, autistic people who gather to celebrate autistic togetherness often aren't inclined to perceive either ourselves or our peers as helpless, powerless people. Even acknowledging that we have significant disabilities and require more support than nondisabled people, most of us expect ourselves and our peers to be capable of managing our own support needs.

If it should happen that someone isn't able to manage, we may or may not recognize that the person needs help. Autistic differences in nonverbal communication mean the person who needs help may not be generating recognizable signals of urgency, and bystanders may not pick up on such signals even if they were present.

If we do notice that someone needs help, we may not know how to help. The variability among autistic people means none of us can assume that another autistic person needs the same things we would need in a similar situation.

Even if we can tell that someone needs help and the person can explain just what manner of help is needed, we may be strained to near our limits just taking care of ourselves. We may not have the spare energy to help someone else. Most NT people, in familiar NT social settings, are at a manageable level of stimulation and energy expenditure. Autistic people, for reasons discussed previously, may be struggling to cope even in an autistic environment, and may simply not have the processing capacity to take on another person's problems.

Being among people who don't rush to offer assistance at the first sign of difficulty, and may not be able to help even if asked, can be a severe shock to people who are accustomed to being able to turn for help to whomever happens to be nearby when they need help. A few people at Autreat have expressed the feeling that no one cared about them and no one was concerned about what happened to them, because no one gave them the kind of help they wanted or needed. The fact is that autistic people are no less likely than NTs to *care about* other people and to be concerned about what happens to them. We just may not be in a condition to be able to *take care of* another person, in a situation where we're not familiar with the other person and his/her needs, and had not expected or prepared to be in a caregiving role.

If someone in an autistic social environment is having an obvious emergency, such as a severe injury or illness, other people will most likely notice and try to help. We may not be able to provide any help beyond calling for an ambulance or police, but at least we can do that much. But for less dramatic difficulties and support needs, such as needing help with personal care, orientation and navigation, time management, making choices, or coping with emotions, one cannot always expect other autistic people to be able to drop whatever they're doing in order to help. Other autistic people are likely to be struggling to keep up with our own self-care and to maintain our own physical, cognitive, and emotional equilibrium. We may not know a solution to whatever problem another person is having. We may be having the same problem ourselves, or we may be having different, but equally difficult, problems. We may simply not be equipped to assist another person, when we're already using all our energy taking care of ourselves.

Some autistic groups exclude people who need personal assistance and support, in order to prevent situations in which someone needs more assistance than other group members can provide. Within ANI and at Autreat we do not exclude people based on needing a lot of personal help. Instead we advise people that if there's any kind of help they need badly enough that it would be a serious problem if no one is able to provide that help, they should make their own arrangements to have a support person who can help them.

In autistic space common autistic needs are understood, and to some extent are predicted and accommodated by the structure of our activities and events. To the extent that different people need different, individualized accommodations, people are welcome to communicate their needs. More, they are expected to take *responsibility* both for communicating their needs, and for participating in working out strategies for getting their needs met (Autism Network International, n.d.-b; Autscape, 2009). This is an opportunity for people to reframe their experience of "needing help" from one of helplessness and dependence to one of dignity, autonomy, and equality with fellow participants.

For people who have never been "allowed" to acknowledge autistic difficulties or to ask for support, this freedom, even encouragement, to express needs and request accommodations can present surprising and previously unimagined possibilities. People may begin to envision, and consider, and experiment with, doing things they'd never thought they could possibly do, or would ever want to do. Trying new things opens opportunities to experience new successes—and also opportunities to experience new failures and frustrations. Often people don't even know what kind of assistance and accommodations would help them. Finding out what works is a process of trial and error, disappointment and discovery.

10. Logistics of autistic space

10.1 Contact a La Carte

When preparing any kind of offering for autistic people, organizers are presented with the challenge of accommodating the tremendous diversity among autistics. Creating something that *everyone* likes and *everyone* finds easy to participate in is an impossible goal. Instead, our approach within ANI tends toward offering a variety of different options, allowing individual participants to pick and choose which ones they want and opt out of the ones they don't want.

This didn't require much organizer attention or planning when we had limited options to choose from, and those options involved minimal interactivity. When all we had was a newsletter and a penpal list, the options were very simple: subscribe or don't subscribe to the newsletter; be listed on the penpal list or don't be listed. People who contacted each other privately through the penpal list were able to negotiate their own contacts without any involvement by group organizers.

Our online forum created much more opportunity for contact and interaction between members. This opened the way for more conflict between the needs and desires of different members. Some autistic people wanted a forum for autistic people only, with no NTs allowed. Some parents wanted to hear from autistic people and from like-minded parents about autism-positive approaches to raising and educating their children, but weren't interested in having their inboxes flooded with large numbers of messages about square-dancing llamas or whatever other fun subject autistic members might start perseverating about. Some autistic people were interested primarily in peer social contact and emotional support. Some were interested primarily in disability rights and political advocacy. Some people (autistics and parents alike) found it emotionally traumatizing to read activist posts containing upsetting accounts of mistreatment of autistic people.

We were able to use an existing feature of the listserv environment to enable individual members to choose the types of posts they did and did not want to receive. The "Topics" feature of the listserv software allowed us to subdivide the forum into various sections, and allowed each member to opt into some sections and opt out of others. The Topics feature was probably designed originally to be used on professional mailing lists, dividing discussions according to technical specialty. But we implemented it for a social purpose, to allow our members to personally customize the kinds of sharing they were interested in: peer discussion among autistics and "cousins" (Cousins, 1993; Shelly, 2003), discussion of interest to parents and professionals about meeting the needs of their children or students or clients, disability rights and politics, and eventually also a virtual party section for times when a lot of fun posts were being sent in near-realtime interaction, and a conference section for list members who planned to attend conferences or other live events and wanted to make plans for meeting each other at those events.

The listowners had to do some extra work to implement the Topics feature so that members could have this degree of individual choice. First we had to learn how to use the feature ourselves. Then we had to teach group members how to use it. There was some initial anxiety and resentment among members over having a new complication introduced to their use of the forum, some people who were certain they'd never learn to get it right, and some people who went overboard criticizing other people's mistakes. A lot of people were intimidated by the instructions for setting up their own Topics options, and the listowners received many requests for individual coaching and assistance.

For the most part, after a brief adjustment period, people caught on and recognized both the logic and the usefulness of the feature. People became more confident and less anxious with practice. People became less critical and more supportive when other people made mistakes. Experienced list members cheerfully helped newcomers learn the ropes. As the Topics feature became incorporated into ANI-L culture (complete with virtual M&Ms. for reinforcement) there was more peer instruction and support, and less reliance on listowners to provide individual assistance.

When we took the step from online contact to physical gatherings, there were even more opportunities for interaction, more intensity and immediacy of interaction, and greater diversity among participants. Participation in written exchanges, whether in print or online, requires the ability to use language. People who cannot read, write, or speak can and do attend live gatherings.

Organizing autistic physical spaces requires attention to participants' orientation and sensory needs, because people are leaving their own familiar personal spaces and coming into a new shared space. It requires the creation of structure for people to maintain their own personal boundaries and recognize other people's boundaries. It requires a great deal of detailed preplanning by organizers, so that autistic participants are not overwhelmed by the need to attend to too many details on their own.

A central dilemma of autistic togetherness is that on one hand, most autistic people are easily overwhelmed by having to make complex social arrangements, such as needing to organize their own travel *and* hotel reservations *and* meals *and* personal leisure time *and* informal social plans with other participants *and* local transportation between conference facility, hotel they're staying at, hotel their desired social contacts are staying at, restaurants, recreational facilities, and whatever else they need or want while attending an event away from home. The retreat structure, in which formal activities (presentations and workshops), informal social and recreational opportunities, meals, and lodging are all available within the same venue, simplifies things immensely for the participants. Often this can make the difference between an autistic person's being able to attend at all vs. being unable to handle the complexity of making all the detailed arrangements. On the other hand, the variability among autistic people means that we require careful individualization of our arrangements, and accommodation of a great variety of different and sometimes conflicting special needs. The concept of "one size fits all" does not work in autistic space!

Autistic gatherings such as Autreat, and other events I'm aware of that have adopted a similar model, balance this tension the same way we accommodate autistic differences online: by providing a variety of different options and allowing people to make their own personal choices. Autreat offers a selection of presentations on topics we know to be of interest to many autistic people. No one is required to attend any of those presentations, and if someone does attend one and finds it uninteresting, it's perfectly acceptable to get up and leave. We offer informal group discussions and recreational activities. No one is obliged to attend any of these, and if someone chooses not to join in, Autreat social rules specifically proscribe pressuring people to participate. There's a pool where people can swim, and posted hours when it's open for swimming, but no herding of participants to the pool for swim times. The general principle is to provide opportunity but not pressure.

Autreat has always been organized as a residential retreat with on-site meals and lodging, partly to minimize costs to participants, but mostly to minimize complexity for people who prefer not to have to attend to each and every detail for themselves. People can register for the "standard" Autreat package which includes meals and lodging, and then once they manage to get themselves to the venue, they can eat, sleep, socialize, participate in formal and informal activities, and take down time when they need it, all without needing to travel off-site or decide all the minute details of their daily routine. But if some aspect of the standard package doesn't meet their needs, they can opt out of that part of the package and make their own personal arrangements. If they prefer the greater privacy and comfort of a hotel room to the shared lodging on-site, they can register for day attendance only, without overnight lodging. If they prefer the 24-h immersive experience of lodging on-site but don't want to share a room with another participant, they can request a private room for an additional fee. If the Autreat meal plan doesn't appeal to them, they can register without paying for the meal plan, and make their own arrangements to bring food with them or go to restaurants near the venue. If they're not up for attending the full retreat (currently lasting four days and four nights), they can register for as many days as they wish to attend.

This degree of individual choice and program adjustability is not built into the standard event-hosting services offered by most conference or retreat venues. In seeking suitable venues for an autistic gathering, the organizers need to negotiate exacting details that the venue staff are probably not accustomed to having to think about. Each time we've searched for a new venue for Autreat (our current venue is the fourth), our checklist of requirements has included such things as allowing participants to opt out of the meal plan and/or on-site lodging, having facilities available for participants to store their own personal food, providing menu information well in advance of the event so people can decide whether or not to sign up for the meal plan, providing complete lists of ingredients for each dish so people with food allergies can find out what they can safely eat (the latter two requirements are constant sources of frustration when venue staff don't realize how important they truly are), having some "modular" food choices such as salad and pasta and sandwich bars for people who have trouble eating "mixed together" foods, and allowing people to take their trays out of the dining hall and eat in their rooms if the dining hall is too noisy and crowded for them.

10.2 Information and orientation

As ANI space has developed from a small group of friends visiting each other's homes, to a larger group of email correspondents in our own online forum, to a four-day retreat where dozens of people share physical space, our approach to the challenge of unfamiliarity has been to provide a lot of detailed information to help people get oriented. Early home visits were typically preceded by email exchanges or (less frequently) phone calls in which people planned the visits and asked each other questions about what to expect in being together. The ANI-L forum has a set of rules and policies that people are required to read before they are allowed to join the forum. People who register for Autreat are sent a collection of files before they attend, providing extensive details about the venue, daily schedule, and Autreat customs and social expectations. There is an Autreat Information forum online where people can ask questions and share information and pre-plan their Autreat experience. There is an orientation session on the first evening of Autreat where non-readers can hear the information that was sent in advance to read, and everyone has a chance to ask questions.

Most autistic people who are capable of formulating questions have frequently experienced the following scenario: We ask for information that we need in order to prepare ourselves for a new experience. Instead of answering our questions, NT people tell us that we don't need to ask these questions at all. We just need to relax and stop being so anxious. The fact is that being able to ask questions, and getting clear answers to our questions, and thus knowing what to expect, are often the very things autistic people need in order to be *able* to relax and not be anxious. Asking a lot of questions about the details of a situation is usually *not* a "maladaptive behavior" that increases an autistic person's anxiety. More often it's an adaptive strategy that an autistic person is using to reduce anxiety or to prevent being in an anxiety-provoking situation in the first place. It's very important for us to have thorough explanations and ample opportunities to ask questions.

Experience has taught the leaders of ANI and Autreat that taking time to answer people's questions is a necessary part of organizing autistic activities. And for autistic people who participate in these activities, learning to identify the information they need, and to ask the right questions to enable their participation, is an important self-advocacy skill.

10.3 Autistic social rules

Autistic people commonly have difficulty with NT social rules for three basic reasons: We may not know what the social rules are; we may know what the rules are but not see any reason to comply with them; or even if we would like to comply with the rules, the demands of social compliance may be inimical to our natural ways of processing.

Autistic social rules develop and evolve naturalistically within autistic space, just as social rules and customs develop naturalistically within any other community. Autistic rules and customs reflect common characteristics of autistic people.

Autistic people characteristically do not pick up on vaguely described, implied, or unspoken behavioral expectations. We don't often learn by "social osmosis." Therefore, in autistic space the rules and the expectations are clearly and explicitly spelled out, sometimes in excruciating detail.

Besides being explicitly spelled out, autistic social rules are also clearly explained. Some autistic people scrupulously, even rigidly, follow rules simply because they've been told that they're rules. Other autistic people care nothing at all for rules unless the rules make sense to them. At both extremes of the rule-following spectrum, as well as for autistic people in between, it is helpful to have clear explanations of the reasons for rules. If we know *why* something is a rule, and we can understand that the rule makes sense, then those of us who require logical explanations are more likely to respect and follow the rule, while those of us who tend toward rigid rule-following are better able to be flexible when necessary. In autistic space, asking "Why?" is a perfectly normal and acceptable response to a rule, and is not considered impudent or disrespectful. Autistic community leaders who create rules must be prepared to explain and justify those rules.

Most autistic people who have never experienced autistic space have never meaningfully experienced "social rules." And until there was such a thing as autistic space, there were no such things as autistic social rules. But once autistic space came to be, autistic people began doing kinds of social processing that we had never done before. As a result of this authentic autistic social processing, authentic autistic social rules began to emerge. Autistic social rules evolved in accordance with autistic processing, and meanwhile autistic social processing was able to further develop because of the facilitative environment created by the autistic social rules. Thus, autistic social rules develop synergistically with autistic processing both online and in real space:

The ANI-L forum was created in the aftermath of heated conflict between autistic adults and NT parents in an NT-dominated autism forum. Some early autistic members were reluctant to allow NT parents to participate, and some early NT members were apprehensive about whether their participation was welcome and appropriate. The social rules we created involved detailed and extensive written rules for the forum; the division of the forum into different sections for discussion topics of interest primarily to autistic people and discussion topics about parenting; firm boundaries defining what kinds of topics could be discussed in which section; and clear identifying signals (using the Topics feature) enabling people to recognize what kind of message to expect before opening a post.

Autistic processing—explicit statements of the group's rules, very detailed explanations of the reasons for those rules, clear boundaries to maximize individuals' ability to pick and choose among available options for contact, and protection of autistic people from adverse forms of contact—produced those rules. NT communities probably would not make such rules.

The autistic rules enabled autistics to have experiences that we couldn't have anywhere else: the development of social connections with autistic peers, and also with NTs who were willing to respect our rules.

These new experiences and connections led to greater familiarity and comfort with social interactions in autistic space, more spontaneity, less "us vs. them" thinking between autistics and NTs, and less rigid reliance on constantly checking the rules.

The original rules may be seen as creating separation and distance between autistics and NTs. From our perspective, what we were doing was *acknowledging differences* between autistics and NTs—as well as between autistics and other autistics—and claiming the forum as autistic space where autistic needs and autistic sensibilities have priority. Through acknowledging our differences, claiming our own space, and accommodating autistic processing needs (e.g., clear and detailed rules), we were able to have the experience of successful and enjoyable interactions between autistics and NTs in a comfortable autistic space. These successful interactions created bridges to traverse the original distance between us. Within a year of its creation, the ANI-L forum had evolved into an integrated community of autistic and NT people sharing ideas as equals.

A similar pattern unfolded with gatherings in physical space. Before the ANI strand in the 1995 parent-oriented conference, ANI members used the ANI-L forum to discuss what we needed to make the physical space comfortable and safe for us. Our online forum provided a safe and comfortable virtual space in which we could discuss our past experiences with, worries about, and possible solutions for live gatherings. Among the things we came up with was a set of "Guidelines for Interacting with Autistic Speakers and Attendees" based on experiences many of us had had at other conferences. These guidelines stated clear and firm boundaries regarding contact. They explained the reasons we need these boundaries.

We were told by one of the NT conference organizers that our interaction rules would be perceived as rude and unfriendly by NT conference participants. We insisted that they were necessary anyway.

That conference was attended by a larger group of ANI members than had ever come together in real space before. With our social rules clearly stated and explained, most of us were able to have a wonderful experience of autistic togetherness—*and* also have some positive real-space encounters with NTs who respected our processing needs and accepted our rules.

Those same guidelines, with only minor modifications, remain in place as Autreat rules. For thirteen years, as of this writing, dozens of autistic and NT people have been able to come together each year with remarkably few serious difficulties, and share a great many positive experiences ranging from simple "fun times" to profoundly life-changing insights and discoveries.

Autreat social rules and customs, which have been adopted by some other autistic-run events around the world, include:

- Accommodation of other people's sensory sensitivities. Flash photography, use of perfumes or scented personal care products, and touching people without permission are all prohibited by Autreat rules. Smoking is limited to designated smoking areas, and if people come to group activities smelling of smoke, they may be asked to go change their clothes. Loud activities, such as music and dance sessions, are assigned to places in the venue that can easily be avoided by people with auditory sensitivities.
- Respect for people's privacy. We do not compile an Autreat participants' directory in advance. People make their own individual decisions about what personal contact information they wish to give out to which fellow participants. We have an Autreat privacy agreement which everyone is required to read and sign. We have an even more restrictive set of rules for journalists or researchers who wish to gather material at Autreat for research or publication.
- Use of badges as visual cues. Many autistic people have difficulty recognizing faces. We therefore strongly encourage Autreat participants to wear their name badges at all times when they're in group space (except in the pool) so other participants can recognize them. People who don't

wish to be photographed at all, even under the terms specified in the privacy agreement, may wear a black circle badge ("photon trap") to signal that they are not to be photographed. Since many autistic people do not generate automatic nonverbal signals to indicate their interaction-readiness, nor pick up on such signals from other people, we provide optional color-coded interaction signal badges. People can use these badges to indicate whether they wish to be left alone, are open to interaction with familiar people but not with strangers, or are having trouble initiating interaction and would like someone else to approach them and initiate.

- Respect for boundaries. Autreat social rules don't require "acting normal" or suppressing "weird" behavior. Actions are determined to be acceptable or unacceptable based on whether or not they infringe on other people's boundaries (Sinclair, 2004). As coordinator, whenever I have occasion to mediate a conflict among Autreat participants, I generally start by identifying and clarifying the relevant personal or group boundaries. I almost always find that once the situation is framed in that manner, people are able to arrive at reasonable and mutually acceptable conclusions.

At Autreat as on ANI-L, we have accommodated autistic processing by providing clear and detailed rules, logical explanations, careful attention to boundaries, and individual freedom to choose among available options. At Autreat as on ANI-L, these accommodations have created a social environment in which autistic people have been able to have social experiences that we don't have anywhere else.

I realized this for myself just before the first Autreat in 1996. While our current venue offers apartment-style lodging with only two people per room, our first venue was a camp facility in which six or more people shared each cabin. The camp administrator had told me that there were two private rooms in the infirmary building, and that when groups rented the facility for retreats, the group leaders usually stayed in these private rooms. I decided it was more important to make those private rooms available for people who needed them for accessibility reasons. I made all the cabin assignments based on participants' completed questionnaires, then called the camp administrator to read her the names of the people who would stay in each cabin. When I read my own name on the list assigned to a cabin, the administrator said, "I thought you were sleeping in the infirmary." Without even thinking about it I said, "No, I'm sleeping with my friends." I was startled and then astonished to hear myself say such a thing. I realized that I would *rather* share a cabin with my friends than have a room to myself. I realized that I actually *had* friends—five of them!—whose company was so enjoyable to me that I'd willingly give up the luxury of a private room in order to spend more time with them. This is not something I would ever have imagined myself choosing, before I had any experience of autistic social space.

I've observed many similar examples of autistic people doing things at autistic gatherings that are contrary to conventional observations of autistic social behavior. During the first few years of Autreat we planned few or no group activities other than workshops. The very idea of "group social activities" was too threatening to people whose past experience of social activities had involved coercion, rejection, failure, or other sorts of unpleasantness. Gradually as people grew more secure in the acceptance of peers and the freedom to opt *out* of activities in autistic space, it became less scary to consider deciding to opt *in*. We also discovered that there were more interesting possible activities to opt into, in a group of people who shared similar interests and processing styles.

Autreat rules are carefully designed to protect people from unwanted social and sensory intrusions. People are expressly told not to touch anyone in any manner at all without permission. People who don't want to be touched can be secure in the knowledge that nobody is likely to touch them (and that if anyone does touch them, they're allowed to protest). With the freedom to consciously choose whether and when and how they want to be touched, some Autreat participants have recently been discussing possible adaptations of the badge system to signal that they are open to hugs.

The Autreat dining room may be the quintessential example of this phenomenon. It's a common stereotype that autistic people in a group dining situation sit alone, as far away from other people as possible. And indeed there are people who choose to sit alone in the Autreat dining room, or to take their meals out of the dining room to someplace more private. We expected that from the beginning. We negotiated agreements with all our venues to accommodate the needs of people who opt to eat alone.

But there are also people at Autreat who not only choose to sit together, but who actually crowd *closer* together in order to fit more people around a table than the table was designed for. I have watched an autistic person bring a tray from the serving area, walk to an empty table, take a chair from that empty table, and carry it to another table where all the chairs were already occupied. I've seen the people already at that table move closer together to make room. I've seen this happen again and again with person after person, until as many as ten people—most or all of them autistic—are happily crammed around a table that was meant to seat six.

11. The future

Our community is still young, but a generation of autistic children has already grown up having experience and familiarity with autistic togetherness. We're a small minority among an NT majority, but more and more of us are reaching out, finding each other, and making connections. Twenty years ago it was almost impossible for an autistic adult make to contact with even one other autistic person (unless that autistic adult was one of the fortunate autistic people who have autistic family members). Today there are numerous autistic forums and web sites and blogs on the Internet, autistic social and support groups in many cities across the U.S. and around the world, and immersive cultural experiences in several different countries. Autistic children today have the opportunity to grow up with knowledge of themselves as autistic people, with awareness of possibilities and resources for them to lead happy and meaningful autistic lives.

We face many serious challenges, as individuals and as a group. But now, for the first time, we are able to face them together.

Acknowledgments

This article is adapted from the author's presentation "Being Autistic Together," Plenary Lecture at Autscape 2005, Radstock, UK, July 27th 2005. The author would like to thank Ruth Lavender for providing her notes from that presentation.

The author would also like to thank the following people, without whom this article could not have been written:

Sola Shelly and Kazumi Honda for endless hours of writing support and facilitation and for technical and research assistance, Phil Schwarz for communication assistance, Kiesa Kay for editing and formatting advice and assistance, Emily and Ralph Savarese for bearing with my very slow progress on this article, and most of all, the members of Autism Network International and ANI-L, participants at Autreat, and the many other autistics, cousins, and allies who have created the reality of autistic space.

References

Adam, P. (2009). *Small Talk. Wellsphere.* http://www.wellsphere.com/autism-autism-spectrum-article/small-talk/590433. Retrieved December 14, 2009.

Aston, Maxine. (n.d.). Cassandra Affective Deprivation Disorder. http://www.maxineaston.co.uk/cassandra/. Retrieved December 14, 2009.

Autism Network International. (2009). *Autreat 2010: Call for Proposals.* http://www.autreat.com/aut10cfp.php. Retrieved December 14, 2009.

Autism Network International. (n.d.-a). Introducing ANI. http://www.ani.ac/intro.html. Retrieved December 14, 2009.

Autism Network International. (n.d.-b). Autism Network International presents Autreat. http://www.ani.ac/autreat.html. Retrieved December 14, 2009.

Autreat Planning Committee. (n.d.). Policy Regarding On-Site Manager's Discretion to Make Executive Decisions about People Attending Autreat. http://www.autreat.com/Decision-making%20discretion.html. Retrieved December 14, 2009.

Autscape. (2009). *Expectations*. http://www.autscape.org/experience/expectations. Retrieved December 14, 2009.

Baggs, A. (2006a). *Lessons on Inclusion from a Segregated School, Version 2 (from memory, not an exact reproduction of the first one)*. http://ballastexistenz.autistics.org/?p=81. Retrieved December 14, 2009.

Baggs, A. (2006b). *My friend, in the disability gulag, without support*. http://ballastexistenz.autistics.org/?p=35. Retrieved December 14, 2009.

Broun, L. (1996). The experience of teaching a congregated class of autistic children. In Education Committee of Autism Society Ontario (Ed.), *Navigating the special education system in Ontario - A handbook for parents of children with Autism/PDD* (pp. 69–70). Willowdale, Ontario: Autism Society Ontario.

Claiborne Park, C. (1968). *The Siege*. Gerrards Cross, UK: Colin Smythe.

Cohen, L. J. (2006). *Autreat. Aspies—A Place for Self-Advocacy*. http://aspies.blogspot.com/2006/05/autreat.html. Retrieved December 14, 2009.

Cousins, S. (1993). Neural connections in Toronto. *Our Voice: The Newsletter of Autism Network International, 1*(3), 8–11.

Foden, T. J. (2008). *Sibling studies: The good, the bad, and the contradictory. IAN Community*. http://www.iancommunity.org/cs/articles/sibling_relationships. Retrieved December 14, 2009.

French, K. (1993). My personal holiday. *Our Voice: The Newsletter of Autism Network International, 1*(3), 2–3.

Frith, U. (1992). *Autism: Explaining the Enigma*. Malden, MA: Blackwell Publishing.

Helms, J. E. (1995). An Update of Helms's White and People of Color Racial Identity Models. In J. G. Ponterotto, J. M. Casas, L. A. Suzuki, & C. M. Alexander (Eds.), *Handbook of Multicultural Counseling*. Thousand Oaks, CA: Sage.

Judd, A. (2009). Death highlights lack of regulation at 'Psychoeducational' schools. *The Atlanta Journal-Constitution, 27,* 2009. http://www.ajc.com/news/death-highlights-lack-of-100352.html. Retrieved December 14, 2009.

Kaufman, B. N. (1995). *Son rise: The miracle continues*. Tiburon, CA: H J Kramer.

Lane, H. (1992). *The mask of benevolence: Disabling the deaf community*. New York, Alfred A: Knopf.

Maurice, K. (1994). *Let me hear your voice: A Family's triumph over Autism*. New York: Ballantine Books.

McNabb, J. (2001). *I Dont Want To Be A Pioneer I Just Want To Be Me*. http://www.autistics.org/library/nopioneer.html. Retrieved December 14, 2009.

Meyerding, J., & with the collaboration of KB, Clark, Patty, and Comm, Marla. (1998). *Why Are We So Unfriendly? Or: Hello, Friend, Now Please Go Away*. http://mjane.zolaweb.com/unf.html. Retrieved December 14, 2009.

Muskie. (2002). *Institute for the Study of the Neurologically Typical*. http://isnt.autistics.org/. Retrieved December 14, 2009.

Schwarz, P. (1999). *The Autism community as a neurological diaspora*.

Shapiro, J. P. (1991). *No pity: People with disabilities forging a new civil rights movement*. New York: Times Books.

Shelly, S. (2003). Cousinhood: Who cares and other questions. In *Presentation at Autreat 2003: Transforming our world, June 29–July 2, 2003, Brantingham, NY*.

Sinclair, J. (1995). Cognitive, communication, and interpersonal deficits of non-autistic people. In *Presentation at MAAPing the future, September 15–16, 1995. Indianapolis, Indiana*.

Sinclair, J. (1998). *Concerns about inclusion from within the disability community, compiled by Jim Sinclair, 1998*. http://www.jimsinclair.org/inclusion.htm. Retrieved June 14, 2008.

Sinclair, J. (2004). The nature and functions of boundaries, rules, and social conventions. In *Presentation at Autreat, July 1, 2004. Philadelphia, PA.*

Smith, J. (2005). *You Have It So Good.* http://www.autistics.org/library/youhaveitsogood.html. Retrieved December 14, 2009.

Smith, Joel. (n.d.-a). Living With Executive Dysfunction. http://thiswayoflife.org/ef.html. Retrieved December 14, 2009.

Smith, Joel. (n.d.-b). You're not Autistic! http://thiswayoflife.org/youarenot.html. Retrieved December 14, 2009.

Spicer, D. (1998). *A Way of Describing Autism.* http://www.ani.ac/dave.html. Retrieved December 14, 2009.

Tustin, F. (1990). *The protective Shell in children and adults.* London: Karnac Books.

Williams, D. (1994). *Somebody somewhere.* NY: Times Books.

WrongPlanet.net. (2008). *What does small talk mean?.* http://www.wrongplanet.net/postt73054.html. Retrieved December 14, 2009.

Zaks, Z. (2006). *Life and love: Positive strategies for autistic adults.* Shawnee Mission, KS: Autism Asperger Publishing Company.

Further reading

Debbaudt, D. (2002). *Section A: A Definition & A law Enforcement Handout. Avoiding Unfortunate Situations.* http://policeandautism.cjb.net/handout.html. Retrieved December 14, 2009.

Sinclair, J. (2005). *The History of ANI.* http://www.ani.ac/History_of_ANI.html. Retrieved December 14, 2009.

Some Cultural Parallels and Some Practical Responses. (1999). *Presentation at Autreat 1999: Telling our stories, August 22–25.* Canadaigua NY.

Van der Klift, E., & Kunc, N. (1994). Hell-bent on helping: Benevolence, friendship, and the politics of help. In J. Thousand, R. Villa, & A. Nevin (Eds.), *The practical guide to empowering students, teachers, and families.* Paul Brookes: Baltimore.

CHAPTER SIX

Self-made design notes for an autistic's office

Marie K. Harder*
Values & Sustainability Research Group, University of Brighton, Brighton, United Kingdom
*Corresponding author: e-mail address: m.k.harder@brighton.ac.uk

Contents

1. Introduction

This chapter is in a book about architecture designs for people who think differently. In particular, those with neurodivergence—conditions where people's brains function in substantially differently ways to normal. Some of the chapters are written by academics, some by architects, and some—like this one—by *bone fide* neurodivergent people. I am an autistic person. I moved towns every year as a kid, so my inherent differences were overlooked and went without diagnosis until I was in the same job for several years and middle aged…and people started to notice that there were some things I simply could not "learn" despite being a professor. (My academic specialism is sustainability.)

People sometimes categorize autistic persons as "high-" or "low-" functioning, but I think that's a red herring, as my own level of apparent "functioning" is more likely to be related to the fact that I am a glutton

Developments in Neuroethics and Bioethics, Volume 5
ISSN 2589-2959
https://doi.org/10.1016/bs.dnb.2022.08.003

for punishment and punishingly-hard work, and I "punish" myself every day trying to "manage" in the world that many non-autistic people like you create around me. I'm always pretty close to "not managing at all." I'd actually love to escape to a desert island (with food!)—or under a duvet all day—but I get up and take the punishment. Whereas maybe an autistic person who finds many "normal" tasks daunting and overwhelming—maybe from a bit of attention-deficit disorder—might feel at the age of two that interacting with people is just too much trouble...become a voluntary mute...and appear to be "low" functioning. So please read on to the end before you pigeon-hole me!

Another point I'd like to clarify is that I am writing this chapter as a contributing "autistic" person, and not as any kind of academic expert on the subject. On the other hand, I have been working so long as an (introvert) academic that I will naturally write in that style, which may confuse readers. My warning to the reader is that, despite the fact that I will inherently reach for authoritative works to provide evidence on some things I say, I STILL do not claim to be an expert. Just a layperson who looks things up. The difference? It is that I am not claiming to give you reasoned or balanced view of the matter: only my personal, biased view, and possibly in an inappropriately overconfident manner.

Having made that clear, I'd like to introduce the two key contextual themes that I will be constructing my chapter content within. The first is about *the importance of meaningful work* in life, and the second is about the power of the *concept of validity*—and in particular unitary validity—which I will explain further.

1.1 Meaningful work

Most people hope to have meaningful work, but it is not a top of priority for everyone. However, there are some reasons to think it should be. Maslow developed a "Hierarchy of Human Needs" which is very well-cited (although less well-proven), which puts self-actualization at the top, well above basic survival needs (Maslow, 1954). He originally purported that higher needs were not very relevant to persons until lower needs were met, but in his later works he wrote that for some people self-transcendence was even higher than self-actualization and these might well be *the underpinning drivers* throughout the other levels (Mazlow, 1993). What I am suggesting here is that for autistic people it seems possible that ordinary life is so hard—since basic physiological needs are rarely met—that there must be something higher which drives those who fight their way out of self-isolation (or at least, the strong desire to avoid other people) and join the workforce. Some reports say that only 20% of autistic people are employed; others that they are more likely to commit suicide. For those who are familiar with why ordinary life for autistic people is stressful, the idea that they might need their work to be meaningful—*must* be meaningful—will probably make sense.

Viktor Frankl established an important field of psychology on the principle that the innate human quest for meaning is so strong, that strengthening it for those in distress could dramatically improve and transform their lives (Frankl, 1959). And there are philosophies and faiths which make this explicit, as illustrated in this quote from the Baha'is: *"Unlike animals…human beings are impelled to express the immense capacities latent within them through productive work designed to meet their own needs and those of others. In acting thus they become participants, at however modest a level, in the processes of the advancement of civilization. They fulfil purposes that unite them with others."* (Baha'i International Community, 1995)

The reason that I labor on the topic of meaningful work, is that I feel its importance is greatly overlooked, for all people, but especially for neurodivergent people. In a world where many people face no employment at all, or increasingly low-paid employment as gaps increase between the top- and bottom-earners in every country, I think it is necessary to keep in mind that meaningful occupation (even if unpaid) is probably a fundamental necessity of being human. Without it, we don't feel human: well-being goes down and suicides go up. Humans seem to *need* a channel to transcendence: to feel they can make some kind of contribution. And of course, contribution is akin to sharing, and sharing means connection to other humans: this is something I imagine many autistics yearn for but cannot easily manage.

My question to the reader is then: what kind of an architect or academic are you? What kind of a health worker or scientist are you? Are you carrying out your work with no regard for the potential it might give neurodivergent people to have meaningful purpose in their lives? Are you setting your career goals in line with some kind of mainstream metrics like how many clients you have, or income brought in, or number of books written…or are you actively considering the contribution of your work towards facilitating other people to have a meaningful life? Because…if you're not bearing that in mind, I would suggest it might not be worth you doing it at all. From my (neurodivergent) point of view. So help me: help me so that I can find more meaning in my work and my living…or call yourself by another name. I can get a "builder" to build an ordinary office; a "GP" to treat me as a neurotypical. I'd very much like you to offer me something more.

But how will you know the way to proceed: what information will you need to seek out and keep to hand, in your journey to be of positive service to neurodivergent persons like me? Here I suggest the guiding principle of unitary validity.

1.2 The concept of unitary validity

Professionals like architects and health workers rely on expertise for the collection, interpretation and analysis of valid data, which can then be synthesized to inform their decisions. But what makes data valid or invalid? In some subjects this is a philosophical question, but here I am going to stick to the very pragmatic approach of *unitary validity* (Messick, 1995; Onwuegbuzie et al., 2007). It is conceptually quite simple: if you want to be certain that something you are measuring is valid, then ask: *is your measurement (data) valid in every sense needed: such that it is appropriate to use to answer your question?*

Here's an example. Standard IQ tests are sometimes used to determine whether children would benefit being provided with educational opportunities which are more academic, or less academic. They usually consist of two test parts, verbal and nonverbal—because psychologists have a construct that these two are different and largely separate intelligences with different assessments—fulfilling *construct validity*. Each question in the test is designed to be an indicator of a different capability: they each have *item validity*. The number of questions across sub-topics properly represents a balance of those which contribute to that intelligence—*structural validity*. And overall, the IQ tests can predict which children will flourish in more-academic opportunities, which is *predictive validity*. If EVERY LINK in the chain has good validity, then there is UNITARY VALIDITY: the data is valid for the specified purpose.

But what if IQ tests are used to decide whether children are enrolled for Advanced-level Physics or Business Studies? Teachers may assume that children with higher IQ's should take Physics, but in fact this involves several breaches of validity for using the data that way. Higher nonverbal IQ might be predictive for Physics success, but not overall IQ which includes verbal: that use of the data would break unitary validity. And higher-predicted success for Physics does not mean lower-predicted success for Business Studies, but teachers may assume this, breaking predictive validity. Because Physics is perceived as more difficult, teachers may be inclined to think higher-IQ scores indicate those children should preferentially be directed towards it—when in fact they might instead do outstandingly well at Business Studies. This invalid use of the data could have been spotted if unitary validity had been maintained throughout.

I make the effort to introduce this concept because it is so relevant to architects and medical researchers who are trying to work towards better design for neurodivergent persons. The fact is, that there is much uncertainty about autism: it is not well-defined in terms of characteristics nor symptoms nor life challenges. Until 2013 there was a separate diagnosis category of "Asperger Syndrome" in the American Psychiatric Association's Diagnostic Manual of Mental Disorders, but the 5th Edition (DSM–V) eliminated it because it could not be well-defined. There is much learned every year in recent years, and new related formal diagnoses produced (such as Social Communication Disorder), but there is currently no simple "checklist" of aspects to be considered by architects, nor provided by medical researchers. What valid information, or data, is there?

Scientific measures and metrics show that autistic persons are known to often have hypersensitivity to sound—or hyposensitivity. They can be oblivious of their visual surroundings—or so strongly influenced that they "shut down" their cognitive functions, or "bolt to safety." They can come from the traditionally-diagnosed young males, or the increasing numbers of middle-aged women. They can appear unfeeling and robotic, yet have traumatic emotional "meltdown" reactions for apparently minor issues, and have far higher rates of suicide and depression. The bimodal nature of this data—everything seems very high or very low, and different people can have different highs and lows—can be considered a boon or a curse for designers. An engineer would probably walk away from a contract to produce a working solution with this type of data, knowing that there can never be a simultaneous solution for both modes, and that the use of any averages would fail everything (Fig. 1).

An architect or designer, however, might overlook the lack of validity of the data for their purposes. They might focus on the *language* of the data, rather than its reality, and allow it to lead them away from useful designs. For example, "hypersensitivity to light" might immediately be translated to the architect as "a chance to justify some really innovative lighting systems." Or, "tendencies to flee or bolt" might be snapped up as justification to design aesthetic "winding corridors to delay evacuation and allow calming down." This type of justification (my_favourite_aspect_of_a_solution to

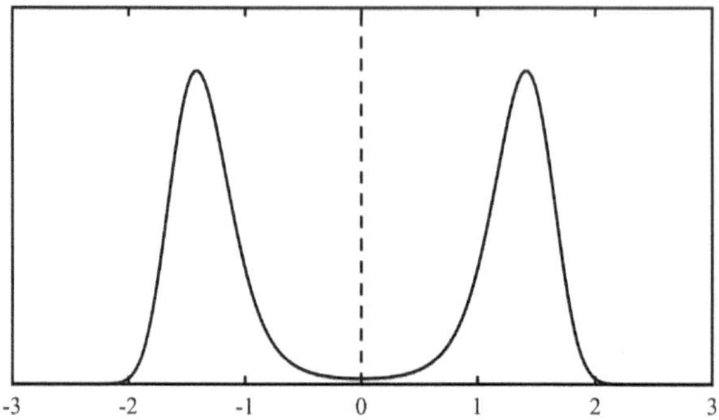

Fig. 1 A bimodal distribution (e.g., representing Neurodivergent characteristics of hypo- and hyper-sensitivity). Note that designing for the average of them all, is useless to (almost) all.

the_nearest_related_aspect_of_the_problem) is not uncommon in many subject areas, especially creative ones, and it is often a pleasure to see the creative responses triggered. In some arenas like marketing, the exploitation of one piece of data ("people want to feel they deserve more") to justify pathways to profit ("why not upgrade today: you know you're worth it!") are also commonplace. The question is, is it ethical for architects and designers, and is it what you want to be part of? If not, then the guiding principle of unitary validity may help you stay on your chosen path.

Unitary validity starts by asking, what is the specific question you want answered, for which you require data? It might be, "How can I use this design brief to justify unleashing my creative abilities and becoming really well known for cool stuff?", in which case, use the bimodal data mentioned above and go ahead and justify whatever you wish. But if your question is something like, "What are the characteristics of the people I am designing for?", or perhaps, "What are their built environment needs?", then the above bimodal type of data, even if authoritative, is not very useful. Ideally, you could just research studies of what autistic persons perceive their needs to be. Unfortunately, there is not likely to be much data there, because of the sad fact that we have terrible difficulties in two-way communication because we cannot easily perceive the mind or mindset or point of view or attitude of another person. At least, not intuitively; not more than superficially; not for more than a few minutes—because by then the "normal" person has shifted to another frame and many autistic persons cannot follow that fast enough. For example, when I stop in the tea room to listen to a colleague tell me about their latest project, I cannot really understand his intended meaning until I have "uploaded" typical vocabulary from his discipline; turns of phrase he has from a rural upbringing in Midwest USA; terms of social reference he collected in Oxbridge; and the fact that he's hungry for his lunch. I have to remember that my facial movements might shift his way of speaking, and that as a woman passes by his change of bravado and gesticulations might not be related to the content of his speech. I notice an illogical emphasis of a sub-topic and frantically search for my lack of understanding so I can keep up with his speech: then I remember his wife is an immigrant from the East, which resolves the conundrum. If he asks me a question I need to have all this knowledge in place so I can then calculate which words to use to express my "thoughts" in a manner he would consider "normal": appropriate not only for the content matter I'm trying to convey, but for that stage of the conversation. (Beginnings and ends of conversations are full of empty phrasing to be disregarded for content.)

Answering a questionnaire is not very different: I still have to construct an idea of what on earth the researcher *really* wants to know. The difference is that they are not there, so I have to guess a lot. And the next question might be from a very different perspective or intention of the author. So, even though I have a doctorate and decades of interacting with people daily, which many autistic people do not have... I doubt you could obtain much accurate data about me or my "built environment needs" from a questionnaire. More likely I would rip it up in frustration.

There is another way to collect valid data for this question. It is qualitative, and idiosyncratic. It requires the questioner to be willing to enter the rich world of detail, in the world of just a few individuals. It is not representative data: it does not earn its validity in that way. It derives its validity from the holistic and relatively pure manner in which the information provided is communicated, with very little interpretation along the way. The questioner then has the full responsibility to interpret the data themselves, for their purpose. This is the approach I offer in this chapter: I am going to give you my personal, one-way communication on how I see the built environment I occupy, with respect to my needs. Then, you get to interpret it, and unitary validity is (almost) ensured. (I've looked up the formal method: it's called autoethnography -see also Clarke, that volume.)

2. The source of my built environment needs

I would like to provide some coherence to the information I offer about how I perceive my built environment. The only place I can find to start, however, is in describing the way that I think. The main message is: I do not think in words. I don't know what to call them: thoughts; clouds; swirls; grunts. But they are not in words. This means that every time I wish to communicate (outwards) in words—to write or to speak—I have two choices. I can either compose the words from scratch, or I can grab the ones nearest to me from recently-seen text or recently-heard conversations and use those as proxies. The latter is quite powerful as a mechanism to survive in a school playground. In some places like the USA it can be considered normal to construct conversation from tag lines and sound bites of other people, and it is generally a convenient way to appear to fit in without too much mental effort on language choice. A major drawback, though, is…that all my mental energy goes into noting, banking, practicing, categorizing sound bites: analyzing where they were used and how to template them for future use. Then, when I need language, I have to do the inverse calculations for determining if they are (at least moderately) suitable for the context, topic, audience, etc. The drawback? That I cannot do my own authentic thinking at the same time. To use proxy-language means to not have time or space or habit of thinking for myself: only of "mirroring" and "reacting" via soundbites. Imagine being at a funeral: I hear phrases, tones, gestures, and draw from my memory bank of films of funeral scenes, and I act and speak accordingly. But I don't have headspace or any perspective space to consider how I actually feel; what I would actually like to express to others; perhaps to realize I don't feel much sadness and isn't that strange because everyone else seems emotional. On the other hand, it's socially safer (as long as I don't use the wrong template), because I'm mostly mirroring the people around me—who presumably are "normal."

Composing words from scratch, on the other hand, is much harder. It's so much effort, that I think I completely avoided it for many years of my life. In order to create a composition of thoughts or words, I have to put my language part of the brain in contact with my actual thoughts. And that is a difficult and complicated process. Because, as I said, I don't do my main thinking in words.

I recently was involved with some studies on Tacit-to-Explicit Translation (Odii, Huang, Bouvrie, & Harder, 2021) that incidentally have started to help me understand what might be similarly happening in my own head. Those studies involved small groups of people who have a lot in common—say a workplace team. They are taken through a process whereby they are asked to think about a tacit experience they all share, such as "dealing with difficult customers." The facilitator of the process asks them to identify something "worthwhile" about that experience, and so all group members start talking "around" the subject. They haven't previously articulated it, so they struggle for words; try some out; exchange them for new ones; make suggestions to each other; challenge the fitting of the word to the mutually accepted meaning...they are actually collectively creating jointly-shared meaning at the same time as they create their shared vocabulary. Fig. 2 illustrates the micro-processes (related to those of Polanyi's Personal Knowledge Theory). Every now and again they achieve some kind of (more-explicit) articulation and they pause; try it out; possibly decide it doesn't fit; and consult together again for a "reinterpretation." Finally, they achieve a statement they all agree on: perhaps something like, "We feel good that we manage the situation of a difficult customer without it getting out of control: even better when they buy something as well." Although they threw away a lot of other statements, those rejected somehow still all contributed to shape the final meaning.

The reason I mention the above study, is that it is useful to illustrate my own way of thinking. My process requires "diving deep" into tacit-type thoughts; "fishing around" and bringing up some ideas for inspection; then modifying them... modifying the way I might express them; reconsidering... a kind of ladder of words and thoughts which have to evolve together in order to get them into any kind of communicable form. All of the above means that I have to submerge in *non-words* before I can build any *words*. This process requires huge mental focus. I guess it is kind of like an

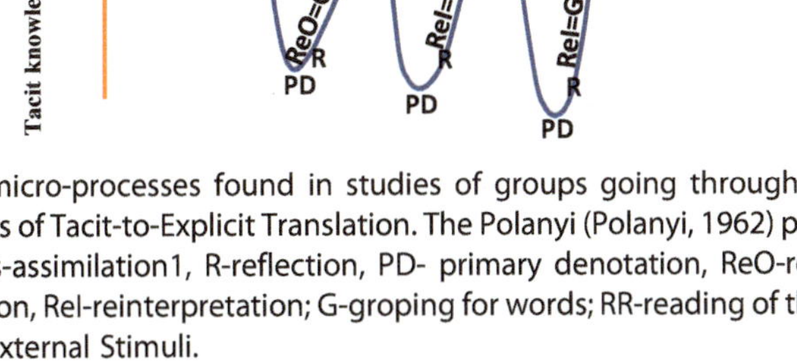

Fig. 2 The micro-processes found in studies of groups going through the WeValue InSitu process of Tacit-to-Explicit Translation. The Polanyi (Polanyi, 1962) processes identified are: As-assimilation1, R-reflection, PD- primary denotation, ReO-reorganization, Ds-designation, Rel-reinterpretation; G-groping for words; RR-reading of the result/ read result. ES = external Stimuli.

internal meditation. But an (almost) all-day mediation: it takes time. And... the built environment around me can completely mess that up. Or (very rarely, and usually accidently), enhance it.

Below I give specific examples. The information I gave above should help you read them with some context: imagine me trying to do some kind of "yoga meditation" with these built environment challenges.

3. Aspects of the built environment which interact with my neurodivergence

3.1 Sound

When sitting in an office attempting to "compose" thoughts and meanings, any loud noise is an intrusion. Constant, and low-volume noise is something I can sometimes cope with, but in many cases stress levels build up without my acknowledging them, and I later lose my temper or show aggression or anger that I did not realize had been increasing while the noise continued.

Since I have described my work approach as "composing"," then try this thought-experiment. Imagine your boss has insisted you write accurately about your thoughts and feelings the first time you rode a bike by yourself: when your dad let go of the bike. You close your eyes: you try to remember the time. Try to picture the road, the bike, the grass: now you're six and you're on the bike and can feel the movement, and the sound of your dad talking to you as he runs alongside you. *How do you feel, how do you feel? You've got to interrogate the situation: that's your task.* You can feel your grip on the cold handlebars; you're trying not to look at the moving ground; and as you look up you realize your dad has... WHAM! Back in your office, a ROARING ROARING ROARING sound tells you someone has just set off the hand drier in a nearby toilet. BUT NO! You're not allowed to indulge in the tingling of your fingers caused by the shock: you MUST IGNORE it, and finish writing this passage for your boss... so WHERE WERE YOU? Ah yes... you're six again, gripping the cold handlebars of the bike, trying not to look at the ground, and DING! DING! DING! A truck is reversing in the turning circle up the road: you can tell without looking, from the warning alarm. You breathe hard a few times to settle your heart rate...and after a few minutes try to immerse yourself again into the past...you're six, and the bike is moving too fast; you can hear your dad talking to you, and then everything goes numb: you're flying...freely flying...not even moving your feet... and the sound of dad falls far away in the distance...and you feel free. *Is it free? Or happy? Or exhilaration? You revisit the scene to explore: which one is* it? hmm... seems more like happy. *Or is it... something nameless? Is there a word for this feeling...?* WHAO! A colleague calls out to ask if you're going to the pub tonight. You try not to contain the startled movements of your arms. Drat! You nearly had it: you nearly had put your finger on that feeling, on the bike. Will your irritation, at being brutally interrupted so many times, erupt?? And even if not...how much time will it take to get back to that "feeling"?

What I've described is quite similar to my situation with sounds in the office area, when I'm trying to work. It's true, I can probably manage: I can probably get through the day and produce some kind of work that will pass as acceptable. But I will not find that type of work meaningful: more likely it will feel like work-for-food: I only do it to pay the rent. You might think: go do something else: your demands on the workplace are too high. But if that's the way my thinking works, then I don't have many options. Either I find an environment where I can think, or... I don't get to think. I literally stumble from one moment dominated by someone else's noise, to another. I don't even have a quiet moment to touch base with myself; to remember who I am. I certainly cannot be satisfyingly creative. So now comes the issue of Meaningful Work (Life?): do I have any entitlement to it? or not? And the answer is: only if the society I live in is civilized enough to understand and try to cater for my office needs; my built environment needs. So now let's get down to the details.

3.1.1 Sounds from the street

Even while working, I can't stop my ears from listening. So, if there is a bus stop nearby, I will hear every stop and start, the doors opening and closing. If one is an express stop and another a normal bus stop, I'll probably be able to tell you the ratio of express to normal buses passing each day. I won't need a clock to tell me the time: I'll know it from the traffic noise. Similarly for the weather: people and vehicles make different sounds in rain or heat waves. I'll probably be able to tell you if there's a football match up the road, from the extra buses. The same applies for trains: if there is a rail line nearby then I'll soon be familiar with the patterns of arrivals and departures. Although these are often tolerable sounds, I am more calm and productive without them. Nearby carparks are less of a problem: I get used to the sound of a three-point turn: the rhythm of cars searching for empty spots. Unless they sound their horns at each other... POSSIBLE DESIGN LESSON: you might not need to do anything in such as case, but be sensitive to potential ND needs.

Traffic lights are worse. This is mainly due to large vehicles with air brakes, and the associated sounds of breaking to bring them to a stop, and then the engine revving to get going again. If there are pedestrian crossing signal noises, then I doubt I would last more than a day before feelings of flight/panic/violence rise. POSSIBLE DESIGN LESSON: don't site my building near noisy pedestrian crossings; avoid traffic lights.

Nearby recycling or scrap yards—places where materials are thrown into piles—these are difficult to cope with. The problem is the irregularity combined with the loud and relatively high frequency sound of metal banging around. I'm like a sleeping dog, trying to focus but at the same time with one ear up to listen for, identify, and furiously try to filter or erase, sounds. While working meaningfully. POSSIBLE DESIGN LESSON: don't site my building near such yards. Actively contest proposed changes in land use.

Road works (engineering work) should not need pointing out. But in some countries like the UK it is almost perpetual: a stretch of road may take months to complete, during office hours, and with no apparent limit on the decibel levels. Jackhammers are commonly used. I might last 2 h. (In Shanghai for 10 years I only saw work take place at night.) POSSIBLE DESIGN LESSON: have 3-layer sound proofing windows and possibly (heat pump) air conditioning if this is a likely scenario. Can't afford it? Why not design some portable, self-fitting ones?

3.1.2 Sounds from the site grounds

One of the most incomprehensible aspects of working (in the UK) for me, has been the lack of professionalism of university estates managers in keeping noise levels to reasonable levels. Every Friday morning, for 4 h, a rotary line mower was used on the grass around my building, followed by air-powered leaf blowers rated at about 120 dB at 10 m. Every one of the 240+ offices and teaching rooms facing it were impacted. This is 20% of my working week. But there are apparently no clear guidelines on noise pollution: nothing that the Health & Safety department could do. (The fact that the grass did not even grow a lot of the year made this feel worse.) If I had a line-manager willing to stand up to the estates department maybe things would have been different but...I have yet to see a university where the estates officers have less pragmatic power than any academic. POSSIBLE DESIGN LESSON: design OUT the estate maintenance noise (artificial grass??): design IN enforceable clauses in maintenance contracts. INNOVATIVE SUGGESTION: autism-savvy architects could set up a side-business providing sympathetic specifications to estates departments.

This suspended disbelief at lack of professionalism has been extended several times to academic managers. Such as one who commissioned the refurbishment—including gutting out of concrete walls—of an eight-storey building, WHILE THE STAFF continued to work in it. The manager purported that leaving one floor clear above and below would be enough soundproofing. The staff had few options: for 2 years, we had jackhammers and similar noise. At one point I lost my short-term memory for several days, and was allowed to work elsewhere. POSSIBLE DESIGN LESSON: do not authorize this unless your conscience has a very thick skin.

3.1.3 Killer sounds in my office

I can tolerate some noises if I have an office of my own, because I can feel some reassurance the feeling that the noises are "out there" and unlikely to suddenly be at my shoulder. All of this is lost in an open plan office. Short of surrounding myself with partitions and attending with an astronaut-style helmet, I would not be able to work meaningfully—or at all—for more than a few hours. POSSIBLE DESIGN LESSON: don't put ND staff in open offices without lots of special measures.

With an individual office (Fig. 3), I can begin to work seriously, but some noises make it completely ineffective. One is the location of nearby audio-visuals, such as lectures or videos running in nearby teaching rooms. Another is noises from nearby toilets, such as hand driers, flushing toilets, water feeds or water tanks, or drainage pipes. Lastly, noise from neighboring offices where people are speaking or in video-conferences with speaker phones. Let me be crystal clear: it is not just a possibility that such sounds will be a problem: it is highly unlikely they will not be a significant problem. POSSIBLE DESIGN LESSON: use excellent soundproofing between offices (including well-fitting doors). Design-in water supply and drainage pipes which don't intrinsically make noise. NEVER authorize air hand driers.

3.1.4 Daunting sounds in work rooms

I used to have a high-level meeting every 2 months in a specific room sited near the CEO's offices. My job involved a lot of nuance: it was essential that I recognized subtle opportunities and challenges revealed from participants during the meeting. However, the conference room had an automatic air pumping system, which had an irregular beating-hum with varying pitches. I realized I was extremely stressed after the meetings, but only after a year did I realize that I was actually counting the days, 42 days in advance,

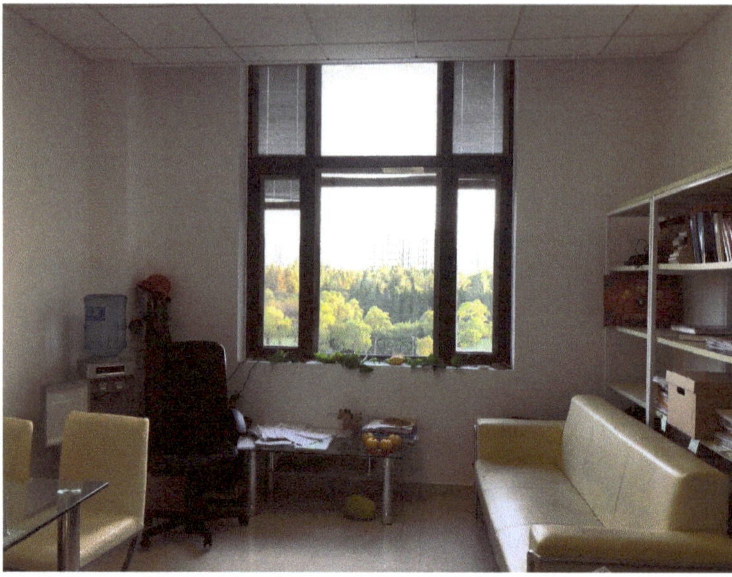

Fig. 3 My office in China. I open the door: the room is waiting for me. A bubble of space: waiting for me. Nothing moves, nothing stirs, nothing has any volition: until I command it. Everything in its place. Everything there to serve. No sound, no Others, but my being. The Master has arrived: the Conductor settles. The engines whirr: the knowledge machine begins, in the silent womb of the room. As I work, I draw meaning like cotton candy out of the air, and it silently forms.

of when the next meeting was. The problem is not the noise, of itself: it was that if I was concentrating hard on the meeting, then my filters to the noise couldn't work, BUT WORSE: in trying to go back and forth between sound sources, my brain was somehow exhausted. It reminded me of childbirth when my husband tried to chat to me between final contractions: the stress of trying to CHANGE FOCUS from blocking pain and comprehending chatter, was immense. I tried sitting in the conference room when it was empty to get used to the sounds: I could not. Finally, when I realized I was waking 2 h early worrying about the meeting, 4 weeks in advance—then I made a plan to informally persuade some mid-level estates staff to get the air pump serviced. The impact on me was incredible: the subsequent meeting was easy-peasy. But funnily, several other people commented at how much nicer the room was without the annoying sounds: it had actually been bothering several people. POSSIBLE DESIGN LESSON: don't design-in mechanical equipment that you cannot vouch for: do NOT let your designs be corrupted by maintenance-reliant elements. Perhaps you should include mandatory servicing?

3.1.5 Terrifying toilets

Have you ever seen a film with a scene from a heavy-metal concert or disco, where it's impossible for people to hear each other? That's what a toilet hand drier sounds like to me. Solution? Save up toilet needs until you go home? Or maybe... find a toilet for disabled people—at least while you're there no-one else might come in and activate it. POSSIBLE DESIGN LESSON: don't use noise machines: recognize them for what they are.

3.1.6 Fire alarm freeze

In UK schools and public buildings it seems there is a norm to have fire alarms of incredible volume. For me this means that every 2 months when there is a fire alarm test, I will be subjected to a audio beating which feels like a stick. WHY DO THEY HAVE TO BE SO LOUD? For you, it's a chuckle. For me, I'm in such a state that I actually start to slide down the wall into a sitting position with my ears covered, rocking. Sometimes it takes me a few minutes before I can remember that I am supposed to be moving out of the building. My desire to face a wall and head-bang, to ease the pain of the sound, is very strong: I have to keep repeating to myself something simple like: move out/move out/ move out or I will forget everything else besides trying to fight the noise. Again, childbirth pain comes to mind. Luckily, in other countries that I work in, this bestial practice does not exist: just simple alarms. POSSIBLE DESIGN LESSON: don't use moronic volumes for alarms.

3.1.7 Quiet pods

Many years ago, it was common to have soundproofing "hoods" over public telephones situated in corridors. Those hoods could be reconsidered for a new purpose: "sound havens". Whether in pokey corners in corridors or in modern looking "pods," it would be useful to me to have a place where I could stick my head just to have some noise protection for a few minutes. Or... perhaps I need to wear noise-canceling headphones everywhere instead. (But those never seem to do the job...) POSSIBLE DESIGN LESSON: provide occasional quiet pods for recovery? AND/OR: provide a hook for my headphones. (joke).

3.2 Lighting and signage and loud colors

3.2.1 Lighting

The second-worst intrusion in the built environment for me, is lighting—if it is moving. In that case, it is almost inescapable. This usually includes slight flickering in fluorescent lights. It doesn't seem to matter much if they are far away; sometimes having them in the far background when I'm (trying) focusing on something nearby, is actually worse. I don't know how to describe the problem: I guess it is similar to the normal room light being switched off and on all the time: it seems other people don't notice but I do. It seems some people don't mind a police car light flashing outside, but I do: it's like a constant vibrating drum that resonates with my head instead of my chest. It's very tiring to endure—but not as bad as sound. And I have learned to relax quite well while covering my eyes with the nearest jumper/ pillow/ shoe… But if I have to keep my eyes open, it would stop me from being able to work. If it is from my computer screen, I would not be able to use it. Strong lights also seem a problem, but mostly when I am stressed or am having to deal with noise as well. Online video meetings have lights that are painful, but endurable. POSSIBLE LESSONS: don't use flickering or moving light sources (inside or outside). Or very strong ones.

But what if you wanted to make me feel comfortable: ready to sit down and effuse creative outputs? In that case, in this modern age, the answer is pretty simple: give me a remote control and install multi-effect lighting, like LED ones. Then I can adjust them as I wish! Why mess around with complex daylight tubes when LED lights can produce something similar—very even and adjustable frequencies and luminosity. But I do suggest…a wall hook to place the remote control so it isn't lost…

3.2.2 Signage

I don't recall if signage has been reported as a problem. But certainly, for me, it is. There I am walking down the hallway trying to daydream, and someone shouts, "EXIT! EXIT! FIRE EXIT THIS WAY!". Of course, there is no one shouting, it's just the sign, but it feels like someone is shouting. FIRE RISERS; CONFERENCE ROOM; DO NOT ENTER—it is impossible to walk down the corridor without being commanded to pay attention to incessant trivia. Every 10 m. Every day. INSTRUCTIONS FOR EVACUATION…it never ends. And in my office, it is inescapable (although I often hide signage…). DO NOT OPEN THIS WINDOW OUTWARDS!!! One of the things I love about working in China is…they don't have many advertising boards. Anywhere. And the written language is in pictographs: it's not phonic. So the Chinese word for fire, looks like a small dancing flame; the word for EXIT! looks like a route out: a kind of trident. Not a yob: silent. Peaceful. Dignified. Respectful.

3.2.3 Loud colors

If you've ever been outside on a very bright day, say midsummer California, then you'll know it hurts the eyes. Regardless of the beauty of the day, your head will hurt. Well, that's what clashing or bright décor does for me. I prefer to close my eyes and not look at all. I once had a living room with raspy ochre orange walls, and several weeks later realized I always became very grumpy there. I painted it lilac and everything became calm. I appreciate some neurodivergent persons will say they like stimulation: then let them put up posters. I cannot repaint all the walls in my life! My office has cream and glass furniture, and cream walls. Thus, they blend into each other: as close to invisible as I can get them. If even the struts of the shelving had been brown, I would have a constant audio playing in my mind every minute I looked up from my papers, saying, "brown, white, brown, white, brown.....are they regular? Check it again. Are they straight? Check it again. Brown, white, brown, white...." But to work meaningfully, I need quiet.

3.3 Office frills

Let me move to ways that my built environment can enhance my life. Ok—you say "working life," but let's remember: autistic persons don't necessarily have fulfilling social lives, so those who have fulfilling work, would possibly consider any workplace enhancement to also be a major life enhancement. I'm curious whether you, the reader can anticipate what I'm about to say: I doubt it. Let's take a tour of the office I'd love to have.

3.3.1 Privacy and solitude: a lock, and no light showing

If I am to immerse myself in thought (the land with no words...tacit space), then there are times when I'd like to be sure no one is going to knock on my door. It's hard to focus otherwise. So... I'd like a lock on the door, and a way to prevent people knowing I was there—like covers on any windows to the hallway. Maybe an adjustable sign on the door saying Don't Disturb'? Do you know when I work in the office the most? At night. I love to be there until 1 am. There is much, much, less noise, and less chance of interruptions. Therefore, I need a building with 24-h access, preferably with safe passage to wherever my car or scooter is parked. In China I can walk home safely anytime of the wee hours: in the UK I need to phone someone to say I am about to leave the building, so my path can be tracked in case I am mugged. Therefore, my built environment should be designed to keep me safe while working alone.

3.3.2 A den for the long haul

Immersion into my tacit zone takes time. It takes me 1–2 h usually, to settle deeply into work (assuming I am not required by circumstances to constantly interact with PeoPle, **noise**!!! or FlAsHiNg LiGhTs and SIGNAGE !). Then, I prefer to work in that tacit zone for a long period. It's like floating underwater: it's peaceful; my thought processes are such gentle interactions compared to people talking; there is an ocean of wonderful ideas to explore. Even when I am focused on bringing ideas from there to "the surface," that process of "composing" thoughts into language can be very calm and satisfying when I am not disturbed. It reminds me of losing oneself playing a musical instrument: there is no time, no place; only the music and exploration of it. This is not only the case for creative work. I can also zip off large amounts of boring work in this mode (Fig. 4).

But…this means the natural pattern of am enjoyable work session would be about 12–16 h long. I'm quite happy with the length of time but…I might need some snacks/food; some variety of clothes layers for changing

Fig. 4 Views from my office window in China. I am immersed in my ocean of thoughts: all is well, all is well. My feet walk the floor; turn at the window: all is well, all is well. While my mind wanders in depth; my eyes pause out the window: all is well, all is well. Trees upright; buildings set back; lampposts in rows; layers and layers aligned. All is well. Thoughts continue to weave.

temperatures; some variety of furniture for different sitting/slouching positions to avoid aches. This means that my "dream" office would have 2–3 different types of seating including a couch long enough to nap on (!); and easily adjustable desk; places to store a variety of shawls/ scarves/ hats/ socks/ small blankets. Places to store dry food/ UHT milk/ bowl/ spoons. Absolutely need a kettle and mugs and places to store various teas and coffees. The basic idea is to have everything there so I don't have to go out and risk interacting with "people".

3.3.3 A walking place

Finally, I love a walking space. 4–5 m is fine. I usually arrange any furniture so that I have clear passage as long as possible across the room. This is because I often like to walk while reading or thinking. But, since I don't look up much at the time, I need nothing in the way (I can do a turn-around on auto-pilot). I actually often do composition that way: walk with a clipboard and pause to write now and again (or write as I walk if walking slowly). At night, in some buildings, I can walk scores of meters along corridors in a straight line: it's great for writing out complex ideas. There is a catch with the lighting: if the lighting is too spaced out then as I walk the light on the clipboard changes a lot: it's nowhere near as pleasant and I usually give up. In China my office has long, circular walkways nearby in parklands, and the weather is often just right for walking: it's blissful. And then, when I have been writing and walking for about 90 min, I can continue the same deeply tacit thoughts while sitting down to give my whole frame a rest: both in walking and resting afterwards, my body can be disregarded. Very meditative (Fig. 5).

3.4 Social common areas

3.4.1 Well-crafted space-sharing places

I wonder if you anticipated this one? Since I don't really know how to start a conversation "nicely," and I am unlikely to attend non-purposeful social events, then…it is very useful if the built environment nudges people into situations where I can socialize in other ways. One way is to have long tables in cafeterias and cafes, so it's "normal" for strangers to sit next to each other. Then, I can just listen in to several conversations and join if/when it seems to work out. Or I can just observe: I require a lot of data about how different people interact in order to function, and that's a good way to collect

Fig. 5 Circular paths outside my office. Clipboard, paper, pen, sneakers, walk, walk, walk… Head up, head down, where's that thought, write it down, Walk, walk, walk, walk… Walk slower, walk faster, it doesn't really matter, Mind is disconnected, feet do what they want. Eyes on the clipboard, mind in the mist, if the turning is missed Then the next one will do just as well. Walk, walk, walk….

it. The seating arrangements have to be quite subtle: if I join and a group is already there, they should be able to politely signal they wish to ignore me, or just continue their chats with tacit acknowledgement that I am now a passive listener. If too many or too few seats are available, then it's hard to make that concept work. On the other hand, an isolated seat with a view of the room allows a social observation data collection session during a tea break.

3.4.2 Non-interactive common areas

If I'm just after a functional need, like a purchased cup of coffee, then I probably won't feel like interacting with the café staff. There are shop set-ups which facilitate this, so that I don't feel obliged to make full eye-contact; full interaction. I can just browse quietly; pick out my own selection and put it on the counter with the money (or digital payment on phone in

China!). I really don't want to do the big calculation of: 'what phrase do I have to say as a 'hello' to the cashier; do I need to remember their kid's name; is this the same one I saw yesterday; which tone should I use; how many microseconds is ok to hold eye-contact... Often I want to just put the sandwich and money on the counter and keep looking at it until the money is taken...

3.5 Wrap up conclusions

It's probably true to say that any special considerations made to improve the built environment influence on "special" people, would also make good sense for "normal" people. It seems to boil down to civilized considerations of sound, signage and office space protected for its purpose: reading, composing, and thinking. However, that's only true on the surface. "Normal" people won't notice very much if half of the civilized considerations are missing. But if architects could string together ALL the apparently "normal" desires of neurodivergent users, then they would have a chance of producing a built environment that truly enabled them to have a meaningful life. Versus a miserable, unproductive and unsatisfying one. The difference? Don't just design out one or two noisy features: systematically eliminate them. Don't just specify a few signs to be pictorial: look into it for all. Write into the specs for the fire alarms that they have reasonable decibels. Design-out leaf blowers, hand driers, traffic lights for trucks; ventilation fans with sounds. Get a holistic picture, and implement the implications in a comprehensive manner. This is not a checklist exercise: this is not a place to tweak a little, the features you were going to put in regardless. You need to achieve the Whole.

So am I high-functioning? And just unlucky, that much of my life I am surprisingly stressed out and explicitly uncomfortable in standard built environment which is sub-optimal for my marginal problems? Or, am I low-functioning but then super-enhanced by having obtained working spaces that maximally nurture my potential, if only by accident? I was nearly fired when my communication skills fell apart and my behavior was mistaken for non-cooperation and loss of ability—at a time when I worked in a building being gutted and refurbished for 2 years around us, and leaf blowers and hand driers made regular appearances. Yet when I changed to a campus with a very quiet office with Chinese signage and white

Fig. 6 Teaching Halls in China. Time to learn; time to be serious. Everything ordered, ready to file. Knowledge in; knowledge out. See the patterns, see the patterns, see the patterns. Columns straight, regular rows. Bridges and arches and balconies Plenty of volume to hold all the thoughts All the thoughts, all the thinking, all the talking, all the meaning. Plenty of volume to hold all the thoughts.

furniture, and gardens to walk in—I became a senior professor and produced five times more outputs at several levels' higher quality, year on year (Fig. 6). I thus propose that my "level" of autism depends on the "level" of accommodation or disregard of my built environment and line managers.

3.5.1 Plenty of order to file them to knowledge

And what can you do if you cannot accommodate me through building design: or if your budget runs low? Well, bear in mind that advances in mitigating technology is overtaking what architects can usually provide. Maybe we just need excellent headphones. Why not make up a pack of materials to retrospectively 'outfit' an office for people like me: wall coverings to dramatically improve soundproofing; third-glazing units with adjustable fittings; door-stripping kits; special office carpet and ceiling tiles; replacement signage. In other words, do not assume you will be the savior: you are in competition with technology. Maybe pre-fab sound pods in the car park are the best solution! Remember the concept of Unitary Validity: your solutions are not valid until you can justify the entire chain of decisions for the purpose. And the purpose for your designs is to make space and place for neurodivergent persons like me to make meaning in their life (Figs. 7 and 8). I wish you strength to achieve it.

Fig. 7 Detail in the ground. Lined up Lined up Lined up yes. Nice oval ends, oval ends both sides. And the Man-made cement meets the Earth-made earth. So cleanly, tidily, naturally OOPS. Who let that plant shoot grow here? Haha! Nature always wins.

Fig. 8 This is my place. Oh! Didn't you know! They made this for me. I am unknown until you've seen this as me. This is my place! This is my place! It's ME! Sunlight spots on the smooth flat floor. Strips and crosspaths, diagonals. Look up, look up: hah! What a trick! Its weaving and wefting and basketing light. Not a tent! Not a roof! But a Den, clever Den. Welcomes wind sun and air. No fight to exclude. No fight to partition. No separation of Me. From the Rest: just Us. It's a Place, in a Space, in the Rest: it's One.

References

Bahá'í International Community. (1995). *The prosperity of humankind*. Haifa.

Frankl, V. E. (1959). *Man's search for meaning*. Boston: Beacon Press.

Maslow, A. (1954). *Motivation and personality*. New York: Harper and Row.

Maslow, A. H. (1993). *The farther reaches of human nature* (1st ed.). Penguin: Arkana.

Messick, S. (1995). Validity of psychological assessment: Validation of inferences from persons' responses and performances as scientific inquiry into score meaning. *American Psychologist*, *50*, 741–749.

Odii, B. C., Huang, Y., Bouvrie, N. D., & Harder, M. K. (2021). Cycles of meaning-making crystallization in the WeValue InSitu process as clear contributions towards transformative learning. *Journal of Cleaner Production*, *304*, art. no. 127024.

Onwuegbuzie, A. J., Witcher, A. E., Collins, A. T., Filer, J. D., Wiedmaier, C. D., & Moore, C. W. (2007). Students' perceptions of characteristics of effective college teachers: A validity study of a teaching evaluation form using a mixed-methods analysis. *American Educational Research Journal*, *44*, 113–160.

Polanyi, M. (1962). *Personal knowledge: Towards a post-critical philosophy*. London: Routledge & Kegan Paul, Ltd. pages 84,110–111.

CHAPTER SEVEN

Equalities design: Toward post-normative equity

Natasha M. Trotman*
The Wellcome Collection Hub, The Wellcome Trust, London, United Kingdom
The Helen Hamlyn Centre for Design, Royal College of Art, London, United Kingdom
Kensington Gore, South Kensington, London, United Kingdom
*Corresponding author: e-mail address: natasha.trotman@network.rca.ac.uk

Contents

1. Introduction

I am an international Equalities Designer and researcher. My practice explores extending the frontiers of knowledge around mental difference, non-typical bodyminds, ways of being and marginalized experiences. It seeks to challenge and reframe normative notions of equality and diversity, and to move toward post-normative equity in recognition of a wide range of social and structural dynamics and contexts. I use a contextual equalities lens that considers the bio, psycho-social fractals and intersections that inform articulated outputs to meet a brief and/or need; as well as the broader, more nuanced situations that neurodivergent, multiply—marginalized designers like myself operate within.

My design journey started in a conventional way with a BTEC Foundation in art and design; I then went on to complete an undergraduate degree in graphic and media design. During this time, my interests and approaches diverged from what was considered design. I arrived at the

Developments in Neuroethics and Bioethics, Volume 5
ISSN 2589-2959
https://doi.org/10.1016/bs.dnb.2022.08.004

Royal College of Art (RCA) in London UK and although I already possessing a broad suite of skills, I struggled to find a good "fit." I worked on various student projects, live briefs and then expanded out into the creative terrain with an MA in Information Experience Design.

Once I graduated, the real lifelong design challenge began. Theory and practice began to collide; a paradoxical change theory occurred in me, a term coined by Beisser (1970). Through this creative journey, I have realized that the most profound change transpired when I've allowed myself to become what and who I am, on my terms, not when I have attempted to be what and who I am not. The journey is not over, though; I am still becoming, still emerging. Once I began working in the field, the depth and breadth of the difficulties, and lack of non-normative inclusion-based offerings, was vividly brought into focus. I acknowledged that exploration and mining for a truth that goes beyond the data were necessary. I spent some time at this, but the nuances and scaffolds that created or upheld the challenges for multiply-marginalized groups became increasingly apparent. Non-normative production and consumption was being deferred, decentered, and disenfranchised by design.

The numbers told a part of the story: women, people of color and disabled and neurodivergent creative people were still being excluded. The fundamental design challenge was within the creative terrain, which for me then became fertile ground for knowledge production and dissemination from marginalized visionary thought leaders. I began to investigate the inequalities within the profession itself. This was through learning from the work of Dr. Elizabeth (Dori) Tunstall, for example, who has pioneered areas such as respectful design (as informed by Indigenous principles) (Tunstall, 2021) and the "decolonization" of design education (Tunstall, 2013). She discusses the interconnecting threads across philosophy, academic design research (and psychology, respectively), the importance of hybrid practices, and presented me with much to think about on alternative approaches to normative structures and systems.

Using the frame of post-normative equity, I have since been developing Equalities Design tools, methods and approaches that interrogate and queer normative thought patterns in order to create alternative bodymind explorations and experimental outputs. These can be contrasted with solutions that have been presented as inclusive but are in fact simplistic design responses—such as sensory playgrounds based on people who are visually hypo-sensitive but leave out other bodymind variations such as hyper-sensitive and tactile sensory seekers. There is a danger with well-intentioned, assumptions-based inclusive design when it lacks contextual complexity. This can lead to the creation of inclusive guidance and projects that focuses on the needs of one group, with a defined and limited set of needs, and which have been decided by "experts" rather than through participation of the diverse individuals affected.

In some cases there are agile and diverse design teams, combining access consultants and inclusive architects to ensure both the neurodiversity paradigm and social model of disability undergirds all outputs, providing (non-normative) equitable solutions to (historic) pain points for the multiply marginalized. In my practice, the aim is to discover and map out all those aspects that fall within the exo-system (see Bronfenbrenner, 1981, 2004) that is, the effects of a setting or settings on not just its active participants, but on their lives and actions beyond that setting. These are the things that may not be conventionally understood as "design" but deeply inform and impact designers; and shape who gets to be a designer and what is ultimately designed. I am still in the process of mapping the impact of such processes on both the designer and on the designed, but the statistics that I will expand on later in this chapter help to make a case for post-normative accessible and inclusive offerings for neurodivergent and multiply-marginalized bodyminds; and where such bodyminds are not just users but also creators.

Lisa VeneKlasen and Valerie Miller, educators and activists, ask us to consider notions of power and the mechanisms through which various dimensions of power operate (VeneKlasen & Miller, 2002, 2007). Here, this is about excluding/privileging agents and organizations through existing structures such as normative, sanctioned rulemaking and by enforcing hidden power to those belonging to dominant groups, deputizing those in the dominant group to uphold and enforce normative practices (hiring, outsourcing, materials, awards). The visible and invisible realms of design and its actors should therefore be considered. VeneKlasen and Miller discuss power as "both dynamic and multidimensional, changing according to context, circumstance, and interest. Its expressions and forms can range from domination and resistance to collaboration and transformation" (VeneKlasen & Miller, 2002: 39). Forms of power include: Power Over (authority, and also a duty of care); Power To (able to act and actualize); Power with (collective action, creating change), and Power Within (dignity and empowerment).

This leaves me contemplating the identification of the markers for neuronormative induced pain points in the existing exo-system and how best to pause or transform them through non-normative making and place holding; reworlding and building in alternative standards and approaches through practice. My lived experience has resulted in mapping out where I thrive best; it's in the in-between spaces, in flux as the liminal and unknown spaces, unfamiliar areas where the rules are least likely to be enforced, providing a space for alternative meaning-making and shaping. I seek out the in-between and unnamed areas to provide a space for the under-served and to enable those locked out of design(ing) to have an opportunity to weave their meanings where they are, on their terms. Sometimes this looks like reshaping or bringing the words or forms of meaningful exchanges into existence free from established "baggage" and constraints.

A significant shift in my practice began to emerge due to the events during the 2020 pandemic that saw the global community pause at the death of George Floyd, gasp in horror at the Do Not Resuscitate (DNR) scandals and ongoing malpractice concerning neurodivergent and multiply marginalized people. Does design need a normative step change to foster a deeper, more prosperous and more contextual understanding of and around multiply-marginalized needs and challenges? This answer is of course yes, and this is what I am trying to address through practice.

Previously in my practice I looked at alternative lenses and frames of reference and how to articulate these experiences to the mainstream; naturally progressing to interrogating and unpacking the intersectional scaffolds that Kimberle Crenshaw examined, and by bringing the term intersectionality (which she coined in Crenshaw, 1989) beyond the borders of the US where she is based. I also began to consider the two-sided coin that informs the majority of designed outputs, the parallels of neuro-normativity and hetero-normativity explored by Dr. Nick Walker (2021) and colleagues (Walker & Raymaker, 2021). Walker invented the term neurodiversity paradigm and neuro queer (queering: customizing subverting, abstracting); I am still exploring where I and my practice sit within this further future landscape. These profound and challenging realities reveal the contextual equalities that need to be considered when trying to engage with the problem of those being left out or mis-represented; and how best to design in dignity that moves away from the deficit or pathology paradigm. This also requires a moving away from "accommodations" and "reasonable adjustments" frameworks, that require individuals to argue for "special needs" and to explain an entire personal medical history in order to get a dimmer switch installed in an office space. Inspired by Urie Bronfenbrenner's ecological systems theory (Bronfenbrenner, 1981, 2004), the Design Exo-System acknowledges Western frames of reference that undergird neuro-normative, sanctioned, valued and accepted design. I started to consider how best to address these challenges through my practice; I began mapping out what a four quadrant

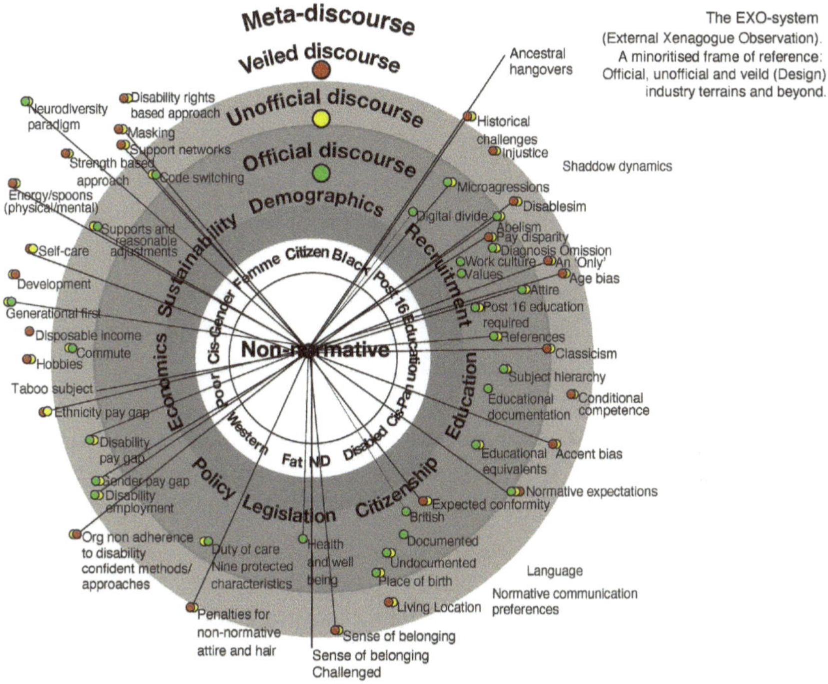

Fig. 1 Diagram of design EXO-system (external xenagogue observation). A minoritized frame of reference. *Created and drawn by the Author.*

Design Exo-System involved: performing the racialized normative, the abled (performance), the authentic self, and the gendered normative (Fig. 1).

The following factors play out in the Design Exo-system; the designer is not usually an active agent within this system, but the previously mentioned factors are shown to profoundly effect the marginalized designer's career ranging from entry to the field, through to development and overall trajectory, because of a variety of reasons. My multiply-marginalized experience designing as well as using the services, offerings, and designed objects created with neurodivergent and disabled persons like myself in mind led me to realize contextual and intersectional analysis are at the core of designs that are a good "fit." This means something that goes beyond being purely functional by also taking into account form and context. Contextually robust methods and approaches—within designed offerings and developing outputs, place-shaping and more—would ensure that articulated outputs for neurodivergent D/deaf and disabled people do not only benefit those it's intended for but can also be created by the multiply-marginalized for the multiply-marginalized—an authentic radically inclusive virtuous circle.

2. The (racialized) normative (dominant culture)

Current creative initiatives, that is the many socio-economic projects, programs and schemes based on normative arcs of development, diversity, access and pathology paradigms, are problematic and must be revisited. For example, these approaches cannot be simultaneously diverse and inclusive while being ageist and ableist. Non-normative intersectional analysis needs to form a part of the equation when organizations create such initiatives. Currently, entry to stated offerings based on the traditional age and career axis is common, unthinkingly focused on those who fall within the 25–30-year-old age range; and leaving out all who fall outside of this normative career development arc and "traditional" career timeline. This "aging out" of the profession then misses the barriers experienced by older people. How might we create consistent embedded support for multiply-marginalized people, irrespective of age, color, gender etc.? Many may be returning to a primarily normative field (or may have left for this reason), or need support with progression for various reasons. Multiple and customizable routes for progression are required for those who fall outside of the traditional, normative career development arc for various reasons.

Organizations that state they aim to achieve racial and/or neuro equity via their schemes and programs are often well-intentioned. However, often the criteria fall within normative and often binary identities. Those who grant funds to such initiatives and schemes also need to apply non-normative intersectional analysis to ensure the terms and conditions provide equitable and positive outcomes for all. This will also help to embed change (similar to when funding bodies' terms and conditions for monetary support are challenged on their underlying age exclusive and inequitable practices). Additionally, initiatives that seek to address diversity and access need to apply intersectional analysis at an organizational level to illuminate and remove in-house normative ageist and ableist mechanisms; such changes will provide fertile ground for new policies and offerings for those who fall outside of the traditional career trajectories. An intersectional approach that draws on civil and human rights-based methods may be needed to tackle the ongoing ageism and ableism that plagues inclusive initiatives, schemes and offerings. By insisting on our rights and dignity and creating responses and solutions that erode such inaccessible practice, we can "level the playing field' so that people with disabilities can access jobs, education, health and other services. A rights-based approach is about removing physical and social barriers; it is about attitudinal adjustments for policymakers, employers, teachers, healthcare professionals, and even relatives. A rights-based approach aims to ensure universal design, accessible technology, and coordinated public programs and service. This approach requires the government to provide the resources necessary to implement these goals and enforce penalties for those who refuse to cooperate." as defined, for example, by *The Southern African Federation of the Disabled* (SAFOD, 2022). Using a rights-based approach, we need to begin to consider how a multiply marginalized designer navigates across the axis of ableism, racism, sexism and disablism that can impact the design(er) singularly and simultaneously.

These observations gave rise to uncomfortable conversations and exchanges with my neuro-normative non-disabled peers. I ask them (and you) to consider—how might contextual equalities through design illuminate the root causes of inequalities and historic challenges? This does not require an immediate answer, but it serves as prompt for recognizing and responding to the fact that normative comfort/s and associated designs may be coming at a cost for non-normative bodyminds. I acknowledge that not every creative or scientific sector is ready to interrogate and decentralize normative methods and approaches and instead pivot toward creative, unofficial discourse and non-normative creative hybridity and multiplicity. So, I also began to use this process as a window to peer into the field, and as a mirror to look into myself and ask fundamental questions concerning the challenges faced by those with multiple protected characteristics and how inclusive and diverse the field is at ground level. To arrive at a genuinely inclusive and equal terrain, the absence of non-normative bodyminds needs to be acknowledged and considered, to inform embedded equalities focused approaches in the built environment and beyond.

3. Equality, ethics and disability justice

The basis of Equalities Design, as I have already outlined, is centrally about equality and human rights. Until the eighteenth century in the Western World, human beings were largely perceived as "naturally" unequal. This attitude collapsed with the advent of the idea of human rights, which assumed a natural order in which all human beings were equal. Against the ideas of Plato and Aristotle and the classical formula for justice (in which an action is considered "just" when it offers each individual his or her due), a rights-focused, egalitarian meaning emerged in which everyone deserves the same dignity and respect. Here, equality means each individual or group of people is given the same resources or opportunities; while equity recognizes that each person has different circumstances and allocates the resources and opportunities needed to reach an equal outcome. Meanwhile, if conventional ethics has centered on positions and actions, within disability activism and scholarship, the focus is more often on disability justice, that is, on situations and consequences (see chapter "Disability studies, neurodivergence and architecture" by Boys). Disability justice as a concept has been developing since the 2000s, aimed at integrating disability as intersectional with other forms of discrimination across race, class and gender, etc. (Piepzna-Samarasinha, 2018; Sins Invalid, 2019).

Although much change has happened, some challenges endure for true disability justice. Accountability, responsibility, and intersectional analysis all need to be strongly considered personally, professionally and collectively. When providing instructions concerning navigating spaces as one's "whole self," these factors need to inform the advice given as a singular approach may erase the lived experience and risks posed to those with intersecting or multiply marginalized identities. From my point of view, the mechanisms for all to turn up as their "whole self" do not currently exist.

In my own practice this has clear ethical implications; based on principles of accessible participation, consent, anonymity, confidentiality, safeguarding, risk analysis (internal and external stakeholders), dignity, access analysis, communication, justice, autonomy, agency, transparency, and trauma-informed methods and approaches. These approaches are informed by the social model of disability, the UK Equality Act 2010, the CRPD rights-based approach and the 12 pillars of independent living outlined by Disability Rights UK (Spectrum, 2022). These disability-centered principles can be applied to the human spectrum of needs, abilities and across age ranges; however, some additional protocols may need to be engaged with if working with children and young people in supported settings. There are various legislations depending on the age of the people being engaged. The best standard practice is ensuring that enhanced DBS is obtained, and that safeguarding, GDPR requirements, a trauma-informed approach, autism awareness, D/deaf awareness, and awareness of appropriate professional boundaries (e.g., in relation to appropriate touch) are completed before commencing. Before any engagement, receiving information regarding access needs, adjustments, and likes/dislikes is also recommended for those leading and assisting with the engagement; the granularity of consent must be communicated and consent provided; a granulated approach breaks down the elements of the brief or engagement, providing choices to opt-in or out at any stage. Being aware of sensory profiles can also be helpful, as each person is unique, with their own preferences and access needs. If this is shared, this will enable the preparation of an appropriate setting. This also ensures that engagement does not leave any negative post-session experiences.

4. Dignifying the non-normative through design

In the next sections I explore what such Contextual Equalities requires—how multiply-marginalized groups can be recognized and dignified by design. In Design Council (2018b), the UK Design Council report *The Design Economy* confirmed what my personal experience of actually living as a multiply-marginalized bodymind in the design world felt like; the Design Council report found that the UK Design Sector is made up of 78% of designers who identified as male with 13% from ethnically diverse backgrounds. The design sector employs a slightly higher proportion of ethnically diverse people than the broader UK economy (13% in comparison to 11% of the wider UK workforce). But racially diverse designers are least likely to be in senior roles, accounting for only 12% of all design managers. In 2019 UK government statistics show that only 5.1% of small and medium-sized businesses in the UK were led by a majority of people from an ethnic minority.

How can my practice speak to this? How can I use a Design Exo-system framework to examine the nuances? Having provided feedback on interview processes as a sitting panel member, sitting on the other side of the table or screen has shown that routes to entry are still entrenched in normativity, something I understand the sector is working on. However, both statistics and my lived experience leave me with questions: where are all the non-normative, gender diverse designers? How best to speak on this through my practice? I began to look for projects that would enable me to embrace the hybridity and multiplicity of my practice. Embarking on work as a developer on the Wellcome Collection's Social Justice curriculum (Burn, 2022) enabled me to map out, with the organization, particular design aspects that had not been identified or considered as design before my intervention.

5. The gendered (normative)

The design world is still "performing" (normative, dominant culture) gender (representation) in the sector. The UK's design landscape is majority male-identifying. The 78% figure towers above the broader workforce in the UK, which is 53% male. In the UK, women generally form the minority in six of the eight design subsectors focused on in the Design Council (2018b) report. Yet, women are increasingly going into design education. Contrast this with up 63% of students studying creative arts and design at university level in the UK identifying as women/femme.

What is happening to women designers, students, and recent graduates at the axis of theory and practice in this instance? How does this impact design? I began to question the products, services, and built environments I engage with daily, particularly the aspects I found challenging. I was then left with one question: Was the offerings I have been engaging with made for "Reference Man," and if so, who is He? According to the International Commission on Radiological Protection (1974), he is in fact a hypothetical combination of normative human physical and cognitive characteristics determined by consensus of the dominant group:

> *"Reference Man is defined as being between 20–30 years of age, weighing 70 kg, is 170 cm in height, and lives in a climate with an average temperature of from 10—to 20 0C—He is a Caucasian and is a Western European or North American in habitat and custom." (p. 4)*

And he is still around informing designs today; the above characteristics may be used by professionals across sectors in designs, estimates and offerings; an example of a design informed by and made for Reference Man can be seen in personal protective equipment (PPE) throughout the pandemic, with "standardized" formats for items such as face masks. The TUC (2017) looked into this pre-pandemic, finding that ill-fitting PPE was high in the emergency services, with only 5% of female workers feeding back that PPE did not negatively impact their work or the carrying out of their day-to-day duties, a very telling response; additionally items that needed to be worn on other areas of the body such as hi-vis vests and jackets were also deemed unsuitable. Sociolinguist Tatman (2016) has shown how Google's speech-recognition software was 70% more likely to recognize male speech accurately; in addition Automatic Speech Recognition (ASR) systems make more errors for Black speakers than white speakers (US). This places dominant culture and subcultures (and the associated ideologies) within the Design Exo-system. So where else is Reference Man sneaking in and what does that mean for neurodivergent and disabled designers and service users like myself?

6. The abled (performance), disability and neurodivergence

Disability and neurodivergence were not accounted for in the UK Design Council's (2018b) report. Thinking through the Design Exo system enables us to map out the aspects, events and dominant ideologies that marginalized designers do not create, uphold or actively engage in. This has a profound impact on the marginalized designer's professional development and longevity in the field, and acts as a form of erasure. Dominant value systems do not account for the non-normative or marginalized; for instance, well-intentioned activities during Autism Awareness week or Black History Month create moments of engagement instead of long-term shifts. My attempts to engage with this through my practice involved working on a two-year research project for the Wellcome Trust and The Royal College of Arts (RCA) Helen Hamlyn Centre for Design on a two-year research project called *Design and The Mind* that focused on co-creation and engagement with neurodivergent, disabled and D/deaf people and groups using the Wellcome Collection Hub in London. The central goal of *Design and the Mind* was to explore ways to address issues surrounding cognitive, physical and digital access to the Wellcome Collection Hub and its resources for a variety of users, by framing this as an integral and positive part of design and research processes. The project focused on new methods of collaboration through interviews, workshops, drop-in sessions and experimental co-creation and engagement methods, to map out key user insights into the space and written inclusive briefs to improve visitor experiences.

This led to articulated outputs such as toolkits: *Ready, Get-Set, Go!* (Fig. 1) for use in the Wellcome Collection Hub and *Find It* (Fig. 2) for use in the wider Wellcome and Wellcome Collection; as well as the creation of a public-facing 5-day Program in the Wellcome Collection Hub: *Our Brains and Us*, co-programmed with the Institute of Imagination (Trotman & McGinley, 2018) (Figs. 3 and 4).

I began to foster conversations aimed at illuminating the lived realities around intersectional non-normative existences during this time. There is a growing appetite for such "unofficial discourse"—that is, non-normative ways of communicating—around diverse experiences; resulting in the mapping out and holding space for those who identify as disabled and neurodivergent. As well as critically and creatively interrogating ableist

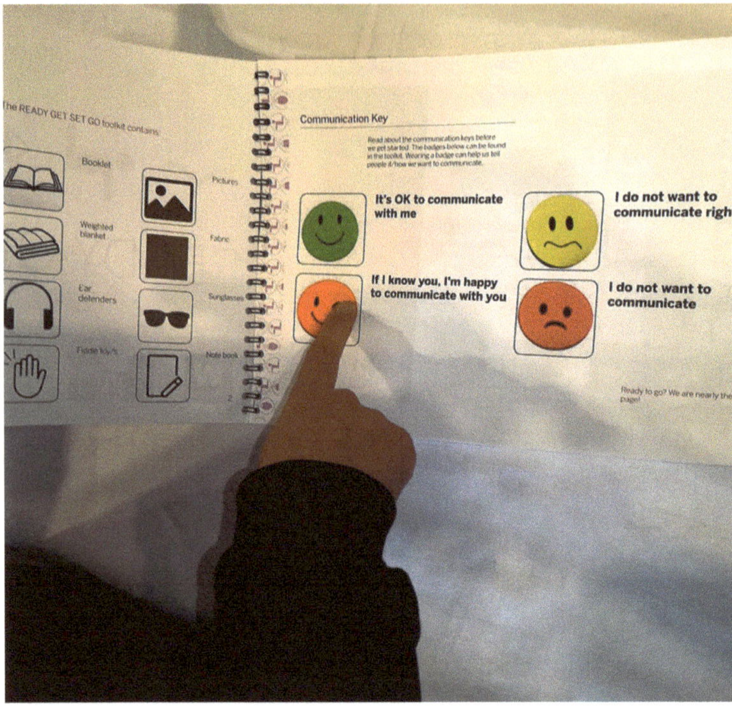

Fig. 2 Natasha Trotman (2018) *Design and The Mind Toolkit: Ready, Get-Set, Go! A "Sensory Vocabulary Kit"* developed out of a co-creation process with neurodiverse people. *Photo by author.*

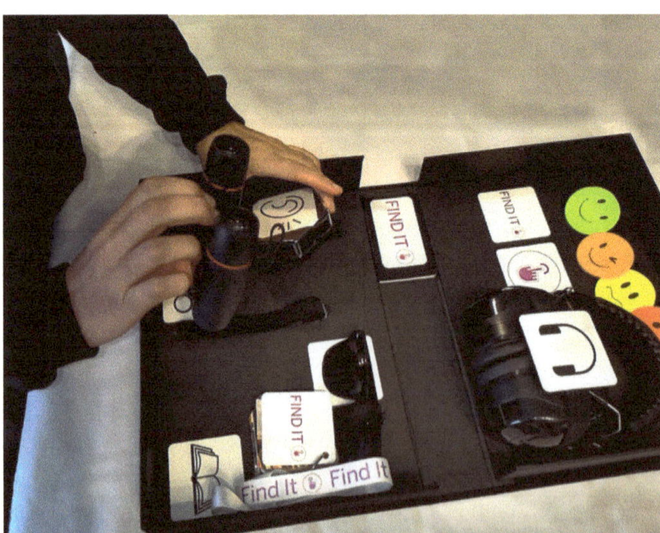

Fig. 3 Natasha Trotman (2018) *Design and The Mind Toolkit: Find It! Photo by author.*

Fig. 4 Wellcome Trust, Natasha Trotman and The Institute of Imagination (2018). Participants co-create as part of 5-day program *Our Brains and Us* at the Hub, Wellcome Collection, London. *Photo by The Wellcome Trust.*

stereotypes and the impact of Reference Man, this is also about unraveling dominant western culture:

> *Not that nothing exists of differences, but that what matters are the systems we use to make sense, to make human societies intelligible. The system we bring to those differences, how we organize those differences into systems of meaning, with which, as it were, we could find the world intelligible (Hall, 1997, p 10).*

For example, who am I beyond the borders of the UK? How might I try to make sense and meaning of the worlds within worlds existing simultaneously through my practice? This led me to embark on one of my biggest challenges to date, taking on the magnificent challenge of being commissioned by the British Council to be part of the Saudi Design Festival in Riyadh and adding my creative voice, holding space as an unapologetically non-normative bodymind in new terrain; with new terms of engagement and encountering unexpected triumphs and challenges that have helped to build my suite of varied terms of references. I learned that marginalization and the embodied psycho-social dynamics related to it/them are dependent on the terrain the person in question is navigating. I have learned that some marginalization/s travel (well) and some do not; this enabled me to broaden my frameworks and concepts and confirmed that context is critical.

When my multiple identities collide, and I am multiply-marginalized all at once, I revisit the Design Exo-System and quadrant consciousness outlined above [(Performing the (racialized) normative, the abled (performance) the authentic self and the gendered (normative)]. This is to consider my position in this current space and time within these scaffolds, as well as my own chronological timeline within design as a discipline, since this too is based on a non-normative arc of development. My particular trajectory is often at odds with the dominant ideology's notions of career millstones and age-profiling career development or lack thereof. When I began to question my peers on whether they felt they had fulfilled their professional aims made some years ago, my multiply-marginalized neuro-kin brought up experiences of the developmental (mythical, youth-obsessed) time-sensitive arc and the feeling of being stymied by barrier based "treadmills" once past 25 years old. Some described feeling like continually emerging but never quite securing consistency, or not being able to access a nourishing terrain to thrive long-term in the sector. Some described needing to "hack" the normative systems when reasonable adjustments were deemed unfeasible. Their various shared, illustrative examples illuminated the repeating intersectional barriers experienced. This included material penalties endured for those unaligned with normative structures and conventional developmental arcs, having to "adapt" by way of code-switching, and long-term masking, among other shared pain points. More strings in the bow of contextual equalities were added—illuminating the desperate need for intersectional analysis, methods and approaches in the sector from recruitment to articulated outputs and beyond.

7. Contextual equalities

For me, Contextual Equalities sits beneath the umbrella of Equalities Design and recognizes that intersections are nuanced, complex, and permeating marginalized designers' internal and external worlds. Sociologist Patricia Hill Collins unpacks some of these pain points in *Black Feminist Thought* (Hill Collins, 1990/2009), particularly what she calls the matrix of domination. I seek to build on this discourse, particularly how age/normative developmental arcs and supporting social constructs intersect with disability and race in the Design Exo-System, the design sectors and academia.

Through lived experience I recall many so-called inclusive initiatives undergirded by neuro-normative value systems and hierarchies that seek to focus on what "value" can be extracted from neurodivergent people. Part of my experiences is encountering situations where there's only enough space for me to be Black *or* disabled *or* working class. Offerings are presented as inclusive but are powered by the currency of singularity; and normative hierarchies and value systems reign supreme. I avoid such spaces. Along with the quadrant of consciousness and the Design Exo-system, academism should be mentioned, as it locks those with marginalized identities out of "official discourse"—subtly but impactfully—in regards to knowledge production. That is because neuro-normative terms of engagement, entry and value systems only accept particular modes of research and knowledge production as valid, valuable and noteworthy.

The matrix of domination is a theoretical approach investigating the intertwined structures and mechanics of oppression in terms of race, gender, class, and other social categories faced by marginalized or othered people. It theorizes power in four domains: structural, disciplinary, hegemonic, and interpersonal. Within my practice, I endeavor to communicate that there is a shared goal, and that we are all "on the clock" due to the age profiling culture in the field. You do not need to be disabled to be negatively impacted by the Design Exo-system, or its underlying ableism. Activist Talilia Lewis defines ableism as "a system that places value on people's bodies and minds based on societally constructed ideas of normalcy, intelligence, excellence and productivity." They put it to us that these "constructed ideas" are all "deeply rooted in anti-Blackness, eugenics, colonialism and capitalism" (Lewis, 2020). These forms of systemic oppression lead people and society to determine who is valuable and worthy based on a person's language, appearance, religion and ability to (re)produce, excel and behave satisfactorily, resulting in real-world material gain or loss.

The social model of disability is a way of viewing the world, developed by disabled people, which says that people are disabled by barriers in society, not by their impairment or difference. Barriers can be physical, like buildings not having accessible toilets. Or they can be caused by people's attitudes to difference, like assuming disabled people can't do certain things. This was developed to challenge medical models of disability that framed it as an individual problem in need of a cure.

However, while identifying systemic barriers, attitudinal barriers, and social barriers that create difficult or impossible terrain for individuals with disabilities to navigate, the social model doesn't include my Blackness or non-normative "age," the extra layers of complexity I have to engage with and navigate. In her book *Design Justice*, Sasha Costanza-Chock asks the reader to consider how design simultaneously distributes both benefits and harms via penalties and/or privileges based on our location within the interlocking system of oppression (Costanza-Chock, 2020). That being said, when race, gender, disability, nationality or place of birth and age (non-normative chronological development) interlock this can ultimately result in further isolation, pain points and unmet needs. The new discourse must seek to critically engage with the dominant structures and the nuances of the multiply-marginalized responses to this. All the ingredients for Inclusive contextual design(ers) are there. Intersectional and contextual design analysis must be woven in, lest we be made bystanders in our own lived experiences; for example, where pain points are being sanitized through being relayed and addressed only though official discourse and its "representatives."

To tackle this requires framing design in contextual, equalized spaces that create post-normal equity. If the objects, services and places we access and use are customizable, they will be made for a broader range of people. The hope is that dignified design will not demand declarations of a person's confidential information to be permitted to engage with the offering. Instead, it is about designing in multiplicity, kindness and access, no longer ignored or overlooked, but as central to a re-worlding with contextual equalities Design at the very core.

References

Burn, S. (2022). *The social justice curriculum at Wellcome collection. (Feb 15).* https://stacks.wellcomecollection.org/the-social-justice-curriculum-at-wellcome-collection-cceeaf76bd72.

Beisser, A. (1970). The paradoxical theory of change. In J. Fagan, & I. L. Shepherd (Eds.), *Gestalt therapy now* (pp. 77–80). New York: Harper & Row.

Bronfenbrenner, U. (1981). *The ecology of human development: Experiments by nature and design.* Harvard University Press.

Bronfenbrenner, U. (2004). *making human beings human: Bioecological perspectives on human development.* SAGE Publications.

Costanza-Chock, S. (2020). *Design justice: Community-led practices to build the worlds we need [ebook].* The MIT Press. Accessed 10 June 2021. Available at https://design-justice.pubpub.org/.

Crenshaw, K. (1989). *Demarginalizing the intersection of race and sex: A black feminist critique of antidiscrimination doctrine, feminist theory and antiracist politics. Vol. 1989* (pp. 139–167).

University of Chicago Legal Forum. Issue 1, article 8 https://chicagounbound.uchicago.edu/cgi/viewcontent.cgi?referer=&httpsredir=1&article=1052&context=uclf.

Design Council. (2018b). *Does design have a diversity issue?.* [video] [Accessed 11 June 2021]. Available at https://vimeo.com/291498463.

Hall, S. (1997). *Race, the floating signifier.* Media Education Foundation. Mediaed.org/transcripts/Stuart-Hall-Race-the-Floating-Signifier-Transcript.pdf.

Hill Collins, P. (1990/2009). *Black feminist thought: Knowledge.* Consciousness and the Politics of Empowerment Unwin Hyman/Routledge.

International Commission on Radiological Protection. (1974). *Report of the task group on reference man [ICRP Publication] No. 23* (p. 4). Oxford: Pergamon Press.

Lewis, T. (2020). *Ableism 2020: An updated definition. [online] talilalewis.com.* [Accessed 11 June 2021]. Available at https://www.talilalewis.com/blog/ableism-2020-an-updated-definition.

Piepzna-Samarasinha, L. L. (2018). *Care work: Dreaming disability justice.* Arsenal Pulp Press.

SAFOD. (2022). http://www.safod.net/safod-content/cid/37/about-us/ (Accessed 20 April 2022).

Sins Invalid. (2019). *Skin, tooth, and bone: The basis of movement is our people: A disability justice primer* (2nd ed.). Sins Invalid.

Spectrum (2022) https://spectrumcil.co.uk/wp-content/uploads/2015/04/SPECTRUM-12-Basic-Rights-2013.pdf [website] (Accessed 6th June 2022), for more details see: https://www.disabilityrightsuk.org/independent-living-0.

Tatman, R. (2016). *Google's speech recognition has a gender bias July 12.* https://makingnoiseandhearingthings.com/2016/07/12/googles-speech-recognition-has-a-gender-bias/. (Accessed 20 March 2022).

Trotman, N., & McGinley, C. (2018). Design and the mind. Engaging and collaborative workshops for the neurodiverse. In G. Craddock, et al. (Eds.), *Transforming our world through design, diversity and education* IOS Press.

TUC. (2017). *Personal protective equipment and women.* April 25 https://www.tuc.org.uk/sites/default/files/PPEandwomenguidance.pdf.

Tunstall, E. (2013). Decolonizing design innovation: Design anthropology and indigenous knowledge. In W. Gunn, T. Otto, & R. Smith (Eds.), *Design anthropology between theory and practice* (pp. 232–250). London: Berg Publishing.

Tunstall, E. (2021). Cultural respect, not social responsibility—The seven principles of design anthropology. In E. Resnick (Ed.), *Developing citizen designers.* New York: Bloomsbury Academic Press.

VeneKlasen, L., & Miller, V. (2002). Power and empowerment. *PLA Notes, 43,* 39–41.

VeneKlasen, L., & Miller, V. (2007). *A new weave of power, people and politics: The action guide for advocacy and citizen participation.* Practical Action Publishing.

Walker, N. (2021). *Neuroqueer heresies: Notes on the neurodiversity paradigm, autistic empowerment, and postnormal possibilities.* Autonomous Press.

Walker, N., & Raymaker, D. M. (2021). *Toward a Neuroqueer Future: An Interview with Nick Walker Autism in Adulthood. Vol. 3.* Mary Ann Liebert, Inc., Publishers. No 1 https://www.liebertpub.com/doi/10.1089/aut.2020.29014.njw Accessed 11th April 2022.

Further reading

Design Council. (2018a). *The design economy 2018.* [online]. [Accessed 11 June 2021]. Available at https://www.designcouncil.org.uk/resources/report/design-economy-2018.

Design Council. (2021). *Design economy 2021.* https://designeconomy.co.uk/ (Accessed 8 March 2022).

GOV.UK. *Ethnicity facts and figures [Website]*. Accessed 8 March 2022. Available at https://www.ethnicity-facts-figures.service.gov.uk/workforce-and-business/business/leadership-of-small-and-medium-enterprises/latest.

Media Education Foundation. (2022). *Race, the floating signifier featuring Stuart Hall* (p. 10). Mediaed.org/transcripts/Stuart-Hall-Race-the-Floating-Signifier-Transcript.pdf (Accessed 20 April 2022).

CHAPTER EIGHT

Dialogic Drawing

Yeoryia Manolopoulou*
The Bartlett University College, London, United Kingdom
*Corresponding author: e-mail address: y.manolopoulou@ucl.ac.uk

Contents

1. Introduction

This chapter examines dementia and architecture while exposing some of the gaps in our knowledge and experiences when we try to engage with brain health and social health challenges not only as built environment professionals but also as an entire society.

I will take as a starting point excerpts from the interdisciplinary dialogues conducted for *Losing Myself*, a project about dementia and architecture that I created collaboratively with Níall McLaughlin between 2015 and 2016, and from these excerpts I will develop a twofold discussion. Firstly, I will discuss the lessons that we took from these particular dialogues—specifically, these were done with a dementia campaigner, who herself lives with dementia, with a health care policy expert, an architect specializing in dementia design, two cognitive neuroscientists, an anthropologist and two neuropsychologists. I will connect these lessons with more recent research and policy developments on inclusive design and on designing for the mind more broadly. Secondly, beyond this, I will discuss the

Developments in Neuroethics and Bioethics, Volume 5
ISSN 2589-2959
https://doi.org/10.1016/bs.dnb.2022.08.005

potential of dialogue itself in architectural thought and practice. Without a doubt, it is dialogue as a creative method that generated the collaborative ethos and interdisciplinary content of *Losing Myself*. dialogue allowed us to include in our own perspectives as architects the perspectives of others. In the last part of my essay, with reference to philosopher Mikhail Bakhtin's concept of dialogism, I will propose that dialogic drawing is one of the most vital design methods that we have as architects if we wish to create genuinely inclusive projects and practices.

2. Dialogue

As architects, we simply do not know enough about dementia. In a commitment to acquire new knowledge on the condition, between 2015 and 2016 Níall and I conducted a series of interviews with experts in the health, cognitive and behavioral fields and with people and carers affected by dementia across the UK and Ireland. These were critical dialogues that sought to examine and empathize with the experiences of people living with Alzheimer's disease and explore the associated challenges that we have in creating and sustaining inclusive communities. What findings from neuroscience, anthropology, art, healthcare and policy can help architects design for people living with dementia and, more broadly, for all of us? How can we build dementia-inclusive cities from the outset and also improve all stages of building, including the adaptation, management and maintenance of the built environment? Central to our project was to interrogate how the human mind comprehends space and the implications of this for architecture. While most architectural literature produced on the subject of dementia focuses on 'best practice' guidance, we questioned reductive design guidelines and asked, first of all, how the brain works. What spatial capacities do we have that we might lose because of dementia? We hypothesized that a more informed and nuanced understanding of how the individual constructs an understanding of space should prove invaluable to all aspects of architectural design.

Taken collectively, these interdisciplinary dialogues offer fascinating but also diverse and at times conflicting layers of content about facing life with dementia and spatial cognition more broadly. They were conducted as part of our wider project *Losing Myself* which formed the sole representation of Ireland at the 15th International Architecture Exhibition in Venice. Our response to the brief of the 2016 Biennale—*Reporting from the Front*—was an immersive installation that attempted to reflect on our own experiences as architects working to 'improve the quality of life while working on the margins, under tough circumstances, facing pressing challenges.'[a] Most of the dialogues are available as podcasts on our open-access website where we also provide summaries of what was discussed as well as a record of all our other research, drawing and collaborations.[b]

As I am revisiting this project, re-listening to our 6-year-old podcasts and tracing newer publications by the dialogue participants and other experts, I am realizing that it is in the edges of these dialogues and outputs, in the unfinished and contradictory bits, that I am finding most meaning—and most questions. Perhaps this is because it is impossible to understand fully the disordered and incomplete sense of self. It is also because recent times have radically transformed our understanding of the world, our social and environmental fragility, and the tenderness of the human condition in more universal terms. It feels more urgent than ever to do more, as an entire society, to understand the interdependence between brain health, social health and environmental repair and to improve all three as a whole. To contribute to this challenging task, as architects, we have no other way than learning how to learn and practice *dialogically*. According to Bakhtin, 'a culture undergoes "dialogization" when it becomes relativized, de-privileged, aware of competing definitions for the same things. Undialogized language is authoritative or absolute.'[c] It is inspiring to think that architecture can undergo a similar process of dialogization to free itself from authoritative practices in order to become more inclusive.

---

[a] This brief and the theme *Reporting from the Front* were set by the curator of The 15th International Architecture Exhibition, Alejandro Aravena.

[b] This chapter will refer only to fragments of some of the dialogues that we conducted. The full list of the dialogue participants with their positions at the time is: Professor June Andrews (Director of the Dementia Services Development Centre, University of Stirling); Professor Sabina Brennan (Co-director, Neuro-Enhancement for Independent Lives Research Program and Director, NEIL Memory Research Unit, Trinity College Dublin); Professor Sebastian Crutch (Neuropsychologist, Dementia Research Centre, UCL); Ms. Kay Doherty, Mr. Desmond Donnelly and Ms. Aisling Guckian (The Orchard Alzheimer's Respite Centre, Dublin); Dr. Sophie Handler (urban researcher, The University of Manchester and Age-friendly Manchester); Professor Tim Ingold (Chair, Social Anthropology, University of Aberdeen); Professor Kate Jeffery (cognitive and behavioral neuroscientist, Institute of Behavioral Neuroscience and Department of Psychology, UCL); Ms. Sandra Keogh (Person in Charge, The Orchard Alzheimer's Respite Centre, Dublin); Ms. Lesley Palmer (chief architect, Dementia Services Development Centre, University of Stirling); Ms. Helen Rochford-Brennan (Chair of the Irish Dementia Working Group and Vice-chair of the European Working Group of People with Dementia); Professor Hugo Spiers (Cognitive and behavioral neuroscientist, Spatial Cognition Research Group, UCL). To listen to the podcasts of most of these conversations, please visit http://www.losingmyself.ie/pages/category/dialogues/

[c] This is how the scholar Michael Holquist discusses Bakhtin's concept of dialogue in Holquist M. (Ed.). (1982). *The Dialogic Imagination: Four Essays by M.M. Bakhtin* (p. 427). University of Texas Press.

3. The brain is hidden

Helen Rochford-Brennan, Chair of the Irish Dementia Working Group and Vice-chair of the European Working Group of People with Dementia, was diagnosed with early-onset Alzheimer's at 61. When we met her in 2016, she talked to us about misconceptions about dementia and pointed out that when she books airport services to assist her with navigation difficulties, she is typically presented with a wheelchair despite her lack of physical disability: 'it's my brain that is slow, not my feet!' Helen talked to us about the need to eradicate the stigma around dementia: 'the general public do not see us… we are cast aside.'[d] She insisted that people living with dementia, like herself, can still contribute to society and have active lives as citizens in their communities.

We have heard many times from carers and researchers that we need to see people living with dementia as people first and foremost, not as patients. It is different when we hear people who themselves live with dementia saying it. More recently, in her Opening to Alzheimer Europe in October 2019, Helen said:

> *I would… advise you to look at the whole person and in particular our social health…. We want to be viewed as people with hopes and dreams and autonomy. Even as our dementia progresses we can still make choices and we can maintain hobbies and interests. We want to be connected in our communities. I am more than a set of symptoms I am a person. No matter what the project or healthcare situation my personhood must be honored.*[e]

For Helen, helping people with dementia to be part of their communities and *to contribute,* continuing to be doing things for others as active and valuable citizens and family members, is essential. 'I don't want to be different,' she stresses, maintaining personhood, belonging and connection is paramount. Reflecting on how her sudden resignation after her diagnosis led to losing her job and with that to losing 'of everything,' she proposes that adapting someone's job in the same organization should help them stay in the same social environment to not lose connections with people. The brain is hidden,

[d] When referring to the individuals we met and had conversations with, I will use their first name to remind the reader of the discursive nature of dialogue. In all other cases, the authors will be referred to with their last names.

[e] Rochford-Brennan, H. (2019, 23 October). Opening for the 29th Alzheimer Europe Conference 'Making valuable connections'. https://alzheimer.ie/wp-content/uploads/2019/10/Opening-Address-2019-FINAL-17.10.19.pdf

people don't see it. What is critical is finding ways to empower people with dementia, and all people facing a change in their abilities, to continue to lead purposeful lives and be active and significant members of their community.

In our conversation, Helen suggested improving signage and the use of landmarks to support people's sense of direction and navigation in the city, and she asked for more spatial organization and physical and visual clarity in streetscapes and public buildings. A greater collaboration between architects and graphic designers to develop successful signage and way-finding strategies is needed. Having roads always changing because of building works and unsigned diversions to pedestrian paths, hiding bathrooms behind other spaces and having automatic doors that are not sign-posted, are all examples of unhelpful situations, Helen complained. She also asked for a reduction in noise everywhere in the city and for making a room for peace and quiet always available inside buildings.

4. Staying at home

As the dementia design principles evolve with new research, we now know that color-coding in a building and creating special sensory rooms are no longer advisable practices. Instead, object placement can help one's orientation and navigation while small and personal practical solutions can enable someone to continue to do the activities they like. Helen questioned color-coding and insisted on the familiarity of surroundings. She has a small moveable unit where she puts all her cooking essentials and carries it with her in her kitchen when preparing food. This minimizes confusion and issues of not remembering what is inside each cupboard. She stressed that we should not institutionalize the home and help people to remain at home enjoying the identity and culture of their own places. She advocated for making small adjustments in order to stay at home for as long as possible.

Staying at home for as long as possible was a recurrent theme across most of our conversations. We discussed it with June Andrews, the Director of the Dementia Services Development Centre at the University of Stirling, who argued that governments, societies and families, and people's entire legal and social infrastructure, should support people with dementia to continue do what they have always done and to stay at home, unless it is no longer possible. No changes should be made in the home except from only increasing light levels and removing trip hazards. 'Our memories of our houses are not about what we see most of the time…it's more of a proprioceptive memory,' our internal awareness of our body in space, our sense of self-movement and body-position.

We need to remember, according to June, that care needs are not linear. People think that life with dementia means first living at home independently, then needing daycare, then respite care and finally total care. But this is not so. Daycare and respite care can become counter-productive because removing people living with the condition from their own familiar surroundings can cause to them enormous and irreversible distress. Thankfully, in the UK, June shared, most people with dementia (75%) live at home and the time in a care home is limited and continuously shrinking (in 2016 at about 18 months). This piece of research suggests that the quality of the home environment and the surrounding neighborhood environment should be the central task of governments and built environment professionals. Creating a safe environment to walk in and increasing opportunities of passive supervision so that the community can collectively support an individual should be one of the key priorities of future architecture and urbanism when dealing with the sea change that dementia causes to our societies. In the world, around 55 million people have dementia and as the proportion of older people in the population increases, this number is expected to reach 78 million in 2030 and 139 million in 2050.

In our various dialogues, we also discussed the value of assistive technology devices (such as locator devisers and movement sensors) to help carers and family members to supervise people with dementia remotely and to deal with the fear of wandering.

5. The human infrastructure

Shifting our attention beyond just physical design to holistically improving the human infrastructure in cities and public buildings should become a major project for all policy and decision makers, designers and managers of the built environment. Designing for better mobility should be addressed but because the city is a continuously changing physical organism, which can become very challenging for memory and navigation, we need to consider the social support network around the person. To create inclusive communities, we need a culture change: to educate everyone in the wider public, people such as bus drivers and workers in supermarkets, how to detect and help individuals who might have difficulties with navigation and memory, so that they can be supported to keep their independence for longer. As Helen has argued more recently, we need to move away from dementia-friendly to dementia-inclusive concepts.[f]

Unsurprisingly, the World Health Organization (WHO) has recently developed a toolkit for how to make society dementia-inclusive so that people with dementia and their carers should 'fully participate in society and have a place in it.' People living with dementia should be 'empowered to live independently, free from stigma, discrimination, exploitation, violence or abuse.'[g] This toolkit, which WHO argues can be implemented by communities worldwide, is grounded in international commitments such as the United Nations Convention on the Rights of Persons with Disabilities (UN CRPD), the 2030 Agenda for Sustainable Development and its Sustainable Development Goals (SDGs) and Universal Health Coverage (UHC) where all individuals should receive the health services they need without suffering financial hardship.[h] Unfortunately, there is a huge gap between realizing WHO's intentions and the bleak reality that many individuals like Helen experience. In her campaigns, and based on her experiences across her country, Ireland, Helen refers to the very many people living with dementia who remain unsupported and lonely. She advocates for more funding, research and services, and for finding ways to improve human connections for all people facing a life with dementia—especially those in rural areas and those who are less able to use technology. She advocates for ending loneliness and isolation and for creating a better human infrastructure of care based on educating every single member of the community—from children to seniors—about dementia.

As architects, we need to learn to anticipate spatial and sensory differences across people beyond the more obvious aspects of differences in physical disability. We need to consider hidden conditions such as confusion, disorientation, lack of balance, impaired memory, sight and hearing, differentiations in visual, spatial and tonal understanding, and spatial insecurity. Notably, the British Standards Institution (BSI) has now developed a framework of standards for dementia-inclusive communities 'where the

[f] Rochford-Brennan, H. (2019, 23 October). Opening for the 29th Alzheimer Europe Conference.

[g] World Health Organization. (2021). Towards a Dementia-Inclusive Society: WHO Toolkit for Dementia-Friendly Initiatives. WHO.

[h] World Health Organization. (2021). Towards a Dementia-Inclusive Society: WHO Toolkit for Dementia-Friendly Initiatives (p. vi). WHO.

participation of everybody, including persons with dementia, is facilitated and encouraged.'[i] Moreover, it is currently working on the creation of a new fast-tracked guide called *Design for the Mind—Neurodiversity and the Built Environment*, expected to be published in the summer of 2022. This design guide recognizes the urgent need to address the challenges and differences in neurological and sensory processing across all people. It is intended to consider 'neurodivergent, neurodegenerative and other neurological conditions which may affect sensory processing and mental wellbeing' and to provide information for planners, architects, managers and decision-makers in the built environment for how to make public places more inclusive, 'in particular by reducing the potential for sensory overload, anxiety or distress.'[j] These are small positive signs of government guidance in a context, however, where the numbers are alarming: at present, one in 14 people over the age of 65 and one in six people over 80 live with dementia in the United Kingdom.[k] During our conversation for *Losing Myself*, June stressed to Níall and I an additional recommendation for improving public policy that should focus on finding ways to give people information earlier on in their lives so that they can make better provisions and prepare for the possibility of living with some form of dementia later in the future.[l]

6. Thoughtful design

Lesley Palmer, the chief architect of the Dementia Services Development Centre at the University of Stirling, repeated to us that we have a responsibility as architects and built environment professionals to design the entire cityscape with dementia in mind: "we should make all public spaces and buildings dementia friendly from the word 'go'." The design of dementia-inclusive neighborhoods would take away pressure from the health system. It would also take away pressure from the families and caregivers because their cared-for individuals would feel more autonomous and have less destressed behaviors.

[i] BS ISO 25552: 2022. (2020, 6 October). *Aging societies: Framework for dementia-inclusive communities.*

[j] BSI. (2020, 6 October). Project launch: First building design standard for sensory and neurological needs. https://www.bsigroup.com/en-GB/about-bsi/media-centre/press-releases/2020/october/project-launch-first-building-design-standard-for-sensory-and-neurological-needs/

[k] BS ISO 25552: 2022. *Aging societies.* https://shop.bsigroup.com/products/ageing-societies-framework-for-dementia-inclusive-communities/standard

[l] BS ISO 25552: 2022. *Aging societies.*

We discussed the next wave of design regulations that will focus on designing inclusive environments for people with a broad spectrum of differences in cognitive ability. Lesley described dementia as one of over 600 neurological impairments that the British Standards Institution tries to address holistically and argued for a universal model of good design. Design recommendations can be similar for a number of conditions. For example, the recommendations for dementia are very similar to those of autism, she argued. Dementia design is not complex. It involves understanding contrast, light levels, ease of accessibility, quality of air and legibility of plan. Just increasing light levels, for example, benefits people more than medication and prevents or delays bringing them to hospital. A consideration of circadian rhythms in relation to daylight is vital: 'at the age of 75 years a person requires three times the light that a 45-year-old needs to elicit the same circadian response.'[m] Therefore, exposure to morning daylight (which has the high blue content of the visual spectrum) is crucial while at night darkness should be achieved.

Safety when living with dementia is to be understood in an open mind, so that it does not restrict pleasure and wellbeing and the person's wish to stay autonomous for as long as possible. Lesley expressed frustration that balconies are, for instance, considered a risky building element, despite the fact that there are only two recorded instances in the country of individuals with dementia jumping from a balcony: 'we're depriving everybody of daylight, for fear of someone jumping.' We need to develop better collaboration between architects, commissioners and end-users, and good methods of post-occupancy evaluation to empirically access the relative success of projects rather than rely on anecdotal information.

Critically, we should ask if a deeper consideration of the challenges of dementia can create creative opportunities for changing more radically the future of architecture. In their edited collection on architecture and dementia, the architect Eckhard Feddersen and architectural journalist Insa Lüdtke describe dementia as a chance to rediscover and *renew* architecture. They try to ask elementary questions about space and perception such as how a space can afford a sense of safety, security and identity. Such questions, they argue, are not just about improving spaces for people living with the condition but questions that attempt to create 'a better quality of life for all.' Architects and interdisciplinary teams should not respond to dementia through 'special' buildings but seek to raise deeper questions about

[m] McNair, D. (2014). Light: Perception and Health. In Feddersen, E. & Lüdtke, I. (Eds). *Lost in Space: Architecture and Dementia* (p. 107). Birkhäuser Verlag.

perception and space with the aim to achieve 'a wider creative renewal of the built environment that can benefit us all.'[n]

Feddersen sees dementia as 'an opportunity to shift our focus back to the fundamental sensory experience of architecture' and shows various examples of buildings that he thinks appeal to all the senses, with Alvar Aalto's Villa Mairea in Noormarkku, Finland, as the most compelling in my opinion. He also writes about giving multiple options to inhabitants as single options constrain the person. A person should choose if they want to be on their own or part of a company, for instance.[o] Lüdtke, in turn, discusses the house as a 'dialectic boundary in which we live' and in which we form habits. The house should give us options and autonomy. She refers to the typology of the roman courtyard house that groups living spaces around an internal courtyard as a much-desired model of space which builds a spatial continuum of inside and outside conditions and fulfills the desire for intimacy as well as for connection to the outdoors. The flow and simultaneity of spatial zones in the traditional Japanese townhouse is another compelling model of space for those who 'need to feel safe but not confined.'[p] Lüdtke explains how the Japanese townhouse transitions from the street to the interior through layers of sliding screens and incorporates the neither indoor nor outdoor *engawa* as a seasonal and inclusive type of space that provides a duality of uses, such as sun warmth in the winter and shade in the summer.

A common misconception is that dementia design has to be something different from the overall design. This is not the case, as Lesley highlighted to us. Dementia design should not come in addition to other design considerations. It should be understood as organically and inherently included in the first principles of an architectural or urban project, simply through things like the making of an accessible and legible plan. Thoughtfully designed buildings and cityscapes will benefit those living with this debilitating condition as well as all of us. 'The provision of green spaces, active travel and play, walkable neighborhoods, improved biodiversity and air quality could all help the fight against dementia.'[q] These are aspects of simply good design, not 'special' design. Dementia-inclusive and well-designed spaces are not mutually

[n] Feddersen E. & Lüdtke I. (2014). Rediscovering Space: Dementia as a Chance to Renew Our Approach to Architecture. In Feddersen, E. & Lüdtke, I. (Eds). *Lost in Space* (p. 11). Birkhäuser Verlag.

[o] Feddersen, E. (2014). Learning, Remembering and Feeling Space. In Feddersen, E. & Lüdtke, I. (Eds). *Lost in Space* (p. 20). Birkhäuser Verlag.

[p] Lüdtke I. (2014). Layers of Living: On the Anatomy of the House. In Feddersen, E. & Lüdtke, I. (Eds). *Lost in Space* (p. 94). Birkhäuser Verlag.

[q] Lesley Palmer, L. (2022, 24 March). What is Dementia Design?. In *Being Patient*. https://www.beingpatient.com/lesley-palmer-dementia-design-architecture/

exclusive; they converge and overlap. We need to stop thinking about designing specially, and start thinking about designing thoughtfully for all of us — whether or not we are affected by dementia.

7. Place and memory

To better understand people's spatial needs and behaviors, we felt that we needed to actively engage with interdisciplinary specialists who could teach us about the spatial functions of the brain. We thought that educating ourselves about impaired cognitive functioning and non-neurotypical needs firstly requires an understanding of the human mind—at least on some basic level. How do humans comprehend and situate themselves in space? How do we acquire spatial knowledge and what are the implications of this for architectural design? A scientifically informed and nuanced understanding of how the mind constructs a spatial understanding and a sense of self-positioning in space should prove invaluable to dementia design and all aspects of architectural design.

To explain how we form spatial knowledge, the cognitive and behavioral neuroscientist Kate Jeffery at University College London explained to us the neurological means by which we acquire knowledge. Neurons in our brain fire nerve impulses along a physical line that is called axon. When the impulse reaches the end of the axon, it activates the next neuron down which passes information to the next neuron, interacting with various other active neurons on the way. So, the ear hears something and the eye sees something and the brain combines the information to draw a conclusion: 'that sound must be coming from that object.' 'Much of what the brain is doing is making these associations...much of what knowledge is, is making associations between things in the outside world.' The incoming information is then synthesized in the hippocampus and used by separate parts of the brain differently. The spatial part of the brain uses this information to make a map and the memory part of the brain uses the same map to attach memories to it. This is why we have the capacity to mentally visit a place and arouse memories: 'in a sense that's what memory recall is—you activate the place where the memory happened.'

There is a profound link between place and memory in the hippocampal structure and function of the brain. Referring to John O'Keefe's, May-Britt Moser's and Edvard I. Moser's discoveries of cells that constitute a positioning system in the brain, we have elsewhere explained:

> *Within the hippocampus, and its neighboring structures, neurons exist that lay down memories in their connections while allowing us to map space and navigate it. This internal representation of space is essentially a 'cognitive map': we have 'grid cells' that create lasting matrices with the longitude and latitude of the spaces we experience; 'place cells' that highlight our position in these matrices, helping us to understand where we are; and 'head-direction cells' that work like compasses to signal the direction we are facing.*[r]

Neural activity, therefore, constantly 'maps,' drafts and redrafts, our understanding of the world around us inside our brains. According to Hugo Spiers, another neuroscientist at University College London, "our neurons are modelling the world—our 'place cells' create an internal matrix of what we think the world is. They form a model of the structure of the environment, the geometry of the environment, how am I exactly positioned, how close am I to this wall, where am I?." In this process, and in our ability to continuously map the unfolding of the world around us, memory and place are inextricably and spatiotemporally linked. Overtime, our brains abstract incoming information through overlaying and linking multiple recordings and eventually store 'fact-like perceptions' of the world.

8. From mapping to story making

What happens in Alzheimer's disease is that a build-up of plaques and tangles disrupts and confuses the synaptic connections in the human brain:

> *This has a disproportionate impact on the higher synthetic functions of human cognition, including but not limited to, the suite of faculties that deal with navigation in time. When neural activity in the regions around the hippocampus weakens, memory retention declines. Tragically, one of the things we can then forget is where we are. Becoming lost and having no sense of place is common in Alzheimer's disease and is an immense challenge for architecture...* [s]

The social anthropologist Tim Ingold offered to us an entirely different perspective on why someone with Alzheimer's disease might feel lost and not have a sense of place: 'knowing where you are' is not about being able to locate yourself in a place but about 'being able to tell the story of the place you are in.' Perhaps dementia is a condition that messes up the temporal

[r] Manolopoulou, Y. & McLaughlin, N. (2022). *Losing Myself: Architecture and Dementia*. Bartlett Design Research Folios, University College London. https://bartlettdesignresearchfolios.com/manolopoulou-mclaughlin-losing-myself

[s] Manolopoulou, Y. & McLaughlin, N. (2022). *Losing Myself: Architecture and Dementia* (p. 11). Bartlett Design Research Folios, University College London.

narrative, our sense of time: 'things can end up next to one another that should be in different positions in the story.' Tim questioned the primacy of the brain in constructing spatial knowledge, as claimed by neuroscientists, and discussed mind, body and environment as a whole and shared continuum: a complete system in which perception is 'the achievement of a whole organism in a world, not a mind in a body.' This ecological conception rejects the brain as 'a central control system' and sees spatial perception as driven by human action in which the brain plays just one part. Being skeptical of 'the map metaphor,' Tim argued that 'one is continually creating and re-creating one's knowledge through the process of going about in it, rather than having it stored away somewhere.' This action-based knowledge is based on the human capacity to tell stories in relation to places and requires a sophisticated understanding of time.

The workshops at the Dementia Research Centre at University College London's Institute of Neurology, led by the neuropsychologist Sebastian Crutch, put person-centered stories at the heart of dementia research. They reveal first-hand experiences told directly by people living with dementia or their families and carers. Two of Sebastian's workshop participants with posterior cortical atrophy (PCA), a rare type of dementia, narrate:

> *I'm usually quite okay here but the minute I step outside the door it all goes mad… Life goes a bit strange, yes… even inside the village can be a bit strange but definitely when I go catching buses and dealing with… interacting with people, in general, no, it's not great, not great… I don't know what happens but it goes mad, yes. Not all the time and not every time but I'm much less comfortable and avoid, now, going unless I really need to go into town. (Participant with PCA)*

> *I do struggle a bit sometimes in working out which way round shirts go… If it's all in a big heap, which it generally is, it's just a question of I will perhaps turn it round, sort of, two or three times before I work out where the collar is. (Participant with PCA)* [t]

Such personal accounts reveal that this form of dementia can affect the individual's ability to experience space coherently and predictably. In our dialogue, Sebastian explained to us that with most types of dementia the person's ability to comprehend the world allocentrically declines. Allocentric spatial

[t] Harding E, Sullivan MP, Woodbridge R, et al. 'because My Brain Isn't as Active as It Should Be, My Eyes Don't Always See': A Qualitative Exploration of the Stress Process for Those Living with Posterior Cortical Atrophy. *BMJ Open* 2018; 8:e018663. https://doi.org/10.1136/bmjopen-2017-018663.

referencing requires a sophisticated mental processing whereby the world is assessed through imagining and examining multiple relations between many different objects and positions in space. It is a complex mental function that searches for object-to-object correlations. The human brain understands the world also on an egocentric level which is based on making much simpler self-to-object spatial connections. Through our conversation with Sebastian, we realized that the problem for us as architects is that the architectural plan, the primary tool of our discipline, is mainly allocentric. It is a visual construct 'based on a totalizing map-like view, disconnected from the circumstantial self-dependent ways in which we live in the world.'[u] Breaking away from this limited definition of the plan, in *Losing Myself* we developed a multi-authored method of drawing that intertwined situation and memory by mixing allocentric and egocentric representations, introducing this neuro-biological concept in architectural drawing for the first time.[v]

9. Connecting through art

Through art we can deepen our understanding of dementia and help people living with the condition. I have found compelling accounts about the role of art in dementia research and dementia care in the recent anthology *Arts & Dementia* edited by Ruth Mateus-Berr and L. Vanessa Gruber. This fascinating book presents an array of art-related projects and interventions that have given insights to researchers, health professionals and the general public and, additionally, multiple benefits to people living with dementia, their friends and families. The editors argue that medical treatments and policies on their own are 'no longer sufficient or effective' and that the function of art can open up opportunities for improving the health and wellbeing of people facing a life with dementia, helping them to hold meaningful places in their social environments, and allowing them to take ownership of their individuality and creative abilities.

Art projects, including architecture projects, can fight misconceptions, reductive policies and cultures around dementia, and get people talking about it. They can communicate less known aspects of the condition,

[u] For a more developed discussion about the allocentric and egocentric aspects of architectural representation see Manolopoulou, Y. (2021). Performing the Architectural Plan: Egocentric and Allocentric Drawing. In Mateus-Berr, R. & Gruber, L. V. (Eds.) *Arts & Dementia: Interdisciplinary Perspectives* (pp. 295–309). De Gruyter.

[v] As argued in Manolopoulou, Y. & McLaughlin, N. (2022). *Losing Myself: Architecture and Dementia* (p. 18). Bartlett Design Research Folios, University College London.

evoking to the public empathy for people living with the condition. Through the social function of art, families and friends, and different parts of the community, can connect emotionally and more deeply with people living with dementia. Sebastian writes: 'Again and again we come across people who say: what really helped us to understand this person was connecting through music, or sharing in a response to a picture, or a carer seeing someone who can still create something. It can still be about who they are, not just who they were.'[w]

Architecture is an art with social purpose. *Losing Myself* developed through this broader context and an understanding of the function of architecture as a social art. We learned from multiple disciplines and individuals, engaged with other authors in a social form of drawing, and through our installation and open access website opened up our findings to the wider public.[x] As architects, we will strengthen our chances to contribute to the making of all-inclusive environments, if we first of all create social forms of practicing where we can think and create together with others. We cannot achieve all-inclusive projects with detached, much-objectified and generalized practices that do not allow us to engage directly with difficult, disordered and unpredictable forms of living. Dry statistics and results from consultations, interviews, site-specific and community-specific research procedures are not sufficient. Beyond these processes, we need to develop architectural methods that are inherently dialogic and imaginative.

10. A million-petalled drawing

In the Arsenale in Venice, 16 synchronized projectors displayed a large composite 16-min drawing film across a floor area of 6.2 m × 4.6 m. The drawing projection represented The Orchard Day Care Centre of The Alzheimer Society of Ireland as vividly experienced through its inhabitants rather than as a fixed built object conceived singularly by the architect's mind (Fig. 1). This drawing was produced by 16 architects/drafters who hand-drew on hundreds of layers of trace to describe possible inhabitations of the building while a video camera recorded their drawing activity from

[w] Ball, P. (2017, 11 March). Forgetting But Not Gone: Dementia and the Arts, *The Guardian*. https://www.theguardian.com/science/2017/mar/11/forgetting-but-not-gone-dementia-and-the-arts-research-project-alzheimers

[x] Beyond the podcasts, our website www.losingmyself.ie collates short stories from people who have had personal interactions with dementia. It also records the process of developing our installation through drawing and making in collaboration with others.

Fig. 1 The installation of The Irish Pavilion, La Biennale di Venezia, 2016. *Photograph by Nick Kane. © Yeoryia Manolopoulou and Níall McLaughlin.*

beneath a glass surface, each time for a maximum of 30 mins. Drawing desks with installed cameras, trace sheets and ink pens were provided, and a notional starting location of where each drafter would be placed as an imagined self in the plan of the building was suggested (for example, 'you are located here,' 'inside this bedroom' or 'at the edge of the garden'). But the content of the drawing itself was freely improvised by each drafter who projected onto the page self-reflections and imagined scenes as they wished (Figs. 2–4). The architect can immerse themselves in an introspective process of drawing, imagining stories and the lives of others, confirming in this way that the self is dialogical. Dialogism is determined by the capacity of an even sole author to find other positions of self in their work. To describe the social areas of the building, multiple drafters co-drew on the same trace, joining their free-flowing lines to represent group activities like dancing or gardening together (Fig. 5). In the end, our task was to bring all recordings of the individually created dialogic drawings (such as bedrooms and bathrooms) and the collaborative dialogic drawings (such as music room and gardens) into further dialogue, to join them spatially and temporally, creating a multi-voiced discourse in and through lines (Figs. 6 and 7). A deeper participation in the world of others

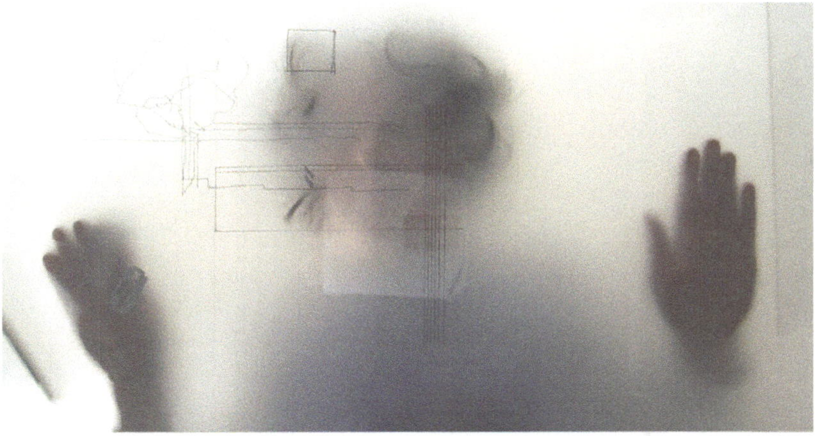

Fig. 2 Early drawing process on trace, video still taken from below and capturing both drafter and drawing inhabiting the surface simultaneously. © *Yeoryia Manolopoulou and Níall McLaughlin.*

Fig. 3 Michiko Sumi drawing a bedroom in the company of family photographs. © *Yeoryia Manolopoulou and Níall McLaughlin.*

Fig. 4 Four drafters drawing on glass tops while cameras record the process from beneath specially-made desks. Design and research material gradually filled the walls of our drawing room. © *Yeoryia Manolopoulou and Níall McLaughlin.*

Fig. 5 Yeoryia Manolopoulou, Níall McLaughlin, Michiko Sumi, and Simon Tonkin drawing collaboratively in an attempt to represent the shared experience of a room of The Orchard Centre. © *Yeoryia Manolopoulou and Níall McLaughlin.*

Fig. 6 Different drawing arrangements on the floor were formed to imagine different plan versions of the inhabited building. © *Yeoryia Manolopoulou and Níall McLaughlin.*

requires empathic observation, conversation and interaction in shared events. In *Losing Myself* a shared and empathic conversation happened in and through drawing.

The projection also contained hexagonal matrices of grid cells to represent the occupants' internal neural mapping mechanisms and blossoming colors and plants to describe the 24-h cycle and seasonal changes in the gardens. In addition, a spatial network of 64 speakers, which played local, regional and global sounds related to the building and Irish culture more broadly, complemented the visual projection (Figs. 8 and 9). Vivid human episodes of a day-and-night cycle of activity were overlapped with environmental transformations happening in the gardens over a full year. The resultant spatiosonic drawing gave an encompassing, yet incomplete, transient

Fig. 7 An overlap of drawn sheets showing the character and texture of the hundreds of drawings that gradually built up the drawing projection. © *Yeoryia Manolopoulou and Níall McLaughlin.*

Fig. 8 Sequential views of a part of the drawing projection, representing an uninhabited evening winter view (left) and its inhabited spring transformation, including the activation of a mat of hexagonal grid cells (right). The Irish Pavilion, La Biennale di Venezia, 2016. Photographs by Nick Kane. © *Yeoryia Manolopoulou and Níall McLaughlin.*

Fig. 9 Detail of the speakers and projectors of The Irish Pavilion, La Biennale di Venezia, 2016. Photograph by Nick Kane. © *Yeoryia Manolopoulou and Níall McLaughlin.*

and disordered sense of space and experience. As already argued, 'a common limitation of the architectural plan is that it represents the building as a static and whole object, neutrally and equally demarcated, from the all-seeing and singular position of the architect.'[y] To address this limitation, we created a denser and more expansive type of architectural drawing that involved parallel and overlapping drawing trajectories by many authors who enacted imaginative co-habitations of the same building. Our interactive multi-authored organism of lines was thought to be analogous to the multiple mind wanderings of the building's inhabitants. It described architecture as a

[y] Manolopoulou, Y. & McLaughlin, N. (2022). *Losing Myself: Architecture and Dementia* (p. 26). Bartlett Design Research Folios, University College London.

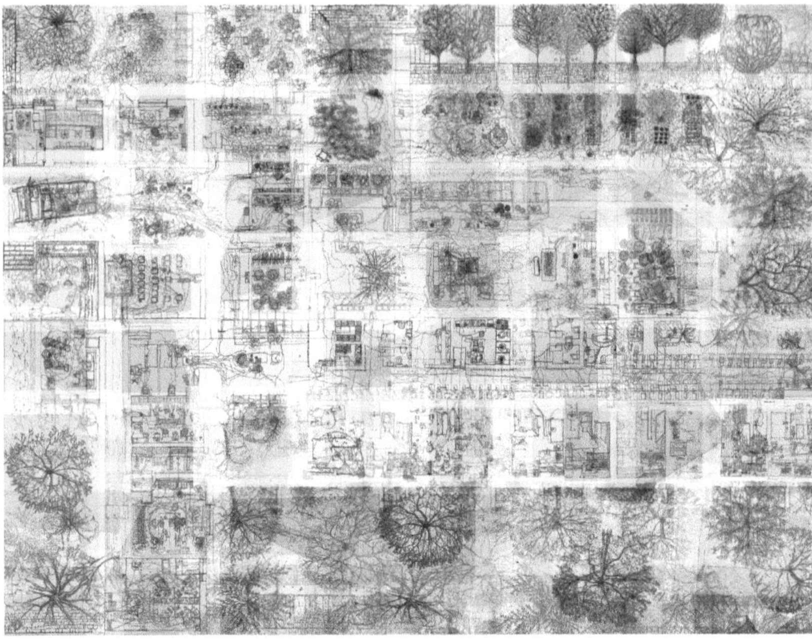

Fig. 10 A video still that captures the stitching together of different drawings in the process of making the new multi-authored and changeable plan of the inhabited building. © *Yeoryia Manolopoulou and Níall McLaughlin.*

polyphonic edifice, not singularly produced, but configured from continuously shifting human perspectives in space and time (Fig. 10).

In his poem The Old Fools, Philip Larkin describes human existence as 'a unique endeavor/To bring to bloom the million-petalled flower/Of being here.'[z] This description was in our minds throughout the research and design process of *Losing Myself* and influenced the dialogic and phantasmagorical nature of the resulted drawing which had a visually blossoming character. Our drawing and soundscape presented a simultaneity of worlds co-existing in the same building where many independent lives constantly meet and separate (Fig. 11). It also evoked something of the co-existence of worlds inside a person's mind—a mosaic of 'lighted rooms' as Larkin calls them. 'Perhaps being old is having lighted rooms/Inside your head, and people in them, acting.'[aa]

[z] Larkin, P. (1973). The Old Fools.
[aa] Larkin, P. (1973). The Old Fools.

Fig. 11 Aerial view of The Irish Pavilion, La Biennale di Venezia, 2016. *Photograph by Nick Kane. © Yeoryia Manolopoulou and Níall McLaughlin.*

11. Collaboration through dialogic practice

Losing Myself is not an architectural conclusion or proposal. It is a project of collaboration that opens-up to dialogic processes of conversing, listening, drawing and perhaps knowing differently (Fig. 12). Lasting societal change requires dialogue and, through this, new collaborative forms of creative interpretation and imagination. Our drawing method embodied what we learned from our dialogues. The plurality of perspectives and research findings that derived from these dialogues led us to decide that we had to define architectural experience and architectural representation in a polyphonic manner. Our learning was dialogic and so had to be our drawing.

Mikhail Bakhtin, one of the most important philosophers of language and ethics, wrote extensively on the concept of dialogue through particularly studying how Dostoevsky's works dialogically visualize the life of human consciousness: characters in the novel are 'respected as full subjects' and speak in ways other than the author's own. Characters are no longer absorbed by a dominating monologue; they are not objects serving the author's plan but 'full subjects,' pursuing independently their ideas. The author shows a kind of disinterest towards the protagonists so that they

Fig. 12 An evening of dialogue with some of our collaborators. © *Yeoryia Manolopoulou and Níall McLaughlin.*

can reach unlimitedly their own 'unpredictable futures.'[bb] Perhaps something similar happened with the making of our collaborative drawing for *Losing Myself.* We let go of control so that others could draw freely in a process that was uncertain yet empowering. Each drawing embodied the efforts, hesitations and determinations of a human consciousness imagined independently by an author. All together, the drawings came to form an evolving web of threads: many minds, represented by wandering lines, intertwining and supporting one another in an open-ended visual discourse which became paused in a particular configuration only for the purposes of the

[bb] Booth, W. C. in Bakhtin, M. (1984). *Problems of Dostoevsky's Poetics* (pp. xxii–xxiii). University of Minnesota Press.

projected animation. Future scores could arrange these drawings in different time sequences and create new configurations of possible spatiotemporal relationships.

For Bakhtin, self-consciousness is determined by a relationship to another consciousness. He writes: 'A person is... wholly and always on the boundary; looking inside himself, he looks *into the eyes of another* or *with the eyes of another.*'[cc] The essence of the person, and of the architect I would add, is in this sense 'unfinalized.' I have found no other thinker who has expressed so clearly how we exist creatively with others and through others. And although Bakhtin has talked mainly about language and literature, we can easily relate his stratified and polyphonic understanding of verbal language to the language of drawing. Drawing can be defined, and practiced, as a dialogic open-ended force, not a finalized object. It can develop dialogically and collaboratively to engage people as 'full subjects' in a process of imagining and conjoining their ideas and futures.

A dialogic learning and drawing process resists forming objectified, absolute and quick, solution-led, approaches to architectural design. It delays authorial design instruction first and foremost in favor of searching for a more open and social imagination. Crucially, this type of dialogic practice encourages a radical and ethical connection between differences and, consequently, helps architecture undergo its own dialogization to contribute to more inclusive futures.

[cc] Bakhtin, M. (1984). *Problems of Dostoevsky's Poetics* (p. 287). University of Minnesota Press.

SECTION 3

Practices

CHAPTER NINE

Allen Kong Architect and inclusive environments

Allen Kong*
Allen Kong Architect Pty LTD, North Melbourne, VIC, Australia
*Corresponding author: e-mail address: allen@allenkongarchitect.com.au

Contents

1. Introduction

Allen Kong Architect (AKA) PTY LTD was established in 1988 in Melbourne, Australia. The practice specializes in community facilities and aged-care and special needs housing, and it has developed a reputation for sustainable and innovative design. This has been recognized in numerous awards, including awards from local government, state, national and international professional associations, and the World Congress for Health and Design. In 1998 AKA projects were the first Australian projects to receive a United Nations World Habitat Award for social housing, and in 2011 the prestigious United Nations Scroll of Honor for Human Settlements.

The design philosophy of AKA is that the quality of life of the individual is first and foremost. This requires that we address their needs, concerns and desires and that we recognize their wider social, economic, cultural and environmental context. AKA's commitment to a design-intensive rather that technology intensive approach prioritizes simplicity and modesty.

Developments in Neuroethics and Bioethics, Volume 5
ISSN 2589-2959
https://doi.org/10.1016/bs.dnb.2022.08.006

Simplicity and modesty are conducive to user-friendly buildings that are reproducible and flexible and that have a long, efficient lifespan.

The philosophy and approach of AKA is a reflection of my own life history and personal journey. I have Chinese, Palawa (Tasmanian indigenous Australian), Scottish and English ancestry, which have disposed me toward a more holistic view of architecture, based on an inalienable understanding of the living spirit relationship and interdependence. This means that environmental sustainability is an essential aspect of AKA's design, but it has also sensitized me to the importance of culture. Architects principally work within their own culture and within the cultural lineage of architectural schools and styles. Architectural schools and styles are as universally pervasive as any colonial styles or ways of thinking. During my studies and training, architectural theory was of interest to me, but I also found that I was drawn to issues of climate, place and culture: because of my non-European cultural heritage I felt I needed to explore how to think about the European culture of Australia. I immersed myself in places of difference to see how architecture fitted with place and culture, but also to explore how I fitted in. I spent almost 2 years in northern Afghanistan building a village, under the guidance Iranian born Sydney architect Syed Sibtain. In the early 1980s I also undertook research in the Antarctic. Via these experiences I came to appreciate the importance of understanding human psychological and physical relations to the built environment, and hence the need to prioritize specific individual and user needs over, for example, conventional aesthetic concerns of architectural styles and schools. In my architecture practice, this has translated into a concern for how design for individual or group accommodation can be rooted in place and culture, while also accommodating those individuals and groups who's psychological and physical needs may be atypical. This has included people with various hearing and sight impairments, cognitive impairments, and also people whose difficult life trajectories have rendered them particularly vulnerable.

In this chapter I will describe three AKA projects that have emerged from this general philosophy. These include the Port Melbourne Hostel, Odyssey House Family Units, and the Wallara-Wintringham- Housing Choices Potter Street Redevelopment, all based in Melbourne, Australia. These projects provide accommodation for people who have drug and alcohol dependency issues, or elderly people who are particularly vulnerable to homelessness. For each project I describe the general purpose, the unique requirements and needs of the users, and how these requirements and needs were addressed in the design response. Before I do this however, the next section will provide some more context to the approach of AKA. AKA, as I will describe, has taken a broadly holistic approach to architecture and its relationship to health. It builds on my formative early experiences of researching design and human wellbeing in polar contexts, and which also reflects non-western cultural traditions.

2. Toward a more holistic approach

Conventionally, design for individual or group accommodation reflects the preconceived cultural idea of the house or home. 'Homelike' generally references the suburban house, and in Australia, that has been an imported mostly English view of the house, and the idealized version of the manor house set in gardens (the bigger the better). More recently, home in urban settings where space is sparse has also been associated with multistorey apartment living. When translated into aged care, this has resulted in multistorey hotel-like accommodations which, for people with atypical needs are the antithesis to healthy living, (except for those individuals who are particularly robust). These cultural ideas and their translation into our built environments raise and important question: to what extent do we pursue the advancement of aesthetic architecture and how can this be balanced with the intertwined real and sometimes subtle but important needs of users and inhabitants?

Architectural training was predominantly via a European lens. There are certainly many challenges to this within contemporary architecture, but the influence of figures such as Le Corbusier on modern architecture remains strong. Le Corbusier's concept of the modular man prioritizes a pleasing architecture aesthetic derived from an idealized view of a human—a 1.83 m tall male (after the English had disagreed with the height proposed by the French) (Ostwald, 2001). The profession is moving away from such theories from early last century, but the tendency to design for 'typical' individuals has remained strong. It likely helps to make sense of some of the failed attempts to provide safe and healthy accommodation for people with atypical needs. A notable historical example with relevance for the work of AKA is Gordon House in Melbourne. In the 1970s, 'old' Gordon House (located in the CBD) was replaced with 'new' Gordon House, which took the form of a budget hotel. The new building (which is discussed in more depth further on) superficially met all the basic needs of occupants: each person had a private room, there were shared bathrooms and toilets at the end of the corridor on each floor, and the ground floor contained dining and lounge areas. However, it gained a reputation as a dangerous and unmanageable place. Occupants were vulnerable and differing needs that were not recognized or accommodated in the design. Elderly residents were reluctant to use the toilets during the night, and some were mistakenly believed to be incontinent as a result.

However, a different design response can emerge if we prioritize the basic needs and human reactions of those particular people who will be inhabiting the spaces we design. A more holistic approach—such as that which has guided AKA—can help sensitize architects to those aspects that are important for health and well-being. We can think of a holistic approach as having a number of important elements that reflect non-western or indigenous perspectives on health and wellbeing (see, for example, Kuwayu, 2022), such as: a recognition of the importance of a sense of connection to country and place for people's identity; recognizing the importance of community and kinship, cultural expression and language (and hence the impact of racism and trauma in health and wellbeing): and recognizing the importance of self-determination and leadership. When informing design, these considerations must also take into account the spectrum of capabilities that characterize people, and hence those atypical capabilities and requirements of users and inhabitants.

The holistic approach of AKA, however, also recognizes the value of scientific understandings of human psychological and physical responses to the built environment. This aspect of AKA's approach can be traced back to my experience with the Melbourne University's Antarctic Program in the early 1980s. As an undergraduate and a postgraduate, I produced a design research and thesis for a design proposal for Davis Base, which is managed by the Australian Antarctic Division. I become familiar with emerging research on physiological and psychological aspects of polar and 'high Arctic' living, produced by researchers such as Leo Zrudlo (who become Dean of Architecture at Université Laval, Quebec), and Des Lugg (joined AAD 1962, 1969 became Head of Polar Medicine, Australian Antarctic Division 2002 became Chief of NASA's Medicine of Extreme Environments Department). From this I gained an understanding of the profound impacts the small and seemingly insignificant change in physical living conditions can having on the health and wellbeing of otherwise extremely robust individuals.

It could be assumed that the best design solution for an Antarctic base would be a large mega-building that simply protects occupants from all extreme external environmental conditions. However the extensive empirical research undertaken by people such as Lugg, Saxton 1981, and Zrudlo suggested otherwise (Zrudlo, 1971). The research generated extraordinary detailed and wide-reaching information regarding group dynamics and health work output, relating to the built and natural environment. It indicated that, in order to maintain the best living conditions for mental and physical health, it was necessary to have a collection of buildings that separated sleeping, living and working arrangements, and that facilitated everyday contact with the external environment. (Gosbell & Holmes, 1982) (Further research and the design response were subsequently published in a report to the Commonwealth Public Works (see Clark, Wellington, and Kong, 1981).

In general terms the experience very much foregrounded for me the value of understanding the psychological and physiological dimensions of human responses to the built environment. More specifically, it has illustrated the necessity for architectural design that connects people with their environment, and according to a more holistic understanding, can also provide them with a sense of place and connection. By the mid 1980s when I began working on aged care facilities, I immediately recognized some parallels between the facilities of that time and Antarctic bases: occupants are essentially taken away from their previous environment their community and placed in a mostly artificial environment characterized by a lack of internal sunlight, (note the certification guidelines from the later 1990s had a category—in bedrooms a resident must be able to tell night form day) and limited sphere of engagement with reduction of easy access to external envirnonment.

3. Project 1: Wintringham Port Melbourne hostel

Designed in 1993 in the style developed in the 1989 McLean Lodge his is a 35-person residential aged care facility located in the area of Port Melbourne. It was the third development in the Wintringham Program to rehouse vulnerable elderly residents from New Gordon House.

New Gordon House—which was briefly mentioned above—was a 10-story building in the brutalist architectural style popular at the time (the 1970s). (Brutalist being a style of aesthetics a term unrelated to any internal spatial arrangement or function) It was located on the south bank of the Yarra River, which at that time was industrial landscape isolated from any residential area. It was also remote: while it was near the city center it was nevertheless surrounded by inhospitable land. This illustrated the value that society placed on those occupants, that they were outcast. Indeed, residents had, in effect, been evicted from what had become a desirable area ('Old' Gordon House was located in an area undergoing gentrification) to an area where no-one wanted to live.

During the mid 1980s, public campaigners such as Bryan Lipman AM (who became the CEO of the Wintringham Program) managed to convince policy makers and public servants that such accommodation was highly unsuitable. A recorded historical account gives some sense of this realization:

> *[A] bureaucrat toured the cement towers of [New Gordon House] with Bryan, taking the slow lift and walking the long barren cement corridors, his senses all the while being assaulted by unfamiliar noises and disconcerting smells. He became increasingly aghast at what he was seeing and told Bryan that he had never visited a prison with conditions like this (Farrelly, 2013)*

There was, in effect, a shift in consciousness: elderly people living in Gordon House were now being viewed as being *aged and homeless*, not *homeless and aged*. In other words, they were now being viewed primarily as especially vulnerable elderly persons, rather than simply 'homeless people'. It was more than just semantics: it permitted access to public aged-care funding which was viewed as a right for all aged people.

People who are aging may have health and cognition issues (Zeisel, Fleming, & Bennett, 2020), and these will be exacerbated by years of living rough. These are often overlain by a variety of other issues such as alcohol dependencies, acquired brain damage by trauma or by substances, and psychological issues. In present day parlance, many of the issues are encountered by vulnerable elderly are understood as being the Behavioral and Psychological symptoms of Dementia (BPSD). One report notes:

> *BPSD…can include symptoms such as agitation, depression, anxiety, apathy, delusions, hallucinations and sleep or appetite changes…. The prevalence of BPSD in nursing homes may be greater than 90%, with 80% having behavioural disturbances and more than half having depression or psychosis. (Masso et al., 2017)*

A confluence of BDSM among residents creates a challenge for designing suitable aged-care settings. Aged care facilities of the late 1970s and 1980s were either 'low-care' and built as hostels (similar to New Gordon House), or 'high-care', built as nursing homes. Typically, a low-care hostel was designed in accordance with minimum requirements of the building regulations. The maximum length of a corridor, for example, was based on the length of travel between points of egress for fire safety and were thus typically around 30 m long. The WC and showers were in a communal block, again responding to the minimum building code requirements.

Such environments were clearly not suitable for residents with complex needs, such as those at New Gordon House.

Around this time, I—and the Commonwealth Department of Aged Care—were developing models of low-care accommodation based on small scale, house-like setting. Meanwhile, Wes Matthews, principal Architect at the Department of Housing and Construction postulated that people who move in to aged care have very little in common with each other, except for their need for care. They are, then, a bunch of individuals. This prompted an exploration of design formats that would recognize this, including AKA's design for the Wintringham Program.

The need for an overall sense of belonging, of community energy, and place, assisted in forming was also engendered through the design process by an understanding of traditional Chinese medicine and feng shui. Moreover, having participated as an undergraduate research assistant to Jan Gehl in his 'Interface Study' 1976 (Gehl, 1977) the urban design concepts of community resonated as being more applicable and appropriate than the concepts of a family 'home'.

A main purpose with our design was to create a sense of 'place' and foster a sense of belonging. In Australia, the notion of homelike aged care was advanced by the architect Brian Kidd AM as a way to encourage the designing of cognitively and emotionally accessible environments. For the most part this revolved around the places where most people who were moving into aged care had lived: the suburban home. The New Gordon House residents that we were designing for, however, were often from itinerant back grounds, having lived life 'on the road', undertaking various forms of casual labor, perhaps on (farms Jack-a-rooing, shearing). Some had been living rough in the city or moving between various night shelters, and not working at all. Some were war veteran, from the Second World War or the Vietnam war. We had to ponder: What would the concept of 'homelike' in these circumstances?

As the example of New Gordon House illustrated, the location of a site and a building's relationship to its surroundings play a major role in facilitating a sense of personal value and belonging.

The site for this Wintringham project was part of the railway reserves and train lines that had previously serviced the petroleum storage areas of Port Melbourne. The large land area was being cleaned and rehabilitated and developed into housing estates. A corner plot of land was chosen by Wintringham as an ideal location for a new aged care facility, as it was on the perimeter of the development and related to the existing parts of residential Port Melbourne. Homeowners in this street had resisted the overtures from the developers to sell their properties, preventing developers a clean slate in designing a whole-new suburb new suburb. Due to this, existing buildings became a critical aesthetic reference point for the new aged care development. This meant that the new project could link with the original Port Melbourne area, as the proposed new buildings would resonate with the familiar streetscape architecture (Fig. 1). This helped to give our new development a sense of belonging. It values the existing context, and hence neighbors are more likely to feel kindly towards the new development. In turn, this can help foster a sense of community and that the new residents 'belong' and are valued and welcome.

It was also important for us to create an environment that felt familiar to residents, so that they could navigate areas with ease and with a sense of safety. Antonsky's 'Sense of Coherence' theory is relevant here. It suggest that people's coping capacity to deal with everyday stressors requires three elements: comprehensibility, manageability and meaningfulness (Super, Wagemakers, Picavet, Verkooijen, and Koelen, 2016). At the time, we were influenced the Richard Fleming's CADE (Confused And Disturbed Elderly) Units (Fleming & Bowless, 1987), which recognized that domestic scale, simple buildings were more appropriate for elderly with cognitive issues than larger infrastructures. The Wintringham Brief was more reactionary identifying the need to be actively anti institutional.

Fig. 1 Port Melbourne Hostel year 2022 Swallow Street Port Melbourne Photo Allen Kong.

The resulting design for Fleming's CADE unit, then, is orientated around a simple 'L' shape. There is a staff base in the corner of the 'L' which affords the ability to see into both legs of the 'L' that contain the living area and bedrooms. Residents' bedrooms open directly onto this single large living area. Residents can see where they want to go, understand what they were seeing, and can easily navigate their way from one place to another.

In the design of Port Melbourne Hostel the core concept of the CADE Units—which emphasize this total visual access—was maintained, however was enlarged to include outdoor areas the shape of the visible space maintained visual access via windows, openings or over gardens through a number of adjacent spaces. This enables a more nuanced spatial division which facilitates visual spatial orientation and enabled residents to engage in self-regulation of social and personal interactions. Wayfinding is supported through the clear delineation of major pathways via the use of multiple visual cues with the building and landscaping.

We explicitly wanted to enable 'self-regulation' of social connections and interactions. When people cannot determine for themselves when to engage with others, negative social interaction can occur. As Edmund Hall famously argued (1969), a person is typically surrounded by an 'invisible bubble' of personal space and which demarcates 'intimate' or 'private' from 'social'. Built space should ideally recognize this, and our design made allowances to cater for those who have larger requirements due to cultural or individual factors. We recognized that the built environment needed to enable people to create and locate their own intimate and meaningful spaces that were comfortable. Architectural staples of ventilation; acoustics; orientation; volume and materials are all relevant in this regard.

It was also important that this concern for space and self-regulation was balanced with the need to facilitate everyday contact with the external environment (Yang, Brandon, & Sidwell, 2003). Verandas were one way that we achieved this. These could also serve as both private and semi-communal spaces. Verandas and external circulation improved daylighting, lighting ventilation and acoustics when compared with more traditional residential aged care facilities (typified by bedrooms in double loaded corridors). Indoor air quality alone, is critical to health Both (Vic EPA 1993) and European lung foundation (Bentayeb et al., 2015) have stated the negative effects of poor indoor air quality in nursing homes. Relevant also to the current Covid-19 era are the considerations for the environmental reduction of cross infection (Anderson, Grey, Kennelly, & O'Neill, 2020) (Figs. 1–3).

Fig. 2 Central walkway Port Melbourne Hostel. *Photo: Rob Reid 1998.*

Fig. 3 Verandas and gardens connection. *Photo: Rob Reid 1995.*

Becauses of the potential of unintended negative responses for people living in a congregate setting with a range of atypical abilities the quality air is even more important. In our design, the separation of bedrooms by external space means that sound and smells are not contained within and transmitted throughout the building. While these may appear to be basic architectural elements, they are very rarely implemented within most residential aged care or group accommodation. Most aged-care settings are derived from basic construction cost-saving principles, resulting in low external wall to floor area ratio. This will in turn lead to usually a double loaded corridor to give the least amount of circulation space to rooms serviced. The rooms are then stacked side to side—again for efficiencies which leads to one window on the eternal wall. Corridors and circulations are a major cost in Residential Care Facilities which generally give no increase to the quality of life to either resident, staff, or visitor. They are 'confused' areas: at times they are public, as any can use them, and at other times they are the second most private space to the resident bedroom. Wintringham Hostel thus represents a major departure from this.

4. Project 2: Odyssey Family Units

Odyssey House Victoria provides long-term and intensive residential treatment for drug and alcohol abuse for individuals and families. It is an 85-bed therapeutic community in the Lower Plenty area of Melbourne, on a bend on the Yarra River. The main building is a converted Roman Catholic Monastery, and the surrounding land is an attractive landscape of trees and bush. It includes several residential and community-based treatment and support services for people with drug and alcohol abuse, and any associated health and mental health, vocational, relationship and family issues. It also provides a treatment program for parents and their young children aged 0–12 years old—Odyssey House is the only drug and alcohol therapeutic community in Australia that enables parents to undergo therapy with their children, and hence there is a high demand for Odyssey House services.

In some respects, the people seeking services at Odyssey House are similar to those residing at Port Melbourne Hostel: both groups are vulnerable and unable secure their own supports. However, there are also important differences. A resident in Port Melbourne Hostel may not be intending and is not expected to make major changes to their lifestyle. A person who enters Odyssey, however, has made an explicit decision to address their substance abuse issues and thus make major changes to their life. As one of the residents at the opening of the Family Therapy building stated: '*I had to change to get my children back.*' Odyssey House thus aims to create and sustain an environment for positive change for those whose lives are affected by drugs and alcohol, providing long term, intensive residential treatment for adults, couples, and parents with their children.

We designed the Family Units within Odyssey house to provide a family specific alternative to the more formal surroundings of the main house (the converted Monastery). The concept underpinning the project was to create an experience that would take people out of their existing situations and create a 'life marker' that would symbolize their change in direction. They enable families to live together in a family environment, reflecting Odyssey's commitment to providing opportunities for rebuilding families and establishing sustainable patterns of day-to-day living. 'Live-in' residents undergo a treatment program that endeavors to change living habits and re-skill people ready to be integrated into the community as families (Fig. 4).

The design of the Family Units supports the therapeutic community philosophy of Odyssey, both in planning layout and building acquisition. The planning and grouping of the units enforce the sense of the individual, their relationship and responsibility within their immediate group and the wider Odyssey community. The individual houses must be shared by residents, giving each a sense of responsibility towards their housemates, whether they be their own family or other residents and their family members. Shared responsibilities include mundane tasks such as housework or keeping bedrooms tidy, to larger responsibilities such as helping to manage

Fig. 4 General view of family units. *Photo: Martin Saunders Photography.*

Fig. 5 Family Accommodation. *Photo: Martin Saunders Photography.*

the overall running of Odyssey house via administration or garden maintenance. This is intended to facilitate a sense of social cohesion and community (Fig. 5).

The relationship between community, the individual and the process of change is reflected in the spatial layout of the Family Units and their relationship to the other buildings in the Odyssey site. As opposed to the strict rigid formality of the Monastery building which forms the main house, the family units emphasize a sense of personal expression within a context of mutual responsibility. The transition between them—and the sense of moving from the public to the personal and domestic, is clear but gradual. The units themselves provide a supportive environment with little presence of obvious security measures, so that a sense of freedom is maintained. The materials of the family units were carefully chosen: handmade essential materials of earth and timber that have carefully molded, polished, and crafted by hand exudes care and support in a personal and individual way. The strength of natural hand polished poles gives the feeling of quiet support. Control of light, ventilation, acoustics and thermal comfort are were also important design considerations, as these elements can agitate sensitive residents experiencing the negative effects of substance withdrawal. The spaces have large and gentle volumes, soft and modulated natural light and very quiet materials (Fig. 6).

Fig. 6 Living room family Unit. *Photo: Maritin Saunders Photography.*

5. Project 3: Potter Street Redevelopment, Wallara Australia, Wintringham and Housing Choices Australia

Wallara was established in 1959 by local council and parents of children with disabilities. It has evolved over the years to become an innovative, sector leading support service for people with 'different' abilities. In the early 2000s Wallara identified the difficulties encountered by individuals who were then in their 50s and 60s and their aging parent careers. Parent carers were needing their own care and after a lifetime of caring for their disabled dependents, and many also found it difficult to relinquish their lifetime of responsibility. Wallara thus sought to find a means to support older people with disabilities and their elderly parent-careers.

In 2007, Wallara completely reinvented itself. The organization restructured entirely from an institutional congregate care facility to a community-based provider of services. It looked for a project partner to provide aged care services, and consequently selected Wintringham to build and manage a 60-bed aged care facility. Wintringham had become well-known for its specialty aged care services for older homeless people (see Project 1 above).

Wallara funded its rebuild by selling a percentage of their land to Wintringham. This would mean that the Wallara residences for individuals with disabilities would be in close proximity to the Wintringham residences for their aged careers. It would also mean that Wallara residents would be able to maintain a connection with the original heritage house that has been their home for a large part of their lives. Wallara's initial thought was that each component would sit on their own but adjacent parcel of land, separated by fences. Instead, after further consideration, the concept was to create a seamless interconnection between the residential components, and to emphasize the positive multi-use of an outdoor space for common use.

The Wallara-Wintringham cohort of clients were recognized as having a wide range of intellectual and psychological conditions. The layout has similar components to the Port Melbourne Hostel however for reasons of staffing, co-ordination and connection to the house cluster style was replaced by a congregate grouping style. This includes an overall connection via veranda covered decks with an interlocking circulation pattern. Staffing and communal dining areas are located centrally and within easy reach. The bedrooms remain in groups of six which provides a sense of place. The Wallara group clusters have their own kitchen and living areas. There is no separation between the Wallara groupings and the 'Wintringham groupings.

One of the prime goals—to allow an aging parent to live in an aged care setting while their adult child could be living in a supported care setting in proximity—was achieved via the design of an indoor-outdoor connection. This supports the feeling that both would be independent, yet only a few steps away from each other: living independently but together. The resulting sense of circulation and separation also enables residents to self-regulate their social interactions and engagement with environmental stimulation. Managers have remarked that the layout means that there is very little need to conduct orientation for new residents. A resident would be shown their room and then left to explore, and soon enough they find a group of residents that they could connect with when it was comfortable for them to do so. Being able to self-regulate social engagement reduces the effect of crowding and increase calmness (Fig. 7).

Verandas are an important element of the design. Indeed, most residents have preferred gathering under the verandas rather than the common lounge rooms. (Indeed, so few people use the common lounge rooms that half have since been converted to other uses such as theater room) Veranda spaces also connect with the smells of gardens which are more likely to have a positive effect (Fig. 8).

Fig. 7 Eunice Seddon Accommodation. *Photo: Martin Saunders Photography.*

Fig. 8 Wallara Family Accommodation. *Photo: Martin Saunders Photography.*

Fig. 9 Courtyard communal area. *Photo: Martin Saunders Photography.*

Maximizing open spaces and reducing corridors has also been important. Corridors, for example, funnel sounds around a building and the source can appear hidden (which residents can find more disturbing). Corridor spaces themselves make a strange and difficult place. Open places dissipate the sound and conversely introduce sounds of nature, wind, rain and birds (Fig. 9). The design also emphasizes the ventilation to reduce the effect of obvious unwanted smells and humidity that can build in enclosed group homes, and more subtle personal and environmental smells that can also create unwanted responses within the residents. The easiest way of management is to ensure 100% clean fresh air (Bentayeb et al., 2015). The combined result of the layout is fundamental to the health and wellbeing of the resident and staff. The maximizing of fresh air, sun light, natural light and connection to gardens provides recognized social benefits such as reduced crowding effect increasing sociability and reducing aggression. Sunlight can improve circadian rhythms facilitating healthy sleeping patterns.

Building materials have also been important, and we aimed to use familiar, soft materials, and bright but earthy colors so as to avoid the feeling of an institution which could invoke negative memories among residents (Fig. 10). We aimed to create a sense of place that would be associated with a domestic backyard BBQ, for example.

Having the variety of residents in close quarters does require specific management procedures, and the unique layout has meant hat staffing procedures have had to be tailored. Some post-occupancy modifications to the

Fig. 10 Friendly familiar material for cladding. *Photo: Allen Kong.*

built form have also been necessary to assist residents and staff and to facilitate effective management. The reception area, for example, was modified to allow the residents to access the reception staff from the outside veranda via a reception window. It was also found that the residents preferred to have their meals in the outside veranda areas and in other dispersed locations around, but this was difficult for staff to manage and observe during meals times. More undercover areas were therefore installed adjacent to the main dining area.

6. Conclusions

Community expectations have changed regarding the accommodation and support for people living with atypical needs, and architects' responses have gradually adjusted. For the most part, architects have been moving towards a more residential style of accommodation and away from anything that is obviously institutional. Newer organizations Wintringham, Odyssey, and organizations like Wallara that were born of the post war 1950s, provide good examples of how the philosophy of care and support for people with disabilities or with substance abuse disorders have changed over the last decades.

In the case of Wallara, some of the families have been through the entirety of their journey with the organization, and Wallara has evolved as the needs of their clients have changed. Attitudes have shifted from simply providing 'care' to providing support for increased self-determination.

For disability support organizations this change in approach from care to support for self-determination has come a change in the style of accommodation. This included the devolution of large institutional models of congregate care to a community-focused approach, involving small group houses carefully situated in a wider community. In Australia, this emergence of different housing models has been facilitated to technology and funding commitments. It is also the consequence of important champions, such as the architect Brian Kidd, who advocated for the creation of more home-like environments for the elderly, and which he demonstrated with the design of the aged-care residential facilities at Pomroy House and Wisteria Lodge in the 1980s (Chau, 2015).

Now, however the style of residential aged care facilities is typically a large-scale, 90 or 180 bed (note that they are usually defined by beds numbers—not residents) residential care facility. No doubt this is driven by economic considerations and the priorities of developers. This is at odds with the preferred style of support for people with disability—to be located within community in either individual apartments or small-scale group homes. An overall observation here, then, is that it seems that the disability support sector is entirely more responsive to the real needs of the residents, given that its approach to facilitating self-determination within the community is backed by an evidence-base.

The work of AKA derives from a deep understanding of the specific needs of each project. These needs balance the sometimes-competing requirements of budget, staff, residents, and environment. AKA strive to maintain a strong, long-term relationship with every client, allowing for a continued collection of understanding, ongoing development, and sustained solutions.

My personal background has been instrumental in AKA's holistic approach to design. My understanding, and passion for community building and defining a sense of belonging is pivotal to the projects AKA design and contributes heavily to our empathic environmental responses.

It has taken decades and an absolute existential crisis for the community and profession of architecture to focus on environmental sustainability. Sadly, the profession has also suffered immensely from a loss of technical ability due to an acceptance of external outsourcing. For the future, it is imperative that students of architecture are exposed to the requirements of inclusive design and learn these skills as a fundamental component of their training. There needs to be greater emphasis on the enabling aspects of design, to further promote a strong and intentional move away from 'skin-deep' architecture. This shift is not only sensible but a moral and ethical imperative as neuroscience continues to reveal the brain basis for differences among people, and society increasingly accepts and integrates neurodiversity into all its layers.

References

Anderson, D. C., Grey, T., Kennelly, S., & O'Neill, D. (2020). Nursing home design and COVID-19: balancing infection control, quality of life, and resilience. *Journal of the American Medical Directors Association, 21*(11), 1519–1524. https://doi.org/10.1016/j.jamda.2020.09.005.

Bentayeb, M., et al. (2015). Indoor air quality, ventilation and respiratory health in elderly residents living in nursing home in Europe. *European Respiratory Journal, 45*, 1228–1238.

Chau, H.-W. (2015). De-institutionalization of aged care residential facilities in Australia: Aldersgate village in Adelaide and Wrintringham facilities in Melbourne as case studies. In P. Hogben, & J. O'Callaghan (Eds.), *Architecture, institutions, and change* (pp. 82–91). Syndey, Australoa: SAHANZ.

Clark, R., Wellington, P., & Kong, A. (1981). *A psycho-environmental evaluation of Australia's Antarctic rebuilding programme*. Melbourne: Habitat Group, University of Melbourne.

Farrelly, E. (2013). *The Wintringham story*. Ellyson Communications.

Fleming, R., & Bowless, J. (1987). Units for the confused and disturbed elderly: Development, design, programming and evaluation. *Australian Journal on Ageing, 6*(4), 25–28.

Gehl, J. (1977). *The interface between public and private territories in residential areas*. The Department of Architecture and Building, Melbourne University.

Gosbell, K. B., & Holmes, I. E. B. (1982). The rebuilding of Australia's 3 Antarctic stations third symposium on Antarctic logistics, Leningrad. *USR Jlne*, 214–225.

Hall, E. T. (1969). *The hidden dimension*. Doubleday.

Kuwayu, M. (2022). *Overview: Mayi Kuwayu—The National Study of Aboriginal and Torres Strait Islander wellbeing*. Retrieved 4 May, 2022, from https://mkstudy.com.au/overview/.

Masso, M., Duncan, C., Grootemaat, P., Phillipson, L., Samsa, P., Fildes, D., et al. (2017). *Specialist dementia care units: An evidence check rapid review brokered by the sax institute (www.saxinstitute.org.au) for the commonwealth Department of Health*. Wollongong, NSW: Sax Institute.

Ostwald, M. J. (2001). Le Corbusier (Charles Edouard Jeanneret), The modulor and modulor 2–2 volumes. Basel: Birkhäuser, 2000. *Nexus Network Journal, 3*(1), 145–148.

Super, S., Wagemakers, M. A., Picavet, H. S., Verkooijen, K. T., & Koelen, M. A. (2016). Strengthening sense of coherence: Opportunities for theory building in health promotion. *Health Promotion International, 31*(4), 869–878.

Yang, J., Brandonand, P. S., & Sidwell, A. C. (2003). Sustainability in residential aged care. In W. A. Kong (Ed.), *The CIB 2003 International Conference on smart and sustainable built environment (SABSE2003) 19-21 November 2003*. Brisbane, Australia: Conference proceedings. Queensland University of Technology, Carnegie Mellon University.

Zeisel, J., Fleming, R., Bennett, K., et al. (2020). Pioneers and Innovators. In *Design, dignity, dementia: Dementia-related design and the built environment*. Alzheimer's Disease International.

Zrudlo, L. R. (1971). *The environmental design implications of Arctic man's psychological problems* (p. 165). Winnipeg, Canada: University of Manitoba. A thesis submitted for Master of Architecture, Report of architect Mark Hall Barrow.

CHAPTER TEN

Not all surfaces catch the light at the same time

Sonia Boue*
Artist and Consultant for Neurodiversity in the Arts, Oxford, United Kingdom
*Corresponding author: e-mail address: soniaboue@yahoo.co.uk

Contents

1. Introduction

Back then I was another person. I mention this to explain any possible duplication of fact contained in *Not all surfaces catch the light* and this introduction. It does all feel very distant. Also, don't we often repeat ourselves when telling our story? Besides, back then there was no COVID-19. Nothing, not even our trust in factual information has been the same since COVID-19, and furthermore I need to arrange the facts in a different order looking backwards. If facts are a bit like objects (which I feel they might be) this makes sense, as arranging ideas like objects is something of a theme (and a necessity). I think this is because what we call 'working memory' is in short supply. When you are a person of 'short attention span'—my mother's take on things—you need objects, I find. Portals of memory, objects can do so much of the heavy lifting where mapping and keeping track are concerned. It has also been said that my kind of people, (or perhaps people like me) are concrete thinkers. I feel this is good information which helps keep me away from abstract cognitions.

Developments in Neuroethics and Bioethics, Volume 5
ISSN 2589-2959
https://doi.org/10.1016/bs.dnb.2022.08.007

Yes, what follows is an exercise in trying to hold onto complex ideas using concrete thinking. For me this is writing neuro-diver-gently. The Neither Use Nor Ornament (NUNO) project is also best understood as the third of four Arts Council England (ACE) funded projects, that enabled me to research how to work as an autistic person in the Arts (Boue, 2019).[a] I discovered that I was autistic in 2016 after hitting a brick wall. I needed to find ways through, for both for myself and a cohort of artists like me. What follows are a series of reflections on reflections. However, rather than turn over my original blog posts for re-examination, I will peer at NUNO from my current vantage point, namely mid-pandemic and informed by project number four, called *Neurophototherapy* and funded by ACE from February–December 2021 (Boue, 2022). This includes illustrating the chapter with work from this more recent project, informed as it has also been by such reflections.

I have taken it as my job to deconstruct aspects of NUNO and expand on the architecture of project space. In conceiving of, designing, embodying neurodivergent space in my projects I have been (lucky to be) able to generate and disseminating findings as I go. These ideas have landed in some far-flung places, but this is probably a matter for someone with a different bodymind (a term I will explain soon) to survey. I can only speak as I go. My thoughts are on the fly, as they say.

2. Project space

A project needs space. What kind of space is the question. How do we make it, and will it also be made of glass, concrete, bricks and mortar? For beings who inhabit the sensory and perceptual in-between, such literal spaces can often be closed doors. The virtual can be both a project space and a haven.

This is complex so my mind takes a lateral turn.

A project also needs people (generally). How people and spaces interact optimally is woven though that which follows. The neurodiversity paradigm

[a] The first project was called *Through An Artist's Eye* and ran from May to October 2016; it included budgeting and curatorial support for me as a newly diagnosed autistic artist. This project was foundational in determining my subsequent projects as I learned more about what I didn't know I needed than I could have possibly imagined. It was a leap in the dark and the first attempt at working out how to navigate as an openly autistic professional with my core team. *The Museum for Object Research 1 and 2* were the follow-on projects, both funded by Arts Council England (ACE).

allows us to articulate inclusive praxis for a diversity of bodies *and* a 'diversity of minds' (Walker & Raymaker, 2020).

I have begun to explore ideas articulated by Dr. Nick Walker who proposes neuroqueer identity and neuroqueering as neurodivergent culture. In a recent interview, Walker unravels the linguistic tangles of neurodiversity to persuade us to think less about brains and more about bodyminds (Dychtwald, 1978). I am most persuaded of this particular point and interested in neuroqueering space, without especially focusing on gender here.

Back to concrete, bricks and mortar. These more literal and often vertiginous spaces (once my bodymind habitats pre-pandemic) feel ever more inaccessible, so that the invitation to revisit the NUNO Project for this chapter is to cross back over a threshold. Some of us have experienced the realities of entirely remote working, and this cannot now be undone. I've relished working from home during the various lockdowns and haven't readapted. Skills have been unlearned and how to relearn them is unclear, with such ongoing uncertainties. Muscle memory is a good example of bodymind. I have been surprised how easily the seemingly fundamental 'habits of life' slip away.

For almost 2 years I have avoided working in public buildings. Neurophototherapy (Boue, 2022) ran from my home office, aka the front room of the house I share with my family (Fig. 1). I'm not sure I can return to public spaces. Agoraphobia apart, the sustained absence of environmental stresses (including in-person peopling) has revealed the toll previously inflicted on my bodymind. I have to admit that important three-dimensional data is missing and Zoom fatigue is real but traveling time and accompanying sensory impacts have vanished. It's not a utopia, but it is new. Why do my thoughts linger on this point? I suppose it is because the pandemic has

Fig. 1 Boue (2022) 'Tell Me How To Feel' and 'Plugged.' Self-portraits from *Neurophototherapy* funded by ACE from February–December 2021. *Photos by author.*

rendered what went before literally impossible. Not only have people moved on, I am unclear where in the literal world I can design and embody neurodivergent space. Could there ever be a further iteration of NUNO? If we were to measure NUNO in spoons (Miserandino, 2003) I would be tempted to say no. Spoon theory (for those who don't know) relates to the quantity of mental and physical energy a person has for everyday tasks and activities. We'll leave spoons there, for now.

The culmination of NUNO did take place in a literal building, but as I analyze that moment, I see it as a gathering of bodyminds. This culmination being an exhibition opening and an events programme, imbued with the spirit of familial celebration. I'm suddenly struck that it could be the coming together of two artist networks at the OVADA Gallery in Oxford on that day, that brings a wedding to mind. The OVADA warehouse space was particularly light and gloriously charismatic, yet it is also an inaccessible building. NUNO was my awakening to physical access as an invisibly disabled person, also highlighted in a review by Caulfield (2019) to center on the opening as a peak moment of human interaction is not to dismiss the significance of the exhibition itself, or the opportunity to make and show work for the artists. The work was a triumph. However, it is the process not the work that is the focus of the chapter.

NUNO as a whole group (the artists and the core team) met for the very first time when our exhibition opened. Yet, this was a fraction of our journey, and NUNO was quintessentially born and raised online. Our funding tagline for NUNO was 'from digital to actual,' which was a reference to our beginnings in blog form. We had mainly connected via countless remote messages and conversations, our bodies geographically apart. I remember the sense of convenience, dislocation, and feeling the physical distance as a barrier at times. Those nearby met in my house and various city cafes. Remember that? These were pre-Zoom days, and I made just two Skype calls throughout. Of course, the journey from digital to actual can't be conveyed in a tagline, or even in a chapter. It is something you have to experience, but which I hope my blog posts allow readers a glimpse. It was compelling, transformational, and profoundly human. I want to say that autistic aversions can be unlearned if the space is right. This was certainly my experience.

3. Head space

We also talk about headspace, don't we? For me headspace operates via 'mind-shape,' which I'm articulating here for the first time (to my knowledge). I feel there must be a term for this. I don't mean literal shapes.

For example, I don't see triangles in my mind. Mind-shape (for me) describes the experience of being in the moment, yet holding an imperative for action and a constellation of cognitions in mind. Its precise relationship to calendar and clock is remote because time and events are also experienced as shapes. Thus, future events can be near or far, big or small. The specifics of calendar and clock (yes, numbers) are not shapes or shape friendly to me, therefore the requirement to focus on specific time and dates can feel like trying to park a barrage balloon on a twig. Shape, in my mind is an amorphous blob, which vibrates and feels warm, creative, and malleable. It can be inordinately difficult to move from one headspace to another, depending on the quality of the shape I'm inhabiting at any given time. This is important because I need collaborators who are good with twigs.

There follows a brief interlude to manage a patch of brain fog.

....

4. Inventory

I feel a need to regroup. Let's lay things out on a nice long trestle table.

(I also need to do this when packing a suitcase. Packing a suitcase, like organizing cognitions can be especially vexing because my bodymind is flighty and rarely settles. Each thought about what to pack and where to find it, triggers a lateral thought. Each action—such as finding something—generates an associated but not necessarily sequential action.) I will proceed in list form.

Concepts

- Spoons = energy and resources.
- Shapes = direction, vision and planning.
- Twigs = precision regarding when, where, how (how many).
- Bodyminds = communication, intersection and connection.

I see the above as an arrangement of 'objects.' Think of an early year's nature table. I find this exercise restful and reassuring. I feel I know where I am.

Learning

- We need to consider diverse body-minds.
- Some people use mind-shape (and other diverse navigation systems) and some people are 'twig carriers.'
- NUNO relied on mind-shape navigation, *and* 'twig carriers.'

I'm glad we began laying this out. I found a gap, or maybe a new list? This often happens when packing a suitcase too. I need to see things to hold them in mind. What I've missed so far is a note about the simplicity, detail and focus of NUNO. I will call this values (it is also an alignment of values with purpose).

Values

- NUNO had one principal value, ACCESS FOR ALL.

As I make my inventory I begin to see the project as an assemblage, not unlike the assemblages I make as an artist. Everything is connected in my neuroverse. I also think another list is needed, and in my neuroverse you can definitely have a list with one thing on it.

Purpose

- Collaboration across neurotypes

5. A note about equality

When I wrote the funding application for NUNO in 2018, the stated purpose of the project was to level the playing field for autistic artists on the project. Equality *is* what's required, but what does this look like in human terms, in a relational sense?

In practice, NUNO's great strength was to interpret all participatory needs as access for everyone on the project. Social bias came out in the wash and 'ordinary' needs could be reframed. How we think and talk about access is key.

There were times when some neurotypical artists on the project needed as much support to participate as some autistic artists. NUNO was a space of learning for us all. This is why I center on mind-shape and 'twig holder' collaboration across neurotypes when I think about purpose today. NUNO was beyond fortunate in the composition of its core team. Relationships and trust had been developed over time, and the project was designed with my access (as the project lead) at its core.

6. End note (part one)

Having laid it all out we've reached the nub of the matter. You could say that my mind-shape navigation was replicated at our project launch in that charismatic warehouse gallery in the center of Oxford. NUNO was a living breathing entity, a warm, malleable and celebratory space pulsating with life. The warehouse still stands, but NUNO passed through its doors in a matter of weeks. What remain are memories, relational bonds, digital traces, art works, writing and professional legacies. When I ask myself what made it all happen, I think it does boil down to relational practice and the social bonds generated through open communication and responding to need. Public funding was vital in making this project not about me, yet gifting the social capital I needed to pull it off. With my Arts Council funding I could open doors for others to come though.

This points in one direction. To create projects like NUNO requires a realignment of power and knowledge exchange. I think this is possible if we are clear that what we seek to generate are sites of learning, and if we make our purpose collaboration. Setting our compass to the north of genuine access is key.

7. Part two: Six curated blog posts

A Delicate Dance

(August 29, 2018)

This post is a series of reflections on autistic 'unmasking' at work on my Arts Council England funded project, Neither Use Nor Ornament (NUNO). 'Unmasking' has become a hot topic for some autistics recently. In reality it is, and always has been, a daily negotiation. What I want to say in this post is that, while it must always be acknowledged that some of us don't have the luxury of choice about 'masking,' others are dancing a daily, if not hourly, delicate dance with their 'masks.' Some have, quite rightly I think, questioned the term 'masking'—here meaning a survival strategy adopted by autistic people. The 'social carapace,' 'role playing,' 'performing neurotypicality,' and 'faking it' are all terms which can be used to describe what we do at any given time, mainly (in my case) unconsciously (until diagnosis that is) (Fig. 2).

Fig. 2 Boue (2022) 'Birdnose' and 'Eye.' Self-portraits from *Neurophototherapy* funded by ACE from February–December 2021. *Photos by author.*

And, before we go any further, 'faking it' here goes beyond 'normal' usage. One fascinating consequence of my identification as an autistic person is that I can now tell when I'm launching into social adaptations that go beyond the 'me' I'm more comfortable with. There must be shades of 'masking,' gradations if you will. Some late-diagnosed autistic writers I've encountered via blogging write about the difficulty in knowing where the line is between the adaptations they've learned and the 'authentic self.' As articulate and deeply intelligent beings, they often conclude that there is no such line. The 'authentic self' as a concept is flawed. Perhaps what we can best say is that we are deeply influenced, as all other humans are, by our nurture—except that in the case of autistics the nurture has so often been the wrong fit. We want—in important ways—to return to nature, to our natural selves. The problem is locating the self in an alien milieu. This search is compounded by the 'loss of self' implied by 'masking' in my view.

My own experience is that second guessing what others want from you as a blanket survival strategy leads to immense confusion in the area of identity. I'm glad to leave this aspect of 'masking' behind me. That is not to say that I don't 'mask'—of course I do—my professional life depends on it (despite my also being 'out' as a professional), and in reconstructing my personal life post diagnosis I am often immensely grateful for the many 'masks' in my repertoire. I am infinitely more comfortable and honest as I move through my days, and as I observe myself slip on a 'mask,' but this doesn't mean I have it made. I don't I still get caught out. I still find certain situations overwhelming. I am still humiliated. All I mean to do here really is observe some changes, and reflect the difficulties inherent in the actions both of 'masking' and 'unmasking.'

This is pertinent to my professional life too.

Central to my work as an artist and creative project lead is the idea of being 'out' as an autistic. My own creative work as such is not about autism, but I am committed to championing autistic arts professionalism (as I see it) within my sector (freelance visual arts practice). The autistic artists who currently work alongside me form part of an experimental project in which we seek to challenge preconceived notions about autistic artists. We are not savants, we are not outsiders, we are professional people doing a great job. Unfortunately this will be news to many! This would seem, on the surface, to be a wholly professional matter. Except that of course it isn't because it involves our 'unmasking'—that delicate and infinitely unstable (because it is constantly shifting) negotiation of the self. This is so utterly personal!

Imagine the 'masks' we use as a cache of theatre props and costumes or a child's dressing-up box. Imagine having to constantly judge each new situation in your day and rummage through the box for the right thing to wear. Imagine tuning in to the voices to find the right voice to use. Imagine studying the gestures to match them perfectly. Blend in, blend in—don't show yourself as you truly are, because you won't be accepted! And all without consciously understanding what is happening or why (before diagnosis).

For some of us this process will have become entangled in our creative work, performers especially, I imagine. For the artists I mentor, I usually suggest tuning in to the inner voice. This is to avoid a tendency to fragment and 'mask' in the face of outside influences. Locating a calm and loving inner voice can be a real challenge, however. This is why I'm currently loving the idea of quiet reflection—the practice of creating spaces in the day for the chatter to die down. Observe the 'masks,' observe the self, practice with and strengthen these muscles of observation (if you can).

Recently I've had cause to think about this key element of the incredibly powerful and exciting *Neither Use Nor Ornament* project. I can't assume anything about where we will all be with our 'masking' at any given moment, and that's a wonderful dynamic to work with when you think about it. What better example of nuance in presenting autism could I dream of than the 'now you see us, now you don't' reality of our lives? In writing this, I realize that so much of the anxiety of 'unmasking' is in the reception we receive—that's the bit we fear most because it can be dangerous (and/or humiliating) for us, and that is unpredictable. *So* privileged am I in my 'unmasking' today, that I often forget the deep, deep root of my social conditioning, in the playgrounds and playing fields of school. I forget that, for many autistics, especially where other minority statuses intersect, 'unmasking' is unsafe and not an option. I plan to work much harder to remember, and to coordinate the project with this at the forefront of my mind.

If we wobble in our resolve it won't be through cowardice—if we choose to 'mask' in specific situations, there will be no judgement at all. I emerge from blogging today with a new image for the project—that of a beautiful, sparkly, multifaceted gem. Not all surfaces catch the light at the same time. Let this be our motto.

I am an artist. I am also an autistic person. I live in sensory peril.

(October 28, 2018)

I am an artist. I am also an autistic person.

The other day someone rudely crashed into my Twitter timeline, expecting me to embrace a sudden urge to include me in a conversation about their work on embodied experience and the sensory world in a wholly unrelated context. I won't say more as I'm not into identifying people, but the stall set out in just one Tweet was honestly not in my area. Not even close. It's a good example of a growing phenomenon of people wanting some of the good stuff, i.e., the benefits of authentic autistic insight, but chasing like puppies at the first ball in sight. Perhaps predictably (and with equal speed) they crashed out again when I made clear my inability to speak to their area of interest. Better to be honest. It's okay—social mistakes emanate from both sides, though it's time to admit to this true fact.

However, what crashed into my timeline that day was privilege and false equivalence. The sensory world and our survival in it is not a drill for autistic people, so please don't come to me with your teaching tools and professional insights. This is mere hobbyism in comparison. I'm deeply interested in sensory survival, because sensory stress disables and ultimately kills autistic people. I'm not being sniffy if I don't want to play with you, but please understand that I'm just not your go-to generic autistic; I'm a professional artist trying to survive and improve conditions for late-diagnosed autistics, and for future generations too.

I am an artist. I am also an autistic person.

I like to say that I'm an autistic professional, not a 'professional autistic'—and I say this with no disrespect to those autistics who do consider themselves professionals in the area of autism first and foremost. I'm incredibly grateful to them for their work—but this isn't my identity.

I am an artist. I am also an autistic person.

The sensory torture of a hospital environment became my reality a couple of weeks ago, and I've been forced to reflect how much activity must be sacrificed to manage sensory stress in my life. It's been a tough lesson, and I've been made aware of my unusual level of privilege with regard to sensory stress at work. In fact—without my realizing it consciously—my working life is organized around sensory stress. I've compensated for it without even knowing it. I work freelance, and increasingly I work from the comfort of home. Hospital was only a series of day visits to support my nonagenarian mother, *but*, as I imagined my own hospital stay, or working in such conditions, I felt the scales rapidly fall from my eyes. Here is a flavor of my long visiting hours, and the level of challenge such conditions can imply.

Another layer of the autistic onion peeled away as I clung desperately to my composure under the cruel lighting of a small shared hospital ward.

Myriad whirls and bleeps monitored patients, during endless hours of uncertainty—in a vacuum of information—as staff in varying shades of uniform darted like fishes, eyes down. I was in foreign waters. Hierarchies of need, codes of conduct, signs and symbols—all had to be absorbed—and so the decoding began. I knew I must measure each interaction carefully. Life and death hung in the air—I am busy, I am busy, the staff blared silently. How to signal that I posed no threat? How to soothe and inch my way toward the person inside the uniform? I know, I *know*, I wanted to say. I will only take a second of your time!

It was a long game of observation over many hours. Snatches of information—disjointed—because it takes many parts to care on such a scale not all of which connect. Time, so much time. Sudden changes. My mother was moved at dawn, a wash bag and reading glasses left behind. A new scene—entire geographies to absorb on my arrival the next day. A ward of four women in varying states of peril. The layout is key—architecture and uniform color signals who's in charge, and who I must woo. Judgements are quickly made, but I am slow.

And now I am in it once more. Reliving it. The vertigo sets in. Tinnitus too. The lighting drills into me and I fight hard to deflect it—I have a long day ahead. A pitiful curtain shields a terminally ill woman as she retches up the awful hospital Friday fish lunch in the bed next to my mother's. She is 2 h at her labor, and her family rush back and forth with cardboard bowls. Can no-one do anything? She needs a private room.

My mother is quite deaf and I encourage her to take her hearing aids off. Television now costs a bomb in hospitals and so the distraction of daytime TV is lost as no-one bothers. Ghost TVs perch on brackets above every bed. Heartlessly we do the crosswords my mother loves. She is losing heart and fearful that she will never leave this place. I'm desperate to keep her spirits up.

Later I encounter the woman's son at the nursing station. Your poor mother, I'm so sorry we did the crossword while she was so unwell. It had felt callous and uncaring, but what to do? You were right to distract your mother, he said kindly. We exchanged stories, which somehow led to a shared history of watching the 1970s TV series Crossroads, famed for its turgid scripts and wobbly sets. It was equally adored and ridiculed in its time. We managed a laugh, but his mother was dying. This was beautiful and terrible. I felt intensely connected.

Hospital time is not the same as in the outside world. To enter is to surrender your agency to both the care and will of others, and to a system.

That system is housed in the kind of environment that I'm sure no-one likes, but has people like me scratching at the walls to get out. There were screamers in that long corridor of rooms and wards. Generally the screamers got their own room. I'd be a screamer for sure.

On my third day of visiting I checked out at 9pm, drove to my mother's house in a complete daze, and resolutely left the lights off. I didn't want any food. I couldn't swallow. I couldn't speak. I ignored my dear sibling and their family and went to bed in the spare room fully clothed, jabbering to myself, it's just a meat factory…. When my husband rang me all I could do was bark like an angry dog. It's often the person you trust the most who feels the brunt of your sensory distress.

I am an artist. I am also an autistic person. I live in sensory peril.

This is my admission to myself and to the world. Mainly, I manage my life, I am happy and I am loved. But it is very hard indeed when I am out of my bubble. I have had many accidents in my life: falling off my bike on a major roundabout is the only one I feel comfortable sharing. Having witnessed several cyclist fatalities in Oxford city I now don't cycle because I know my physical limitations. Accidents of this kind are due either to a sudden onset of vertigo or sensory overload.

This is something I wish all hobbyists to understand: I'm not hostile and I'm not angry. I'm just busy trying to survive.

What does inclusion look like?

(November 22, 2018)

In my day job, I am a visual artist with my own practice, but I'm also a community artist, mentor, trainer, and consultant. I work on various community arts projects as a freelancer, and I also lead my own projects.

I'm writing to share some of my findings after 3 months of working toward an inclusive Arts Council England funded project, where autistic and non-autistic artists will exhibit their work together (in February/March 2019 in Oxford, so there's still a way to go). Our project title is *Neither Use Nor Ornament* (NUNO).

It began with the grand idea to bring together two distinct networks, one predating the other. The longer standing group of non-autistic artists were to show their work in an exhibition that had already been planned.

The newer group of autistic artists would create an events programme to run concurrently, thereby creating a distinct but equal platform while allowing for a cross-pollination of ideas and influences.

In my mind's eye, the group show appeared as a fixed point at the core of the project, and the events programme whizzed around it like a Catherine wheel! I liked thinking about the dynamic interdependence of each element as a metaphor. Could this be a new model in the making? Audiences would certainly gain a sense of contrast, and when we began the project, the two networks were indeed quite separate, their only real point of intersection being me. But would it make any sense beyond my own imagination, and would this represent genuine inclusion? As a visual artist especially, I need to ask myself: what does inclusion look like?

The project (in a nutshell) is really about one person's professional journey toward congruence after a late diagnosis of autism, and their (my) greater commitment to journeying in company for the benefit of a wider group. My project is about making change happen for some of the artists involved. It also seeks to inform arts organizations. Challenging audience perception is important to us, though our spirit is not confrontational. Our first model was what you might call high on visibility. At this point, I didn't know any better. High-vis (or 'Day-Glo diversity'!) could meet with approval from a body like Arts Council England, who we know need to be seen to be doing better on this score. However, as my project progresses, I'm increasingly wary of the Day-Glo approach, which you see quite a lot in the arts right now. Genuine work is taking place in some cases, but I'm disquieted by this trend in diversity signaling.

Inclusion should be an everyday thing, rather than exceptional.

A hegemonic insistence on 'normality' conditions us to believe that signaling 'difference' in highly visible ways challenges perceptions, and therefore creates an instance of inclusion. There are times when this works precisely because our assumptions about who can be a player in society are so rigid. This strategy of 'watching' difference and 'noticing' it (often as 'inspirational') implies a norm from which 'difference' is discernible. This is hidden 'centring' and we must tear off the fetid blinkers of 'normality' conditioning to see it. Losing that fixed point, ditching that norm, and embracing diversity within humanity as the default setting would have us up in arms at the inequities of our (very biased) everyday assumptions.

At this point I refer back to the wisdom of an autistic child I knew, whose logical insistence that if we're all different (as we are), then nobody is special when it comes to educational need (or anything else for that matter). Ghettoization in the name of inclusion within mainstream education can definitely be a thing, and it has marked this young person; as a teenager, their instinct for survival prompted them to ditch all visible support. Not wanting to appear 'different' because it so stigmatized them tells you everything you need to know about being singled out for 'special' attention in 'mainstream' education. Obviously, school pupils turn into adults. Some will go on to wield power and be the decision makers of the future. What will inclusion mean to them? What does inclusion look like? Some of them will also buy lottery tickets—an important source of Arts Council England funding streams. Therefore, if current systems remain, some of these pupils will go on to fund projects like mine. Their adult counterparts of today have indeed funded my project, and ideally they could be among our audiences too.

My project is an attempt to reach out across these invisible fault lines, but the scales have fallen from my eyes. My Catherine Wheel was never going to take off, I was in thrall to a 'neurotypical' hegemony called 'normality.' As our work has gone on, I've listened to the artists on my project and absorbed the effects of high visibility on each one of them—and not all of them want it. Creative practice may seem like a 'safe zone' for the kind of self-expression which extends to autistic 'unmasking'—but how safe is it really? The problem with gaining a professional platform is precisely that you can be seen. Irony! Where invisible disability (such as autism) is concerned, some of us have been conditioned to 'mask' our difference in order to survive, as stigma and discrimination threaten if we show ourselves. Art practices do not exist in a vacuum and art alone cannot dissolve ableism—we need to get real about this. We don't choose to use 'masking' strategies, they occur as an adaptation. I know that it's a relative privilege to 'mask.' Not all of us can do this, but for those who can, it is a right.

Increasingly, privacy is being eroded—we are encouraged to share professional profiles on the very social media that friends, families and colleagues use. It is now almost impossible to control personal information which forms any part of a public persona. This happened to me very recently; the funny look at the non-autism related exhibition opening. *Oh, you're Sonia Boué.* My 'fame' in this instance was an autistic person not an artist. A google-eyed stare is not the end of the world but it's not a great look. I've learned to brush it off, but that's not the point. Invisible disability can demand a calculation at each and every turn. It's exhausting and sometimes the cause of great anxiety.

How much of myself do I show? Where are my safe zones? How often must I pretend and wait for my 'unmasking'? Mainly, we must ask ourselves: what will it cost me? Will it cost me my job, if I have one? Will it affect my mental health? Will I be bullied or abused? Here, of course, my heart bleeds for the autistics who cannot hide. The point is that no one should have to face this. For me, this is one reason why an ideal model of inclusive practice comes without a whizz and a bang. Some of us need to 'mask' our identities while gaining in rightful professional development. In any case, I wonder if great inclusive practice is something you can't necessarily see! Increasingly, I think this could be a truth to live by. Not only is inclusive practice potentially a quieter, more careful and considered game than I'd imagined, but the ultimate goal is that we genuinely don't see 'difference' because we're all included equally. It isn't so much what you see that counts. What matters is the activity that goes on behind the scenes to make a piece of work happen in a manner that's ethical and beneficial to all.

As I move forward with my project I find that my ideas are shifting. Our new model is still forming, and the much-longed-for cross-pollination of ideas and influences is taking place. I'm pretty sure I haven't always got this right, but the learning curve gives a spectacular view. I have a brilliant team and the most wonderful artists on board. The biggest change for me is that I no longer see my project as being one of two parts. Probably, that's what inclusion looks like. I'm grateful for ongoing conversations with the Arts at the Old Fire Station & Crisis Skylight partnership in Oxford, my mentor Miranda Millward, and Thomas Procter-Legg, Headteacher of Iffley Academy in Oxford, in informing aspects of my thinking about inclusive practice. I'm also grateful to Alastair Somerville of Acuity Design for his thinking on 'normality,' in particular his latest writing on *Building a normal world* (https://medium.com/@acuity_design/building-a-normal-world-52fbe7f7d960)

Sharing practice as an 'autistic artist'—some personal observations on ableism **(December 5, 2018)**

I'm still processing.

This is a phrase commonly heard among a particular cohort. The group in question is a network of autistic women I've come to know who've been diagnosed late in life. What I'm processing (to get back to that) is a first-ever experience of sharing my practice as an 'autistic artist.' Previously, I have only ever had cause to share my practice as an artist, period. Let me tell you, there can be a huge difference! For the first time, I understand the fear attached to being labeled.

Perhaps to no surprise, it turns out that 'outing' yourself (to people who neither know you well enough nor have an evolved understanding of autism) closes down the shutters of perception. It can even dictate (it seems) what is considered fair comment—the like of which I don't think would be tolerated for any other minority group in the room. We probably occupy what is currently the last frontier in minority rights. Others will emerge, I'm sure. When I share as the usual 'art me'—Sonia Boué specializing in postmemory work relating to the Spanish Civil War—I feel understood. I never fail to be met with respect and often even a gratifying interest in the many layers of my practice. Hurrah! It is also understood that I have a track record: I'm a professional person who has worked hard and gained significant experience in many areas of practice. So far, so brilliant!

I have always felt included, and certainly never 'othered.' This, I now know, is a privilege. I wish I could say that I was afforded the same respect when presenting my work as an autistic person more recently. Ableism klaxon! With hindsight I can see that it was my fault. Doh! I had tried to broker any misunderstanding of my practice head on. My work is implicitly autistic (because I am) but autism is not my subject. 'Keep it simple' is a motto I try to live by. But I had opened a crack in the door for ableist comment and aggression to pile in (unwittingly, it has to be said). Do I exaggerate? No, not really. Autistic people are subject to aggression and disrespect all the time, it's just that I'm masking and passing usually.

A code of practice?

For me this has highlighted a particular need for a code of practice when sharing our work as 'autistic artists,' which I feel moved to think about more deeply—and process a little more. It shouldn't be needed and perhaps won't be in all contexts, but until we make more progress on autism I'm for being 'share ready' or indeed not ready to share. I think this is about having more boundaries as individuals, but also about pooling knowledge on how to highlight and protect the needs of a community of creatives that is now coming forwards. A great deal of what I encounter in my mentoring and consultancy practice is a gaping hole around 'mindfulness' where diverse neurologies intersect. I'm not talking about a buzzword version of 'mindfulness.' I'm referring to slowing down to a speed at which we can *all* process more effectively. I'm talking about controlling the parameters of our engagement (where we can). This is my ambition for my cohort.

It is an absolute myth that good work happens at speed, or that those who are quick are also more effective. In my mind's eye, I see a giant hand. The palm is out-turned, signaling stop. It is gentle but firm—not a deity but rather a traffic signal. I am secretly enamored of the road sign and street paraphernalia that controls the flow of traffic. I long for tee-shirts with 'stop' and 'go' symbols! No entry! One way! Dead end! All beautifully simple and clear as means of communication. You can't get a license to drive until you've learnt the Highway Code for a very good reason—you'd kill or get killed pretty quickly without it. I like the idea of a nice laid out set of rules for engagement. I like rules. My ideal beginning for sharing my practice as an autistic person would be: *Stop. Look. Listen* (responses on a post-it note *only*).

Now where have I heard that before?

Angel Hair and New Year Resolutions

(December 31, 2018)

Reflections on autistic project design and leadership at the half-way mark #NUNO

A random memory. *Cabello de angel*—sugary threads tucked inside the belly of an *ensaïmada*. Angel hair wrapped in the lightest sweet doughy spiral of my childhood. I shower and reflect on the year about to pass. I think of angel hair. I feel its curious texture between my teeth once more as the white marble staircase to my grandmother's flat flashes before my minds eye. Under the influence of steam, I've visited the bewigged cake shop owner on the street below and am racing up the stairs with my treat. I'm probably 7 years old. In my memory of her, this kindly woman resembled a mature Betty Davis, but underneath her wig (I was told) she was completely hairless. In my imagination, I saw her wig-less at her counter once, but this is surely fantasy.

Cabello de angel means that I'm both nostalgic and happy. Angel hair is all about rewards. The family have been enjoying a peaceful Christmas, and in the gaps between viewing ancient Kodak slides on the viewfinder I gave my mother, and seeing off the remains of the Christmas pud, I've been evaluating my Arts Council England project.

A non-sequitur, I know.

The evaluation had landed in the online portal 10 days beforehand, and I'd only happened on it by accident, as there had been no notification. Not a good look to miss this particular deadline. The second part of our funding depends on it. My boxing day was interesting. I spent the day in a blur playing catch up. Managing a complex project can feel like a big ask sometimes, due to the combined challenges of autism, dyslexia and dyscalculia. It can be scary, for example, when your brain goes walkabout and you know meanwhile that the pesky checklist of vital project tasks won't tick itself. I like the phrase 'buffering,' which I've come to trust as a necessary period of processing. It describes perfectly those periods of time when I simply can't focus on the 'right' details. In such a state it's honestly better to watch an entire series on Netflix than try, but when the stars align there is nothing to match what can be achieved by the converse state of hyper-focus. It seems there must be other states too. States in which we try and fumble. Ones in which we 'do our best.' I often find it hard to remember these in-between places as being anywhere near useful, and yet they must be because I don't think that I've oscillated between the super functional and resting states in a constant loop from July to December. My main impression has been of grafting and trying—without the luxury of time and space to either buffer or hyper-focus in my preferred manner.

It's surprising to me that we've achieved so much, as I write about it for the Arts Council. My project is about making a difference, and it is doing just that, in pleasingly measurable and incremental ways. The 'angel hair' for the artists on this project is not for me to share in any great detail, but for some of us it has been transformational. The opportunity to work autistically has allowed for important developments to occur, the most obvious being our (potentially) day after Brexit exhibition opening!

Other effects will be longer lasting, and relate to vital relationships and networks forming (and consolidating), and further opportunities of work alongside present employment—which will lead to profiles being raised and reputations made. This is the stuff of working lives, but the stuff some autistic artists have been long denied, due to specific challenges in the area of social semantics among others.

So, despite the sweat at times—or more likely because of it—we have some really important half-time outcomes to feel good about. I want to be very un-British and blow our project trumpets loudly! I want to be clear that this is what happens when you begin to work in autistic ways. This is what happens when we are free to design our own projects. This is what happens when we lead.

References

Boue, S. (2019). *Neither use nor ornament 1: An evaluation document for the arts and the inclusion of autistic artists*. Available online at: https://www.soniaboue.co.uk/nuno-pdf.

Boue, S. (2022). *Neurophototherapy evaluation*. Available online at: https://www.soniaboue.co.uk/neurophototherapy-evaluation-pdf.

Caulfield, D. (2019). Neither Use Nor Ornament (NUNO)—Life stories told through the objects that shape experience. In *Disability arts. Online* 26 April https://disabilityarts.online/magazine/opinion/neither-use-nor-ornament-nuno-life-stories-told-through-the-objects-that-shape-experience/.

Dychtwald, K. (1978). *Bodymind*. New York: Jove Publications.

Miserandino, C. (2003). *The spoon theory*. Blog post archived at https://butyoudontlooksick.com/articles/written-by-christine/the-spoon-theory/.

Walker, N., & Raymaker, D. M. (2020). Toward a neuroqueer future: An interview with Nick Walker. In *vol. 3 (1). Autism in adulthood* (pp. 5–10). https://www.liebertpub.com/doi/10.1089/aut.2020.29014.njw.

CHAPTER ELEVEN

Admission: The mycelia of neuroqueer touch in architectures of madness

J.T. Eisenhauer Richardson
Department of Arts Administration, Education, and Policy, The Ohio State University, Columbus, OH, United States

Contents

1. Prologue

In my closet, I carefully store 28 individually wrapped bread slices with dates stamped on the plastic next to the factory name to indicate when the bread should stop being sold (see Fig. 1). I collected these plastic-wrapped bread slices 19 years ago during 28 days confined in psychiatric space.[a] Five of the slices of bread molded inside unopened wrappers, becoming dark fragments and piles of dust. A few years after bringing these slices of bread

[a] In this chapter, I approach "architectures of madness" that include similar spaces that are given many different names (e.g., asylum, institution, hospital, a locked ward, psychiatric unit/floor, etc.). I position the term "hospitalization" critically as it can become a means through which histories and continued experiences of violence, trauma, and loss of agency and rights, become subsumed within a discourse of cure, progress, and care. As a result, such traumas become anomalies, concessions, or a means to blame those experiencing trauma for the injustice they incur from others.

Developments in Neuroethics and Bioethics, Volume 5
ISSN 2589-2959
https://doi.org/10.1016/bs.dnb.2022.08.008

Fig. 1 Left: Moldy bread slice in a wrapper; Right: Self-Portrait (detail), *Admission.*

home along with two hospital gowns and a pair of thin blue hospital-issued pajamas, I created an installation in a local gallery titled *Admission*. The installation included three interrelated multi-media works, titled *Self-portrait, Landscape, and Still-Life*. Each of these artworks integrated found objects mounted on the gallery walls juxtaposed with projections of images of objects. The work titled *Self-portrait* included a section of the gallery wall covered in a grid of white bread slices adhered to a plexiglass backing onto which were projected close-up images of two hospital gown fabrics (see Fig. 4). While these were not the bread slices that I had stored in my closet, the hospital gowns were photographs of the gowns from the hospital. As gallery visitors stood in front of the bread, their bodies blocked portions of the projected images casting their shadows onto the illuminated surface (see Fig. 3). The red amorphous shapes of the hospital gown fabric blurred on the bread's surface appearing like cells on a microscope slide and splotches of jelly. Some visitors, initially aware of their bodies creating shadows, would look down to notice how clothes or skin became illuminated, sometimes covering them with red blotches.

In this chapter, I follow a line of inquiry and curiosity about how I previously framed, storied, and theorized (arguably an academic form of storying) my relationship with these objects, psychiatrization, and academic, psychiatric, and arts institutions. More specifically, I re-turn to my article "*Admission*: Madness and (Be)coming Out Within and Through Spaces of Confinement" published in *Disability Studies Quarterly* in 2009 and what exceeded the representational frame of my *DSQ* article (Eisenhauer, 2009). When I was invited to include my *DSQ* article in this edited book, I considered the possibility of *re-turning* rather than reprinting, particularly as a process of turning over again questions of representation, artmaking, and psychiatrization.

For Barad (2014) "re-turning" does not refer to "returning as in reflecting on or going back to a past that was but re-turning as in turning it over and over again—iteratively intra-acting, re-diffracting, diffracting anew, in the making of new temporalities (spacetimematterings), new diffraction patterns" (p. 168). Inherent to the underlying performative understanding of discursive practices in Barad's work is a critical orientation toward *representationalism*. In particular, Barad (2003) challenges privileging language over matter, and the representationalist belief in "the power of words to represent preexisting things" and "to mirror preexisting phenomena" (p. 802). Rather, Barad proposes a relational ontology that challenges human and nonhuman binaries and attends to matter not as a "thing, but as a doing, a congealing of agency" (p. 822). In what follows, I consider how my relationship with objects and psychiatric spaces through neuroqueer touch and neuroqueer intimacy might elude human/non-human distinctions[b] and the interpretive imperative to locate meaning inherent to representational logic.

My *DSQ* article's structure included a prologue, performative[c] essay, and epilogue. The performative narrative valued juxtaposition and non-linearity to capture the intersection of my own experiences of psychiatrization, the exhibition, qualities of the artworks, gallery visitors' comments, and larger institutional histories and sociocultural perspectives. The article, the artwork, and my own experiences are interrelated and interactive, but cannot be conflated. In this chapter, I include the performative essay from this earlier article and a new prologue and epilogue. I understand the new prologue and epilogue I include in this chapter less as a revision or reflection, but through Haraway's (1997) critique of reflexivity as it informs Barad's characterization of diffraction. Haraway (1997) writes: "Reflexivity has been recommended as a critical practice, but my suspicion is that reflexivity, like reflection, only displaces the same elsewhere" (p. 16). The prologue and epilogue in my *DSQ* article became book ends, providing theoretical and analytical bracing to a structure that may otherwise be perceived as lacking accepted forms of academic "stability," particularly as an article including creative prose and written by a Mad narrator/researcher (Price, 2014). As I wrote this chapter's prologue and epilogue, I reconsidered the function of prologues and epilogues. Perhaps prologues and epilogues "exfoliate" rather than buttress, control, or make "sense" of the narrative "body," and open into reciprocating and generative "transitions" (Manning, 2007, p. 10).

While I made few changes to the original performative essay, some of these edits point to crucial differences. I no longer identify myself with

[b] Trento (2021) proposes "an ecological conceptualization of difference as a continuum, and I propose the dissolution of the human-animal divide, in favor of a mutually inclusive model of difference" (p. 93).

[c] "Performative writing is a method of "writing as doing" where "writing becomes meaningful in the material, dis/continuous act of writing" (Pollock, 1998, p. 75).

diagnostic labels, nor by using the first name and pronouns incorporated in my prior article. However, even Mad, while a reclaimed term, does not capture the multiplicity of my neurodivergent experience or intersectionality. I repeatedly encounter the demand that my body remain categorizable and readable, indicative of others' anxieties about ambiguous bodies. Therefore, in this chapter, I invoke neuroqueer(ing) not simply to establish an identity in opposition to diagnostic labels and their entrapment within psychiatrization, as a type of counter-identification. Egner (2018) characterizes neuroqueer disidentification building from the work of Esteban Muñoz (2009) and Attias (2020) extends Yergeau's engagement with demi-rhetoricity, to consider artmaking practices as neuroqueer rhetoric. Egner (2019) writes:

> *The fluid and multi-definitional term neuroqueer provides members of this community what identity-centered politics cannot: inclusion and value of their unique intersectional selves that are not confined to specific identity categories...[T] hrough practices of disidentification, neuroqueering rejects any form of exclusion and binaries, thus contributing to the deconstruction of typical understandings of identity categories. Employed as a politic, neuroqueer emphasizes the malleability of the meanings and experiences associated with various social locations. (pp. 141–142).*

The question will arise, as I have asked myself, why did I save these pieces of bread? Attempting to find the meaning of this action as being aesthetic, critical and creative will teeter close to the representational logic of pathology and diagnosis. Perhaps, this representational logic intrudes on my relationships with objects as a neuroqueer intimacy, while also refusing my neuroqueer response[d]: *I don't know.* In this chapter, I consider the importance of fostering space for *I don't know,* and what exceeds the impulse toward

[d] Yergeau's (2018) discussion of autism and rhetoricity and demi-rhetoricity challenges the ways in which neuroqueer people are assumed to be "arhetorical or partially rhetorical at best" (p. 50). In addition, to Yergeau's book *Authoring Autism,* Yergeau 's (2013) article "Clinically Significant Disturbance" published in *DSQ,* is an "autie-ethnographic" narrative that critically addresses ToM and integrates Yergeau's account of involuntary commitment after taking their first academic position. Yergeau's engagement with neurodivergence challenges notions of communication limited to the intentional as they expand rhetoricity beyond its exclusion of what is labeled non-communicative and unintentional e.g., stimming, rocking, and more. Attias (2020) builds from Yergeau's discussion of rhetoricity to consider artmaking as neuroqueer rhetoric. Attias' consideration of the body and sensation, particularly in relationship to ADHD, and journaling as a neuroqueer form, create important openings for extending Yergeau's original questions surrounding rhetoricity, communication, and intentionality, to thinking about artistic processes. In this chapter, my consideration of the more-than-representational, neuroqueer intimacy, non-human actants, and psychiatrized institutional spaces does not propose neuroqueer touch and neuroqueer intimacy as communicative.

interpretation, counter-narration, and representation. *I don't know,* can be presumed to reflect a lack of insight and "ignorance." *I don't know* is uttered by the presumed-to-be visually "illiterate" subject, the subject unable to "read," "decode," "decipher the meaning," of visual "texts." Such a visually "illiterate" subject's irrationality, ignorance, and inability to discern the fake from the real, characterized as damaging, sick, "manipulative," images, social media, and so forth becomes perceived as a threat to the future of democracy. However, once again, the boundaries of citizenship become demarcated by rationality and literacy, among other "skills and dispositions" historically used to establish the boundaries of citizenship through racist, heterosexist, ableist, and colonial justifications. The irrational, illiterate subject remains noncitizen, or noncitizen until able to prove one's abilities otherwise.[e]

Academic contexts privilege "insightful" observations and argumentation.[f] Labeling a person as having a "lack of insight" within psychiatrization, and particularly in relation to "psychosis" becomes an indicator of both severity of "illness" and a justification for involuntary treatment. Losing the option to respond, *I don't know* demonstrates one among many forms of trauma experienced through psychiatrization resulting from expectations to disclose, respond, confess, and answer. Intentionality, Bennett (2010) proposes, "is positioned as the most important of all agential factors, the bearer of an exceptional kind of power" (p. 34). When patients are judged to be acting unintentionally, randomly, or uncontrollably subsequent interventions, including nonconsensual medical "treatment," are presumed to "restore" rather than "violate" one's agency. However, in my *DSQ* article, an irony emerges in which I needed to assert my "potential" and ability "to make a meaningful contribution" as an emerging academic labeled "mentally ill," while in my work I questioned the mechanism of "insight" (Jones & Brown, 2012).

I notice in my prior writing how I characterize the bread as cooperative, docile, malleable, glued in nice, neat rows, stored in a box, sealed in plastic,

---

[e] This discussion related to visual literacy emerges in the field of art education as a means to justify art education's relevance. However, visual literacy becomes questioned when notions of "reading" images lose their presumptive authority. Even so, the anxiety of a public, unable to "read" images in a media "saturated" environment, becomes an apocalyptic narrative through which ableist and racist notions of citizenship and competency emerge and proliferate.

[f] Koivisto's (2018) framing of "egress" as encompassing strategies related to resistance in his writing about Blackness, Horrorcore, and Krip Hop navigates important questions surrounding psychiatrization, depsychiatrization and resistance that question the limitations of representational orientations. While navigating stereotypes remains integral to questions Koivisto considers, his proposition that art can provide *egresses* emphasizes resistance imbedded within the work rather than as the intention of its creator.

and turned into a projection screen. In many ways, this narrative asserted my human agency over presumed inactive forms of matter. However, the bread resisted most adhesives. I received a concerned call from the gallery director shortly after hanging the show that bread was falling to the floor. My spouse and I removed the artwork and brought it to our basement to experiment with multiple ways to adhere bread. Bread is layered, organic, constantly changing and can easily release when glued to a surface. It's exfoliative and perpetually changing qualities become incompatible with adhesive. However, when I referred to bread, as a "found object," it also became my possession over which I had the authority to label as "uncooperative."

The opening night of the exhibition, we had a caulk gun hidden out of view, watching for which piece of bread would next fall to the floor. Typically, I may have let the bread continue to fall to the floor, incorporating this unexpected aspect into the work. However, the gallery did not want the bread to be directly adhered to the wall, necessitating the need to adhere the bread to a plexiglass support. Falling bread from a wall seemed interesting, but the surface of plexiglass with the messy remnants of glue, much less so. On a functional level, clear access was to be maintained to a door and set of elevators at the far side of the gallery, which served as the "side entrance" for employees entering the adjacent psychiatric clinic (see Fig. 2). Ironically, the proximity of bread to psychiatry still necessitated it remain controlled and cooperative. This is not to suggest that bread's agency is also located in its abilities or qualities including demonstrating "resistance." Rather, I recognize how my writing about the installation often masked my neuroqueer relationship with objects, while reinforcing a neurotypical distinction between human/non-human.

My experiences and understanding of psychiatrization and psychiatric spaces, continue to change since exhibiting my installation *Admission.* I repeatedly return to questions about psychiatric spaces described as institutions and hospitals (Eisenhauer Richardson, 2018, 2019). While I become frustrated with the romanticization of institutional psychiatrization historically and today including intersecting notions of confinement and creativity and the normalization and perpetuation of institutional violence, I desire a more complex engagement with the experience of inhabiting these spaces (Eisenhauer Richardson, 2016, 2018, 2019). I desire a deeper engagement with the communal, temporal, material, sensory aspects of these spaces as understood by those who have extensive expertise by which I mean those people repeatedly returned to these spaces and who live under the assumption that this can occur again. I desire to understand the relationships

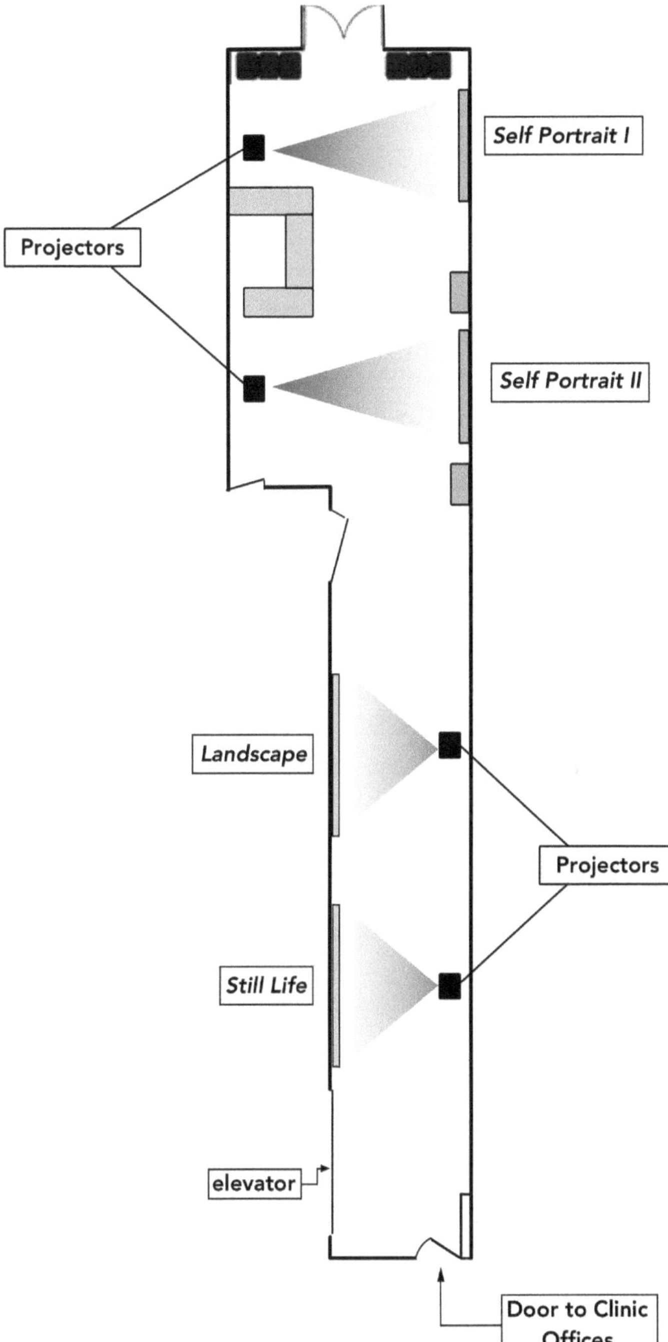

Fig. 2 Gallery layout for *Admission*, 2008.

between 19th century asylums and 21st century psychiatric hospitals because the historicization of contemporary experience justifies comparative constructs of progress such as, "at least it's not like it was," that deny the possibility that contemporary sanism and trauma can still result from psychiatrization. Finally, I desire the possibility for an academic writer to position themself not as an expert speaking for others, but as having expertise and understanding informed by years of returning to these spaces. Regrettably, such repeated returning often justifies one's very disqualification.

Complex mechanisms between dispossession and dehumanization are central to institutional forms of psychiatrization (Goffman, 2007). I continue to wonder about the blurred distinction between human and non-human as well as the distinctive ways in which objects acquire meanings, functions, and agency within institutional space. In the following sections, I include the performative essay from my *DSQ* article. In the epilogue, I further consider my relationship with these objects through *neuroqueer touch* and *neuroqueer intimacy* as relational, distributive and superimposing inside and outside, human, and non-human agencies. Attending closely to a permeable, moving body's interrelationship with mold and spores, I reconsider the space and time of confinement within architectures of madness.

2. *Admission*: An essay, 2009

2.1 Admission

Three works come together: *self-portrait*, *landscape*, and *still life*. I think of many trips to museums walking from room to room, painting to painting, with tired legs and this is what I see, *self-portrait*, *landscape*, and *still life*. They inhabit the same space, but side by side with no peripheral vision. They may actually speak to each other, but we don't hear it. We don't expect it.

2.2 Self-portrait

bread, bread, bread, bread, bread, bread, bread....

I received one individually wrapped piece of bread with every meal (see Fig. 3). Each piece was in a clear plastic wrapper sealed at both ends. I decided to save a piece of bread for every day I was there. I would sometimes wake up in the morning to find someone opening all my drawers, every container, holding the spines of my books and fanning the pages watching for something to fall to the floor. *Please don't take my bread. Please don't take my bread.* She sees my eyes open and says only two words: ROOM CHECK.

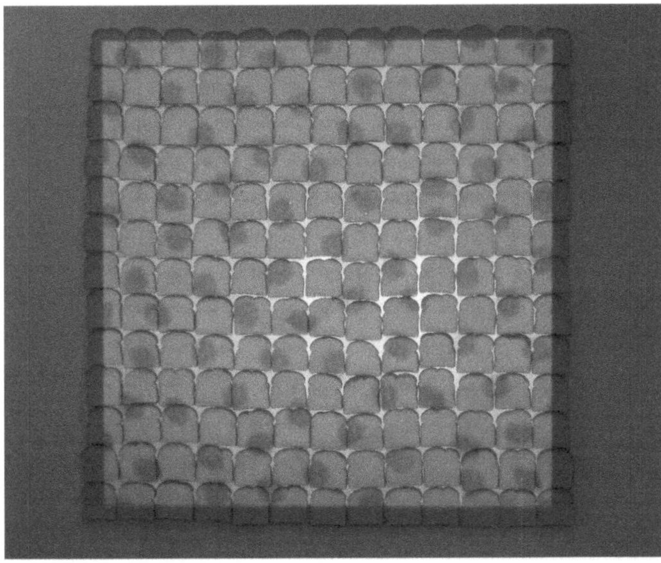

Fig. 3 *Self-Portrait, Admission*, 2008.

My room is deemed safe now. And I know that in 20 min someone carrying a clipboard will look for me again. J.T.? *Yes.* CHECK.

breakfast, lunch, dinner, breakfast, lunch, dinner, breakfast, lunch, dinner.... Can I go home yet?

When I tell the bread story for the first time during a gallery talk the visitors are fascinated with *why*? Why this many pieces? Why bread? Why projected hospital gown fabrics?

They find glimpses of authenticity exciting—*those are the pajamas and gown that I received in the hospital.* And for a moment in this discussion, they are **REAL**...but then they are only light, projections on white bread and now they are bacteria on microscope slides or jelly (see Fig. 1).

The bread that I saved is still in my closet at home and some of it is white, but the moisture has left, and it gets smaller and smaller and some of it molded and became dust without ever opening a wrapper.

One woman asks, *Did any of the hospital staff or doctors ever ask you about the bread?*

No.

She then somewhat jokingly said, *I wonder how they charted that?* and I knew where she came from by her medical dialect. She has likely written in more charts than have been written about her.

She then said, *Think of what could have happened if the staff or your doctor had only asked about the bread.* I imagine the potential of this conversation and realize that she sees this value as well.

These individually wrapped pieces of bread become bodies, preserved, confined, surveyed, controlled, and uniform.

Rows and rows of white bread become padded walls. At the nurses' desk they watch patients put in the isolation room on a small black and white screen. I am continually bruised from bumping into what is no longer visible. (see Fig. 4).

There were no conversations for me with people who held charts and wore white coats even though they asked me every day how I was feeling. I later find out that I am speaking with government leaders in the Department of Mental Health. I think about her point and find an awareness and reflexivity within it and realize my own assumptions about people who wear white coats and write in charts. However, it isn't that simple. She was wearing a blue suit.

A few seats away a man hesitantly speaks up. *When I see that bread I think of my birds. Every morning I take bread outside and hang it in the trees. The birds just make me so happy. I saw a mother bird with her baby this morning and it made my day. I was a line chef for 20 years and have fed many people. I didn't go to college. Now I feed the birds.* Later he introduces himself to me by calling himself a "paranoid schizophrenic" and I knew how he was charted.

Fig. 4 *Self-Portrait* (detail) *Admission*, 2008.

2.3 Landscape

Have you taken your medicine? I have your medicine (Fig. 5).

These almost sound the same.

Have you taken your medicine? I have your medicine.

In the morning and evening, the nurses walk like waiters with trays covered with small cups filled with pills and water. Each of us is a cup filled with pills. *I have your medicine.*

In the morning and evening, my husband will often ask, *Have you taken your medicine?* but I am not a cup filled with pills.

They almost sound the same. But they are not.

Asylum. Asylum.

They sound the same.

The Ohio Lunatic Asylum opened in 1877 and claimed to be one of the largest "insane asylums" in the world. It was demolished in the early 1990s and most of what remains are "asylum graveyards." (Fig. 6).

F 1275, F 5721, M 2481 M 1842. Many of these graves are labeled only by gender and a number.

There are many numbers. **60,000**: A Nazi poster reads "60,000 RM is what this person with genetic defects costs the community during his lifetime. Fellow German, that's your money too."

Fig. 5 *Landscape*, installation view, *Admission*, 2008.

Fig. 6 *Landscape,* (detail), *Admission,* 2008.

T4 euthanasia program. 1939–41, **300,000** the estimated number of disabled people labeled killed through the Nazi T4 Euthanasia program in places called hospitals.

I have your medicine.

1: The United States is the first country to create compulsory sterilization programs in 1907. 1928: The flashing light exhibit by the American Eugenics Society reads "This light flashes every 15 seconds. Every 15 seconds $100 of your money pays for the care of persons with bad heredity such as the insane, the feeble minded, criminals and other defectives." **68,000**: the approximate number of forced sterilizations in the United States most often in places called hospitals.

Do you have my number?

Before the opening of *Admission*, I find out that a donor reviewed my proposal for this installation and was very interested in the work. I am told that she is a psychiatrist and the director of a large residential facility in the

Fig. 7 Shadow of person viewing *Landscape, Admission*, 2008.

city where I live and that she has recently published a book about the Ohio Lunatic Asylum.

I ask the gallery director; *Does she understand the work?* [What understanding did I assume should be known? Can I revise another person's shadow?] (see Fig. 7).

Later, I am given a copy of the book. There is a large stained-glass window on the cover.

I hear voices… stories about forced ice baths, physical restraints, electroconvulsive therapy, lobotomies, beatings, sexual abuse, forced sterilization, concentration camps and euthanasia programs.

One month, two months, three months; one year, two years, three years, four years. My view of the asylum is not through a stained-glass window.

I am told—no, this was a good place. They tried very hard. They treated the patients well. They didn't know anything else. They did what they thought was best. *I have your medicine.*

A mental health worker later said, *When I got home tonight I realized how much I danced around so as not to be cast into the Old Asylum.* In an email another woman wrote about looking at the image of the asylum and now having the power to walk away from that place.

Someone else looked at the projection of the Ohio Lunatic Asylum and said to me, *Isn't the architecture beautiful?*

2.4 Still life

For the longest time, I have had a box of belts in the back of my closet. It is a walk-in closet, a place that I can actually stand in. I tuck away things that other people will never see there. I know that the belts are in the closet even though I have never worn them.

While standing in the gallery a number of people asked me, *Are those real restraints?* (see Figs. 8 and 9).

Lab coats or doctors' coats? Patient or Petri dish?

Hold still! My very active and independent 2-year old daughter rolls around and kicks her legs as I attempt to change her diaper. *Hold still! It is time to put your shoes on. Hold still! It is time to comb your hair. Hold still! It is time to brush your teeth. Hold still! I have your medicine.*

Hold still! I know what is best for you.

In *Madness and Civilization: A History of Insanity in the Age of Reason* (Foucault, 1967), Michel Foucault describes how in the Middle Ages bars were installed at an asylum in Germany so that those outside would be able to view the madmen chained within (p. 68). Throughout history this desire to watch those labeled insane and mad persisted including people paying to watch people in asylums. Within the "art world" the relegation of art by people with mental illnesses into psychiatric art collections or its

Fig. 8 *Still Life* installation view, *Admission,* 2008.

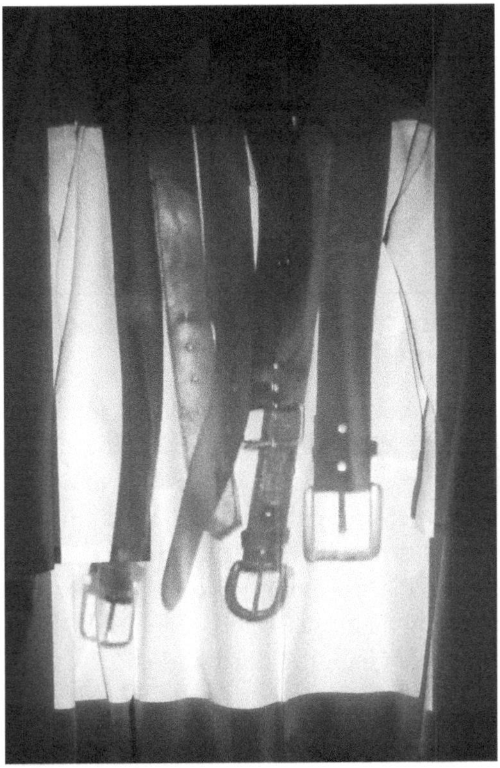

Fig. 9 *Still Life,* detail, *Admission,* 2008.

categorization as l'Art Brut and Outsider Art, can at times produce and/or reinforce a similar kind of pleasure experienced through this desire to look at difference.

2.5 Are those real restraints?

At the opening, the director of Southeast, Inc. stands and reads into a microphone the mission statement for the gallery. I have heard it before, but it makes me uncomfortable: *This gallery exhibits the works of individuals affected by mental illness... Through art, we educate the community....* I imagine the people, particularly those people attending the opening who didn't know this was the mission of the gallery, who are colleagues and students and I feel restrained. Individuals become a blurred mass, the always shifting "us" and "them." I see the title of my exhibition on the wall above his head. *Admission.... this gallery exhibits the works of individuals affected by mental illness....Admission....* Buy your tickets.... *Admission....* Take a look....

During the artist talk, I am asked, *how do you feel when it is time to take it down? Is it hard that you can't keep it?*

My response to her question was fairly academic.

Ephemeral. (definition) Short-lived; existing or continuing for a short time only.

I later realize that it is not my inability to keep it, but rather it is my inability to leave it behind that is hard.

3. Epilogue or "Is this an artwork or something I should be concerned about?"

A number of years ago, I gathered recycled paper and tore each sheet into nine squares. On each small square, I drew the chair in my "hospital" room the same way. I organized the collection of these small drawings into an evenly spaced grid across the desk and adjacent windowsill. One morning, the doctor and his team of residents, medical students, and a social worker crammed into the space of my "room" delineated by a sliding curtain. They began asking their usual series of questions. However, upon seeing the grid of small drawings, the doctor asked:

"Is this an artwork or something I should be concerned about?"

I quickly responded, "an artwork" and after they left my room, put the drawings into a pile and in the drawer, never letting them be seen again. My response, even now, to the doctor's inquiry is *I don't know.* (Although I deeply regret not asking him a series of follow-up questions.) Likewise, *I don't know* why I saved 28 slices of bread.

Toward the end of my performative essay in the prior section of this chapter, I attended to multiple meanings of the term admission, including references to disclosure, spectacle, access, and medicalization. Each of these terms as concepts invoke relationships between bodies, experience, and spaces. Disclosure situates experience through closure and opening, just as admittance recalls the threshold of a door, a crossing, or entering. However, admission can be framed as containment, confinement, and transformation resulting in loss. "Admission," as Goffman, 2007 describes in his seminal work *Asylums* published in 1961, involves "contaminated exposure" through mortification and dispossession (p. 25). For Goffman, admission includes mechanized processes of "trimming" and "programming" the newly arrived patient, "a procedure…characterized as a leaving off and a taking on, with the midpoint marked by physical nakedness. Leaving off of course entails a dispossession of property, important because persons invest self-feelings in their possessions" (p. 18). Dispossession includes the loss of personal possessions, one's personal belongings, but also nonphysical possessions such as one's name. Goffman characterizes "one form of mortification in total institutions as [a] kind of contaminative exposure" which includes:

> *[the inability to] hold objects of self-feeling—such as his body, his immediate actions, his thoughts, and some of his possessions—clear of contact from alien and contaminating things. But in total institutions these territories of the self are violated; the boundary that the individual places between his being and the environment is invaded and the embodiments of self-profaned (p. 23).*

Goffman's critical address of "total institutions" necessitates demarcating the boundaries of a "self" in order to identify its trespass and violation. These include boundaries between self and other, self and environment, human and non-human, and different configurations of time. The "tension between the home world and the institutional world" becomes central within Goffman's assessment of the dominance and control of "total institutions" (p. 13). Likewise, juxtaposing inside and outside of the institution, will inevitably necessitate generalizing social life, behavior, and relationships on the outside.

Goffman's work was integral in bringing critical attention to total institutions and deinstitutionalization. However, I find myself trying to reconfigure the "self" central to Goffman's inquiry, to neuroqueer boundaries, social experience, desires, and empathy. Maintaining humans and non-human distinctions, are entangled with neurotypical definitions and expectations regarding sociality, empathy, intimacy, and touch. Psychiatric hospitals are spaces of *prohibited* touch defined by rules about touching other people, sharing food, sharing objects, and touching oneself. However, concealed physical touching, sharing objects, verbally described sexual intimacy, and other forms of touch often *slip by*. The psychiatric institution is by its own estimation pathological: "adverse" to touch, displaying "deficits in social-emotional reciprocity such as a failure of normal back-and-forth conversation" and "demonstrating an insistence on sameness, inflexible adherence to routines," (American Psychiatric Association, 2013, p. 50).

However, it is within psychiatrized spaces' paradoxical reinforcement of what it otherwise pathologizes that the "improper affiliation" (Chen, 2012, p. 104) of neuroqueer touch and neuroqueer intimacy traces new subjectivities beyond neurotypicality and heteronormativity. New subjectivities and new intimacies necessitate *neuroqueering* touch. Dinshaw (1999) characterizes *queerness as a kind of touch* that works by "contiguity and displacement...

queerness knocks signifiers loose, ungrounding bodies, making them strange, working in this way to provoke perceptual shifts and subsequent corporeal response in those touched" (p. 151). Bonenfant's (2010) writing about queer listening, describes *"sound...as a form of touch"* recognizing that those "invested in heteronormative power" may be averse to "queerness touching them" (p. 78).

The prohibition of touch in psychiatrized spaces fears the fluidity of neuroqueer touch, including its potential to touch without contact, to frame touching or intimacy through the perversion of "aversion" rather than reinforcing established distinctions and hierarchies of "appropriate" and "inappropriate" contact. Psychiatric spaces attempt to contain and control neuroqueer touch given its disruption of neurotypical means to establish distinctions, categories, and separation. I sense neuroqueer touch and neuroqueer intimacy in moments of contact(lessness) with mold, bread, shadows, and histories—with humans and non-humans.

4. Breathing spores: The mycelia of neuroqueer touch

In what follows, I turn toward the *failure* of *What does it mean?* captured by *I don't know* suggesting orientations of unknowing as integral to neurodivergent *ontogenesis* (Halberstam, 2011; Manning, 2007; Trento, 2020). Manning (2007) describes that "sensing bodies in movement are ontogenetic...because they are always in genesis, in a state of potential becoming" (p. 25). I return to molding bread and spores to reconsider the space-time and boundaries of psychiatric hospitals, bodies, and the permeability of "confined" spaces. However, confinement fails to contain, *things slip by* and *hide* and *escape,* challenging the assumption that confinement can absolutely contain bodies and keep them *still.* This in no way brings the door of a locked unit within my reach, nor lessens the traumas incurred when one becomes *out of touch.* For Manning (2007), a *Politics of Touch,* is not "common sense," but rather, "touch is a sensing body in movement" (p. xiii). Manning (2007) writes:

> Politics of Touch *[positions] the senses relationally as expressions of moving bodies. This presupposes a vastly altered concept of time and space. Whereas in the active-passive commonsense model, time and space are located as stable signifiers into which the body enters within a relational model, space and time are qualitatively transformed by the movements of the body. The body does not move into space and time, it creates space and time: there is no space and time before movement (p. xiii).*

When I consider touch as an "act of reaching toward, of creating space-time through the worlding that occurs when bodies move," (p. xiv) that Manning describes, I reconsider the boundaries of bodies themselves in order to reimagine the boundaries that confine bodies within a particular space and time. Such a shift in thinking about bodies, space, and time, challenges the presumed rigidity of physical boundaries of walls and discursive confinement. Manning (2007) inquires: "What is the inside/outside of a skinscape that folds in and out of an exfoliating body?" (p. xxii). My curiosity about mold and spores considers the permeability of inside/outside boundaries and human and non-human distinctions. Spores are not a metaphor. My body circulates thousands to billions of spores each day with each inhalation and exhalation. Spores are part of my body, my cells, and organs. In the biosphere, spores become mycelia, microscopic, networked multiplicities, colonies through the depths of the earth's substrates. They are decomposing-recomposing. People fear mold, find it repulsive, abate it, remediate it. Some molds are biotoxins. Years ago, I stood in line for the only cold shower. I always smelled mold. Each of us stood in this line amidst circulating spores in flowery air-freshened air. Where have these spores continued to take form?

An interest in mold and fungi similarly becomes an expanding network. My relationship with information as endlessly connective, expansive, and detailed, becomes translated to pathology when labeled as a "fixated" or "special" interest, intensity, "excessive… and perseverative" (American Psychiatric Association, 2013, p. 50). (Has an *interest* in fungi and mold become a *biotoxin*?). The purpose and function of things are embodied as curiosities in continuous motion rather than as the sedimentation of meaning producing a spatial rather than linear thought, a kind of architectural rather than linguistic logic (Richardson, 2020). I feel a sense of this movement and space within Deleuze and Guattari's (1987) articulation of the concept of a book and the questioning of meaning in *A Thousand Plateaus:*

> *We will never ask what a book means, as signified or signifier; we will not look for anything to understand in it. We will ask what it functions with, in connection with what other things it does or does not transmit intensities, in which other multiplicities its own are inserted and metamorphosed, and with what bodies without organs it makes its own converge (p. 4).*

I wonder how many spores we breathe while standing in front of a gallery wall wondering what something means. Deleuze and Guattari describe moving beyond meaning and language to metamorphosis, intensity, and convergence. I am drawn again to wondering what mold does.

With each re-turn to spores, to molding bread, I return to movement, permeability, and touch as a non-representational form of thought as sensation beyond explanation in language or what Manning describes as "agrammatical thinking" (xxiii). "Agrammatical thinking brings language to its futurity, to its event-ness, drawing it into the realm of the virtual. The sensing body in movement is an agrammatical invention" (xxiii). The "institution" and "consensual politics" enact "hermeneutic violence" (Flores Ruíz, 2019, p. 207) in order to propel the patient forward within the boundaries of pre-determined "treatments." Explanation as representation forecloses on the movement of curiosity, making I *don't know* a kind of blockage. On the other hand, such curious momentum requires I *don't know* as the productive force, an opening toward that which might lie beyond language, beyond meaning. I *don't know* as asignifying becomes "gesture," as expression rather than representation (Manning, 2007, p. 111). As Manning (2007) writes: "Expression produces momentum. An emphasis on expression leads us away from a signifying subject to a becoming of relation" (p. 111).

The whisper of *I don't know* becomes neuroqueer touch. Yergeau (2018) captures the neuroqueer as "a verbed form...conceived as cunning movements not neuronal states or prefigured genetic codes" (p. 27). Building from Esteban Muñoz's (2009) proposal "that queerness is not quite here," (p. 1), Yergeau traces a neuroqueer futurity within its presumed impossibility in which the curative eradication of autistics as constellations of symptoms and deficits more often positions autism within an "antifuturistic logic" (p. 19). The *I don't know* of neuroqueer touch similarly swirls within its own impossibility including the pathological rendering of autistic sensation and asociality as an "aversion" to touch and "lack" of empathy. However, neuroqueer touch is relational. Neuroqueer touch creates worlds, multiplicities beyond the limitations of touching as contact. Manning (2007) writes, "I reach out to touch in order to invent a relation that will in turn, invent me" (xv). Spores, perhaps, propagate through neuroqueer touch.

In re-turning to *Admission,* I attend to *I don't know* as neuroqueer touch and neuroqueer intimacy as encompassing relationships between human and non-human. My relationship with non-human actants including the bread, hospital gowns, and torn paper squares, is not symbolic, but rather embodied and intimate. Agencies and intimacies beyond human/non-human distinctions exceed the bounds of language, space, and time gesturing toward alternative subjectivities beyond the normative confinement of narrative, institutions, and the representational logic that seeks a definitive response to, "What does this mean?" In Barad's (2012) discussion of touching informed by quantum physics they ask:

> *What if it is only in the encounter with the inhuman—the liminality of no/thingness—in all its aliveness/liveliness, its conditions of im/possibility, that we can truly confront our inhumanity, that is, our actions lacking compassion? Perhaps it takes facing the inhuman within us before compassion—suffering together (p. 8).*

As described earlier, Goffman, 2007 characterization of the psychiatric hospital emphasizes the human becoming an object produced and maintained by the institution. However, I would propose that experiences of psychiatric hospitalization demonstrate the need to attend with greater complexity to our understanding of objects and the non-human. While privileging human agency over that of the non-human justifies violence and oppression, what I have suggested in this chapter's re-turning to saved bread, molding bread, exhibited bread, and shadows on bread, is that asserting normative assumptions about boundaries, bodies, intimacy, space-time, and sociality foreclose on the possibility of understanding the psychiatric hospital, and experiences of psychiatrization beyond neurotypicality.

I challenge the limitations of interpretation and meaning-making captured by the expectation that saving bread should/does *mean something,* I recognize representational logic as also a neurotypical logic. Within my inquiry surrounding neuroqueer intimacy, I encounter others' resistance when I suggest an empathy related to objects, including bread. Inevitably invoking empathy within a human/non-human relationship becomes labeled as anthropomorphic or reduced to symbolic or psychoanalytic explanations when empathy is limited to a particular understanding of "feelings" and experience. Yet, queer intimacies have long been questioned. Chen (2012) reconfigures human and non-human intimacy through the *affects* of molecular toxicity and animacy. Chen identifies "transobjectivity" referring to the "often-different geographies of affective ties to animate and inanimate objects exhibited in autism" (p. 16). Through a discussion of the video *In My Language* by activist and artist Amanda Baggs, Chen (2012) emphasizes relationships with objects that are "*enabled by the absence of attention to human sociality*" (p. 214).

In the original performative essay, in my *DSQ* article, I described asking the gallery director, *Does she understand the work?* I asked this question because I assumed that the psychiatrist, author, and donor requesting to meet with me about the exhibition, likely did not "get" the work. However, I later added the questions: [What understanding did I assume should be known? Can I revise another person's shadow?] My initial question, "Does she understand the work?" positions the "work" of a counternarrative as pedagogical, as teaching someone who "doesn't know" for the purposes of enacting change, including changing existing perceptions, understanding, and "minds." However, I was in many ways continuing the representational logic of the institution in which to understand what a work means and how it structures a critique recirculates the familiar limitations of identifying "a lack of insight." However, bodies moved through this space casting shadows. The shadows enact modes of touching beyond meaning-making (Saetveit Miles, 2019). Projected light touches skin. The beams of light from projected images make visible particles floating in the air, such as dust. Muñoz asked: "What is it to enact a mode of touching that isn't about mastery, that isn't about foreclosure, that isn't about fusing, that isn't about collapsing things ...?" (qtd. in Vourloumis, 2014, p. 124). Where the shadow that is more and less than human touches exfoliating layers of bread, is pure potential.

References

American Psychiatric Association. (2013). *Diagnostic and statistical manual of mental disorders* (5th ed.). https://doi.org/10.1176/appi.books.9780890425596.

Attias, M. (2020). Exploring the implications of Melanie Yergeau's Neuroqueer for art education. *Visual Arts Research*, *46*(1), 78–91. https://doi.org/10.5406/visuartsrese.46.1.0078.

Barad, K. (2003). Posthumanist performativity: Toward an understanding of how matter comes to matter. *Signs: Journal of Women in Culture and Society*, *28*(3), 802–830.

Barad, K. (2012). On touching—The inhuman that therefore I am. *Differences*, *23*(3), 206–223. https://doi.org/10.1215/10407391-1892943.

Barad, K. (2014). Diffracting diffraction: Cutting together-apart. *Parallax*, *20*(3), 168–187. https://doi.org/10.1080/13534645.2014.927623.

Bennett, J. (2010). *Vibrant matter: A political ecology of things*. Duke University.

Bonenfant, Y. (2010). Queer listening to queer vocal timbres. *Performance Research*, *15*(3), 74–80. https://doi.org/10.1080/13528165.2010.527210.

Chen, M. Y. (2012). *Animacies: Biopolitics, racial mattering, and queer affect*. Duke University Press.

Deleuze, G., & Guattari, F. (1987). *A thousand plateaus: Capitalism and schizophrenia*. University of Minnesota Press.

Dinshaw, C. (1999). *Getting medieval*. https://doi.org/10.2307/j.ctv11sn2m8.

Egner, J. E. (2018). The disability rights community was never mine. *Neuroqueer Disidentification. Gender & Society*, *33*(1), 123–147. https://doi.org/10.1177/0891243218803284.

Egner, J. E. (2019). The disability rights community was never mine. *Neuroqueer Disidentification. Gender & Society*, *33*(1), 123–147. https://doi.org/10.1177/0891243218803284.

Eisenhauer, J. (2009). Admission: Madness and (be)coming out within and through spaces of confinement. *Disability Studies Quarterly*, *29*(3). https://doi.org/10.18061/dsq.v29i3.939. In this issue https://dsq-sds.org/article/view/939/1112. (Accessed 23 August 2022).

Eisenhauer Richardson, J. (2016). Exhibiting mental illness: The parodies and performances of Charcot's hysteric patients and bobby baker. In *Visual culture & gender*. Retrieved May 13, 2022, from http://vcg.emitto.net/index.php/vcg/article/view/101.
Eisenhauer Richardson, J. (2018). The art and politics of artists with mental disabilities experiencing confinement. *Studies in Art Education, 59*(1), 8–21. https://doi.org/10.1080/00393541.2017.1401881.
Eisenhauer Richardson, J. (2019). The yellow crayon: Rhetoricity, mental disability, and art education. *InSEA ART Education VISUAL Journal IMAG #8, 1*, 8–15. https://doi.org/10.24981/2414-3332-7.2019-3.
Esteban Muñoz, J. (2009). *Cruising utopia: The then and there of queer futurity*. New York University Press.
Flores Ruíz, E. (2019). Between hermeneutic violence and alphabets of survival. In J. Medina, A. Pitts, & M. Ortega (Eds.), *Theories of the flesh: Latinx and Latin American feminisms, transformation, and resistance* (pp. 204–219). Essay, Oxford University Press.
Foucault, M. (1967). Madness and civilization: A history of insanity in the age of reason (R. Howard Trans.). *Mentor Books*.
Goffman, E. (2007). *Asylums: Essays on the social situation of mental patients and other inmates*. Routledge (Original work published in 1961).
Halberstam, J. (2011). *The queer art of failure*. Duke University Press.
Haraway, D. J. (1997). *Modest_witness@second_millennium. femaleman_meets_oncomouse*. Routledge.
Jones, N., & Brown, R. (2012). The absence of psychiatric C/s/X perspectives in academic discourse: Consequences and implications. *Disability Studies Quarterly, 33*(1). https://doi.org/10.18061/dsq.v33i1.3433. In this issue https://dsq-sds.org/article/view/3433/3198; (2012). (Accessed 23 August 2022).
Koivisto, M. (2018). Egresses: Countering stereotypes of blackness and disability through horrorcore and Krip hop. In R. Sanz Sabido (Ed.), *Representing communities: Discourse and contexts* (pp. 163–180). Springer International.
Manning, E. (2007). *Politics of touch: Sense, movement, sovereignty*. W. Ross MacDonald School Resource Services Library.
Pollock, D. (1998). Performing writing. In P. Phelan, & J. Lane (Eds.), *The ends of performance* (pp. 73–103). Essay, New York University Press.
Price, M. (2014). *Mad at school: Rhetorics of mental disability and academic life*. University of Michigan Press.
Richardson, J. (2020). Against interpretation: Searching Marina Abramović's face. *Studies in Art Education, 61*(3), 207–219. https://doi.org/10.1080/00393541.2020.1778606.
Saetveit Miles, L. (2019). Queer touch between holy women: Julian of Norwich, Margery Kempe, Birgitta of Sweden, and the visitation. In D. Carrillo-Rangel, D.-I. Nieto-Isabel, & P. Acosta-García (Eds.), *Touching, devotional practices, and visionary experience in the late middle ages* (pp. 203–235). Essay, Palgrave Macmillan.
Trento, F. B. (2020). Protocols for a procedural space for failing. *Research in Arts & Education, June*(2), 1–22.
Trento, F. B. (2021). Crip teleportation: The animal that therefore I am—Or I am not. *TECCOGS: Revista Digital De Tecnologias Cognitivas, 22*, 91–104. https://doi.org/10.23925/1984-3585.2020i22.
Vourloumis, H. (2014). Ten theses on touch, or, writing touch. *Women & Performance: A Journal of Feminist Theory, 24*(2–3), 232–238. https://doi.org/10.1080/0740770x.2014.973755.
Yergeau, M. (2013). Clinically significant disturbance: On theorists who theorize theory of mind. *Disability Studies Quarterly, 33*(4). https://doi.org/10.18061/dsq.v33i4.3876.
Yergeau, R. (2018). *Authoring autism: On rhetoric and neurological queerness*. Duke University Press.

CHAPTER TWELVE

Creating autistic space in ability-inclusive sensory theatre☆

Molly Mattaini*
Interdisciplinary Theater Studies, University of Wisconsin–Madison, Madison, WI, United States
*Corresponding author: e-mail address: mgmattaini@gmail.com

Contents

1. Introduction

Ability-Inclusive Sensory Theatre (AIST) is a genre of Theatre for Young Audiences which creates highly intimate theatrical productions in which narrative structures are organized around sensory engagement. Productions are audience-centered and focus on the agency of the child to engage in any manner the child finds aesthetically stimulating. The flexible structure of the productions and the focus on sensory stimulations means

☆ Reprinted with permission from Taylor & Francis from: Mattaini, M. (2020). Creating autistic space in ability-inclusive sensory theatre. Youth Theatre Journal 34, 42–54. https://doi.org/10.1080/08929092.2019.1633719.

Developments in Neuroethics and Bioethics, Volume 5
ISSN 2589-2959
https://doi.org/10.1016/bs.dnb.2022.08.009

the aesthetic quality of the productions is distinctive. Internationally, the field focuses on creating productions for young people with cognitive disabilities but in the US the genre is mostly focused on audiences on the autism spectrum. Due to my own location and experience, this article will focus primarily on US understandings of Autism Spectrum Disorder but will look at an international community of theater companies practicing AIST.

I propose that AIST creates an artistic version of "autistic space" through the audience-centered creation which focuses on the experience of the audience over the messaging of the artist. AIST is built for an audience of children on the spectrum from the beginning and as a result, the form itself is distinctive. The resulting innovative use of materials, adjusted focus of the performer, and customization of the theatrical experience allows children on the spectrum to experience drama in a richer way than they would in a typical TYA performance, even a typical performance with adaptation.

AIST is distinct from "Sensory Friendly" performances. In Sensory Friendly performances (or Relaxed Performances), the rules of audience etiquette are amended; for example, the theatre may allow talking, moving, and eating in the house. Adjustments are also made to the production; for example, bright lights are dulled and sound cues are quieted (Fletcher-Watson, 2015). Sensory Friendly productions are an *adaptation* of an existing work, rather than a creation of a new work. In contrast, AIST builds shows from the ground-up with an audience of young people with cognitive disabilities in mind.

AIST is not interested in theatre as a therapeutic avenue to fix, help, or otherwise adjust individuals on the autism spectrum. AIST can instead be considered a part of the neurodiversity movement. Neurodiversity is,

> the notion that conditions like autism, dyslexia, and attention-deficit/hyperactivity disorder (ADHD) should be regarded as naturally occurring cognitive variations with distinctive strengths that have contributed to the evolution of technology and culture rather than mere checklists of deficits and dysfunctions.
>
> ***(Silberman, 2015)***

Neurodiversity frames Autism Spectrum Disorder (ASD) as a neurological state which is different from, but not inferior to, typical brain structures. Much in the same way as multiculturalism embraces differences as strengths to society, neurodiversity argues that the community of humanity is strengthened by an embrace of neurological variations.

AIST can be connected to neurodiversity both because these productions do not seek to offer therapeutic fixes to autism and because the environment of AIST productions celebrates the distinct ways children with autism perceive the world. None of the companies that practice AIST explicitly call themselves a part of the neurodiversity movement, but the connections are evident. Instead of providing therapeutic resources, the goal of theatre for children with autism is the same as any other Theatre for Young Audiences performance: namely, to provide children access to quality art. AIST expands existing TYA to children with disabilities by creating a piece of theatre that is particularly appealing to a person with autism. This expansion is necessary because both children and adults with disabilities are often overlooked as part of a theatre audience.

2. Social concept of autistic space

This truncated history will focus on the specific label of autism because the legacy of autism advocacy effects the broader culture around cognitive disabilities. The autism community includes a large group of people placed along a complex spectrum of ability and includes many prominent and articulate self-advocates. These verbal advocates have become a voice for a broader group of people who cannot communicate through normed verbal/spoken language, opening up understanding of people with a broad range of cognitive disabilities, whatever their diagnostic label. As a result, the history of the autism community affects all AIST companies, even those focused on serving a broader population than just those labeled with ASD.

The current understanding autism comes from the autism self-advocacy movement which began in the 1990s, when adults on the autism spectrum began to form personal connections at autism conferences which were targeted to parents of children on the spectrum. Adult attendees on the spectrum formed enclaves which subverted the intended purpose of the conferences. Rather than discussing the difficulties of raising a child with autism, these attendees formed neutral social spaces to connect with other Autistics (Autscape, n.d.). Journalist Steve Silberman writes about the phenomenon of these conferences in his book *Neurotribes*, in which he calls the information booths geared to Autistics "little oases of autistic space" where Autistics could take a break from sensory overloads, judgmental looks, and "constant reminders that their existence was a tragic puzzle" (Silberman, 2015, 442). One of those advocates was now notable advocate

Jim Sinclair, a frequent attendee who had grown tired of his conference role of providing insight about the behavior of non-verbal children for parents because he felt like a "self-narrating zoo exhibit" (Sinclair, 2005). Sinclair observed from the enclaves at these conferences that the spaces created by Autistics were different from the public spaces created by neurotypical people. Some of the factors which make autistic space unique are safety from sensory overload, respect for the boundaries of people on the spectrum, and an openness to non-verbal communication.

Out of these social connections made at conferences grew a large community, and eventually Sinclair, Xenia Grant (formerly Kathy Grant-Lissner), and Donna Williams started the Autism Network International (ANI), the first autism organization run by autistic people (Sinclair, 2005). The purpose of the organization is to create spaces, digitally and physically, for people with autism to connect. ANI also welcomed and provided information to neurotypical allies but established clear boundaries for expected attitudes and behaviors from the neurotypical people being invited into autistic space.

From 1996 to 2013, ANI ran an annual conference/retreat called Autreat, which was structured for maximum accessibility for people on the spectrum. Autreat was organized by people on the spectrum, and people with autism were the primary audience of conference presentations. The topics of presentations given at the conference focused around neurodiversity, and neurotypical participants were welcomed on the condition that they not focus the event on their experiences, or "how heroic they are for putting up with us" (Sinclair, 2005). While Autreat featured presentations on how to successfully navigate the neurotypical world as an Autistic, the primary purpose of Autreat was as a retreat, creating a space where unusual behavior was welcomed and celebrated, where participants did not feel a need to suppress autistic behavior simply because their behavior was atypical. Autreat ended in 2013 but its directly inspired UK counterpart Autscape is still in operation, having met annually from 2005. The topic of the first Autscape was "Creating Autistic Space" (Autscape, n.d.).

3. AIST as autistic space

I assert that AIST performances create an artistic version of Sinclair's concept of "autistic space." (Sinclair, 2010) The space created within an AIST show is designed in totality to be pleasing, interesting, and welcoming to a person on the spectrum. This presents a profound effort to carve out space specifically for people with autism. Where Sensory Friendly performances make room in neurotypical spaces for people with autism, AIST makes autistic space which neurotypical people can earn an invitation to enter.

Sinclair's use of "autistic space" described social environments rather than a theatrical experience but a similar sense of safe, autistic space is reflected in both the design and attitude of AIST spaces. Many AIST companies share a sense that the immersive spaces of their shows transcend a simple physical transformation. The language companies use to describe abstract "space" created in their productions is varied yet connected. Oily Cart refers to the spaces within their shows as "wonderland" (Webb, 2018) and Trusty Sidekick calls the metaphysical space "a world of yes," where the goal of the space is to engage and feel comfortable, but not to engage in any particular way (Schmidt-Chapman, 2018). The physical transformation of the theatre space informs a space in which the rules of engagement with art are amended, where children are able to connect with performers and welcome their neurotypical-loved ones into a shared, joyful experience.

However, there is an essential difference between the autistic space that Sinclair discussed and the autistic space that AIST produces. AIST shows are in large part created by neurotypical artists, not by artists with autism. This tension parallels an overall tension in TYA, which strives to create aesthetic experiences for children but is made by adults (the adult/child dynamic also being present in AIST). However, the dynamic is not entirely parallel because while all adult creators were once children, but all neurotypical people do not have previous first-hand experience of being neurodivergent. Adding onto this tension is my own perspective as a scholar, as I am a neurotypical adult writing about this artform with an outsider perspective on the experiences of the young people attending these productions. A scholar who is on the spectrum or identifies as Autistic may provide a more nuanced understanding of these productions. However, I believe that the neurotypical artists creating AIST have engaged genuinely in an effort to create autistic spaces, just as adult TYA practitioners have worked diligently to create aesthetic experiences for children even though they are not themselves children and the field of AIST is quickly turning its attention to authentic inclusion of artists with autism into their development and performance process.

While the concept of creating public, autistic spaces is important beyond the world of theatre, theatre is uniquely equipped to create these spaces because the nature of theatrical performance is to create temporary spaces which reflect worlds that do not truly exist. The habitual creation of temporary spaces positions theatre ideally to contribute to the creation of autistic spaces. As well, TYA companies are in a unique position to contribute because the social capital afforded to children with disabilities is often greater than what is afforded to their adult counterparts.

In the following sections, I will lay out my current conceptualizations of the essential elements which make AIST a distinct art form, in the hope that putting some of these concepts into widely disseminated words will advance an art form which is politically and morally necessary.

4. Building an autistic space: Development and audience preparation

Each company with publicly available documentation of their creative process begins show development with drama education workshops in special education classrooms (Brown, 2012; Performing Lines WA, 2015; Shaw, 2010; Trusty Sidekick Video, 2015). Working in special education environments serves a dual purpose for the company to try out material on the same or similar audience which will be attending the show and serving as a professional development opportunity for the artists to familiarize themselves with the various approaches, languages, and challenges present in the disability community.

Beginning the artistic process in the special education classroom comes out of the tradition of Oily Cart, the foundational company to the AIST field, which developed their first AIST show through a partnership with a special education school (Brown, 2012, 4). Oily Cart's initial company mission was serving very young children, but their work expanded to include children with disabilities in 1988 because of a request from a "Special Needs" school (a term in the United Kingdom for a school that is exclusively offering special education) to bring in a Theatre for the Very Young (TfVY) show. Rather than bringing in a TfVY show for older children, Oily Cart asked the schools to work with them to develop an age-appropriate show which would consider the abilities of all the students at this school. After working extensively with the students at the school, Oily Cart developed a show called *Box of Socks* that would appeal to the broadest possible contingent of students at this school, a challenging proposition considering the age and ability ranges in their school community.

From this history, one can see the artistic heritage of TfVY in AIST productions. AIST emerged directly out of TfVY but with adaptations formed by an audience-first approach. The artistic process began by getting to know the people in their audience, before considering anything else about the concept of the show. This tradition has continued as Oily Cart's professional development workshops and partnerships have passed on their process to other companies, which have then created a development processes of their own with the building blocks of Oily Cart's model.

Once the essential framework of the show has been developed, companies must engage in an intensive audience development process. In addition to traditional marketing material, AIST companies produce carefully curated social materials to prepare all audience for their productions before they arrive at the theatre. This can range from social story (a narrative infographic which rehearses a new social experience(Berrick, 2018)) to Sensorium's mobile app (Berrick, 2018) which includes an interactive map of the production, photos of the cast, thematically related games, and recordings of the music from the show.

Companies have also recently turned their attention to the preparatory materials for neurotypical, accompanying audience members because the social rules of the AIST space can be just as foreign for a caretaker used to managing the behavior of their child in a traditional theatre space as it can be for a child with a cognitive disability. Particularly companions from an educational setting may come in with the expectation that their role in the space is to enforce the social norms of attending theatre, which is counter to the "wonderland" space created in an AIST show. Companies have begun experimenting with ways to establish the new rules of the space as quickly as possible. For example, Sensorium Theatre has developed a sheet of guidelines which are distributed to caretakers in the initial space to introduce caretakers to the idea that their role in this space is not to exert control on their child. One of the responsibilities of AIST artists are to create a space in which caretaker control can gradually, safely, and joyfully be lifted.

The audience development process also requires that marketing, ticketing, and front of house staff be included in the language of the production. In a typical rehearsal process, a theatre company develops a common artistic language to communicate about their show. With AIST productions, a new social contract is being created within the performance space that involves both the performers and the attendees, meaning the people involved in every level of the production to be introduced to this language before the show begins.

5. Entering autistic space: Immersive environment

I have come to visualize both the physical and extra-physical spaces of AIST as concentric circles. The choice of circles calls upon a concept from Sensorium Theatre that the transitions into and out of their productions work to maintain a "circle of care" (Sensorium Theatre, 2018). The circle of care around the child is constant and includes parents, professionals, and other loved ones. During an AIST performance, the circle is expanded to include the performers, and then during the transition out the performance, performers gradually remove themselves from the circle, shrinking the circle back to its initial size but always maintaining the strength of the shape. The following sections are inspired by that imagery (see Fig. 1).

AIST productions acknowledge that a child's experience at the theatre begins with their arrival into the initial physical space of the theatre, rather than their entrance into the performance space. Social assumptions about what it means to enter a theatre space cannot be relied upon with an audience who is unlikely to have attended theatre previously due to their age and ability, and because people on the autism spectrum usually require

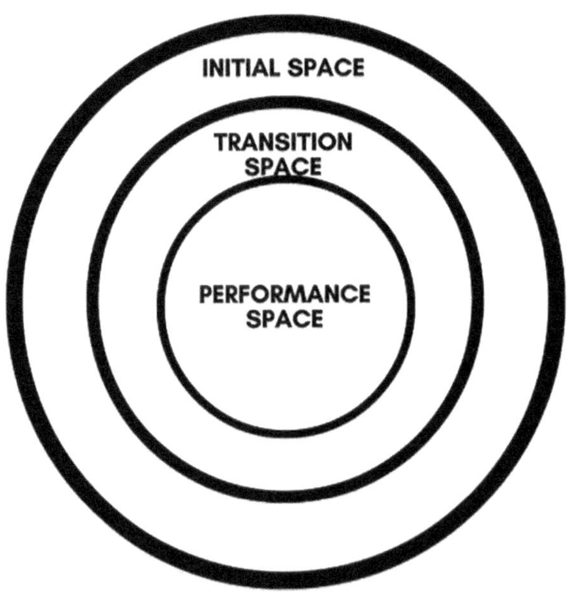

Fig. 1 Spaces within an AIST performance.

more overt social cues to understand the dynamics of a new interaction. As a result, the establishment of what constitutes the theatrical space must begin immediately.

The initial space is the first space a child encounters when entering the theatre. This is usually a lobby, although the physical realities of the building call for companies to think broadly about how an audience enters a space initially—such as with Trusty Sidekick's *Up and Away* in which two musician characters greeted incoming audience in an elevator and then rode with the audience up the lobby space while strumming stringed instruments. The initial space usually includes multi-sensory activities which have an aesthetic or thematic connection to the coming performance, and performers meet the audience member in this initial space before accompanying them into the theatre. The activities and interactions in the initial space begin the establishment of the social rules of the AIST space, for both the young people on the spectrum as well as the neurotypical people accompanying them.

When the show is about to begin, performers lead the audience into a transition space. Transitions can be particularly difficult for children on the autism spectrum, both because changing one's focus requires difficult cognitive work and because any new space threatens overwhelming sensory inputs. To lessen this anxiety, transitions happen slowly and move through some sensory-neutral spaces. In the book *Oily Cart: All Sorts of Theatre for All Sorts of Kids*, Tim Webb compares this transitional space to the "airlock" on a spaceship—a decompression point which marks a temporal and spatial change, and which the audience is welcome to return to at any point (Brown, 2012, 21). This transition space is sometimes established with a clear structure, such as the "observatory" space in *Up and Away*, in which the audience is introduced to the performance space—a room filled with faux hot air balloons—through a wooden "observatory" structure outfitted with screens which reveal the balloons at the narratively appropriate moment. But a transition space can also be an unstructured, less stimulating open area on the outskirts of the performance, such as with Sensorium's *Oddysea*. *Oddysea*'s airlock space is simply the peripheral space of the theatre outside the main action which takes place within a tent. This transition space is inhabited by characters called "Peris" (short for perimeter person) who interact with children who may be sensory-avoidant by providing less intense versions of the multi-sensory materials being used in the performance. Children are free to remain at the peripherals throughout, move on to the main action, or oscillate between.

The initial and transition spaces work to move audience into the central location of the performance space. The performance space is usually based in either a black box theatre or a school auditorium (depending on the company model) which has been transformed into an immersive space which evokes the setting of the show.

The aesthetics of the performance space are essential, because of the connection that people on the spectrum often form with physical objects and locations. For example, in Temple Grandin's memoir *Thinking in Pictures*, she speculates that for people with autism, emotional bonding happens at a stronger level with routines and objects than it does with other people (Grandin, 1996, 140). This highlights the importance of the immersive space of an AIST show. A child in the audience may connect more with the environment created for the AIST show than with the characters within the space. These spaces therefore deserve great care and detail, as the connection with the space may be the basis of the entire experience for the child. This connection to immersive spaces presents an opportunity to go beyond "sensory friendly" spaces that merely avoid triggers, to also include aesthetically compelling elements which contribute to the experience of participation in the theatrical event.

In addition to the importance of the contents of each of these three spaces, artists must also consider how the thresholds between these spaces are crossed. Agency over the transitions between spaces becomes a way of respecting the boundaries of the audience member on the spectrum. In *Oily Cart: All Sorts of Theatre for All Sorts of Kids*, Claire de Loon, founding member and designer for Oily Cart, discussed creating the physical space for their production *Blue* which considered the needs of children who were reticent to enter a new space. She describes each detail of entering the space, with each design choice being made out of a desire to give the child agency in the way they entered the experience:

> *... there was a very wide opening and you could actually watch the show from outside, and the young people could shuffle in gradually as they became less afraid. This only applies to a small minority of the audience but you have to cater to them too. They wouldn't be able to participate at all if you didn't take these things into account. We did two other things for those young people. The first sight you got of the set was seeing what was inside from the outside; it was just the shadows of the actors inside the installation. That was a gradual introduction to what was inside, to demystify it and take the element of surprise out of it. So the young people got an introduction to what was coming next. There were also little windows in the set that audience members could look through. So if you weren't even comfortable with being at the large opening, you could just peer in through one of the little windows.*
>
> **(Brown, 2012, 50)**

De Loon makes it clear that moving into an AIST space requires number of physical and temporal thresholds, which the child must cross with full agency.

6. Enjoying autistic space: Objects and sensations

Within an AIST show, the narrative structure revolves around the relationship between the audience and physical objects. Other narrative elements such as plot and character are still present, but they are secondary to the aesthetic episodes in which the performers and audience appreciate the multi-sensory objects. These aesthetic materials come first, and through joint connection with the object, the object becomes a channel for human connection. This is in contrast to neurotypical theater which primarily explores human relationships through an audience watching the performers form a connection onstage and includes material aesthetics which inform this story of human connection.

The plot of AIST productions are generally organized around the discovery of multi-sensory objects. Some examples of multi-sensory objects include the "asteroid" props in *Red Kite, Blue Moon*, which were music shakers that the children could both hold and hear (Shaw, 2010). Oily Cart brought two productions to the 2011 Manchester International Festival which featured scent-scapes created in partnership with the perfume company Seven Scent (Brown, 2012, 29) Many productions use soft pillows and spritzes of water, and all AIST productions are underscored with music.

Sensory objects are an important factor of Sinclair's concept of autistic space. In Sinclair's article about the formation of ANI, he discusses how he built a friendship with other Autistic adults, Kathy Grant and Donna Williams, through the co-appreciation of objects:

> *I watched her [Williams] go into fits of ecstasy while arranging colorful objects and looking at them through a kaleidoscope … And while she was engaging in this activity of arranging objects and looking at them through her scope,* she kept insisting that Kathy and I look at them too. *Of course, being autistic I'm not supposed to understand things like this, but to me, that looked suspiciously like a person wanting to share a pleasurable activity with her friends.*
>
> ***(Sinclair, 2005)***

Sinclair's framing of this interaction places the object as the central channel through which he is able to connect with Grant and Williams. Williams describes the same interaction in her memoir *Somebody Somewhere*, describing her time with Sinclair and Grant as feeling like finding a "lost tribe"

(Williams, 1994, 186). Williams explores her relationship with objects earlier in her memoir as well, describing how as child she felt an emotional presence from objects: "things never thought or felt anything complex but they gave me a sense of being in company … people were always third-person; they imposed upon an already present sense of company" (Williams, 1994, 67). From this anecdote neurotypical artists can infer that relationships formed within autistic space are more deeply connected to shared material experiences than relationships created in neurotypical space.

These are a few of many examples I have found through reading first-person narratives of people on the spectrum which provides compelling evidence that the strong relationship of many people with autism to the material world presents an opportunity for creative connection through the channel of objects. AIST productions have capitalized on this connection by arranging the narratives around the discovery of objects with interesting sensory elements. While all AIST shows are organized around some semblance of a narrative, dramatic tension is generally secondary to the use of objects.

Essential to this use of object is the agency that a child has over interacting with the object. Performers always ask for the child's consent before manipulating an object which will touch the child, either by verbally asking consent or through nonverbal modeling.

7. Connecting in autistic space: Circles of focus

Because AIST is an audience-centered form, the influence of each person in the room during a performance is more profound than in a typical performance. Performers are constantly reacting to the audience's reception, meaning the audience's presence has a direct impact on the structure of the performance. The audience's reaction to each aesthetic moment dictates how the aesthetic moment is performed. How the audience is constructed is therefore an essential element of the dramaturgy of these performances. The performer's focus is given primarily to the child on the spectrum/with a cognitive disability but the performance is impacted by all audience members, even those on the peripheries (see Fig. 2).

Just as the physical space was conceptualized as a series of concentric circles, I propose that the influence and focus of the audience can also be conceptualized as concentric circles. The most crucial audience relationship is at the center—the participating audience member (child on the autism spectrum) and the performer—and forms around co-appreciation of an

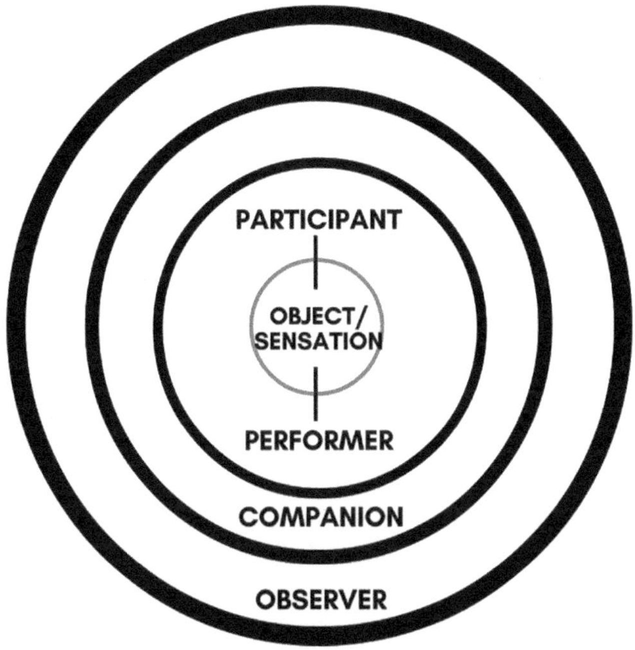

Fig. 2 Audience relationships within an AIST performance.

object or sensation. The less crucial spheres of influence move outward from the center, but from those outward circles, peripheral audience members are able to influence the core relationship. Because AIST is targeted to young people, the space must be inclusive of companions, which can include parents, teachers, paraprofessionals, friends, and other family members. While the dynamic of a sibling versus a teacher is quite different, for the sake of simplicity in this overview article I have grouped anyone that has a previous relationship with the participating audience member into the category of "companion," language which I borrowed from the ticketing structure of the Big Umbrella Festival in 2018. In addition to companions, some performances may also welcome observers, or any person who does not enter the space with a previous relationship with a participating audience member. This can include donors, other theatre professionals, academics, journalists, families with neurotypical children, and curious members of the public. The likelihood of observers attending depends on a number of factors which cannot be fully explored here. However, many AIST shows include these four roles: performer, participant, companion, and observer. Therefore, it is essential when producing an AIST show to understand their dynamic and place within an AIST space.

The relationship between the performer and the participant is at the center of the performance. AIST performances always have a low ratio of performers to participating audience (ranging from 1:1 to around 1:3) which allows for personalized attention. The performers keep focused on the audience, without pretense of a fourth wall, and involve the child by learning their name and asking consent before offering the child an object. Tim Webb emphasizes that the focal point is always the audience reaction, saying, "The Oily Cart has to watch the audience members far more intently than they watch us" (Brown, 2012, 8). Trusty Sidekick calls the process of a performer observing and responding to the participating audience member "empathetic improvisation" (Schmidt-Chapman, 2018).

Shows are built around a flexible structure, but how the participant reacts to each episode of object/sensation dictates the experience within that structure. Empathetic improvisation allows performers to make constant decisions about how to adjust individual moments within the performance structure with the goal of maximum joy for the participant. Because the audience is given previews of the structure through social stories and visual schedules, there is a sense of balance between an anticipatable structure and genuine acceptance of however the participant wishes to engage with that structure. Even if the child chooses to engage with the performance at a distance throughout, their physical presence on the peripheral of the performance does not make their experience of peripheral importance to the performer.

While the performer/participant relationship is central, the inclusion of the other audience members is also integral to AIST performances. For companions, attending an AIST performance is a chance to experience public celebration of the joy of their loved one. Through seeing the performer/participant connection, the world of the person on the spectrum is opened up to their neurotypical-loved ones.

Attending an AIST show can be a challenging experience for companions who are used to acting as caretakers to the children they accompany. In addition to the preparatory materials for neurotypical audience, some of the performer attention is devoted to communicating with caretakers that non-normative behavior is welcomed in an AIST space. This is often accomplished through clear, positive reactions to the child's engagement as well as sometimes through direct communication with the parent.

The most outward circle of observers creates the most ethically fraught relationship in the space.(Brown, 2012) On one hand, inviting neurotypical people who are not already a part of the disability community has positive implications for social integration. It prevents AIST shows from becoming ghettoized. On the other hand, observers might taint the experience with a neurotypical gaze which offers judgment of the way a participant engages or offensive wonderment at the capabilities of a child they assumed was incapable. To combat this potential, companies have also begun to experiment with preparation materials and strategies to ensure positive influence from observers.

The audience conception of concentric circles is my own, although the physical arrangements of spaces are often in literal circles. It is up to every company how to incorporate the members of the audience who are not direct participants. One way in which the company offers a profound communication about who the show is for is through the physical audience arrangement. What kind of "seat" an audience member is offered, the physical location of a seat, and whether that seat fits with the other seats in the space, all communicate to the audience to what degree they are being welcomed in the space.

Some companies opt for more flexible seating (such as floor cushions), which has the benefit of on-the-fly adjustment, freedom of movement around the space, and open inclusivity of every member of the audience. However, seating without clear boundaries can also have the downsides of audience fidgeting disrupting other audience members, and lack of a clear home to establish a place within the performance. Structured seating has potential for physically addressing audience needs (Oily Cart has designed a number of chairs over the years that allow for movement while maintaining support), but can also be more restrictive, and can negatively communicate to the audience who are not participants that they are actively not welcomed in the space.

8. Conclusion

While Sensory Friendly performances are a shrewd use of the resources of TYA productions that have already been developed, the adjustments made in sensory-friendly performances do not offer the same level of aesthetic experimentation as an AIST production. In fact, adjustments made for Sensory Friendly performances sometimes limit aesthetic considerations by focusing solely on avoiding triggers. In contrast, AIST shows delight in the potential of sensation to create an aesthetic experience.

Based on the personal narratives of adults on the spectrum, using the sensory object to create a channel of connection with a child on the spectrum is a rich avenue for artistic exploration.

Three categories of companies currently produce AIST. The first, those who produce work exclusively for children with disabilities, includes Bamboozle Theatre in the UK, Sensorium in Australia, Frozen Light in the UK, Seesaw Theatre at Northwestern University, Yellow Finch Theatre in upstate New York, Bluelaces Theatre Company in New York City, and Jumping Jack Theater in Pittsburgh. In the second category, TYA companies with an established arm that produced multi-sensory work, is Oily Cart in the UK, Replay Theatre in Northern Ireland, Chicago Children's Theatre, and Trusty Sidekick Theatre in New York City. The third category, TYA or other theatre companies that have produced one multi-sensory show, includes The Rose Theatre in Omaha and Bricolage Production Company in Pittsburgh. Academic environments have also fostered AIST work, including productions from Gretta Berghammer at University of Northern Iowa and the immersive "Imagining Autism" project at the University of Kent. In April of 2018, Lincoln Center hosted The Big Umbrella Festival, the first international festival of work created for children on the spectrum, which included productions of *Light Show* from Oily Cart, *Oddysea* from Sensorium Theatre, and *Up and Away* from Trusty Sidekick Theatre in partnership with Lincoln Center. This represents a complete list to my knowledge but may be missing some information.

Looking ahead, the most pressing issue for AIST companies is the incorporation of artists on the autism spectrum into the development and performance of AIST shows to create authentic autistic spaces. Many AIST companies are currently working on authentic incorporation of people with cognitive disabilities into their processes. The newly founded Pink Umbrella theatre in Milwaukee hopes to train adult artists on the spectrum and include those artists in the creation of multi-sensory works for audience members with autism.

AIST companies are looking to expand their work, both by producing more and licensing shows to interested TYA companies. The largest obstacle to this expansion is the scale of AIST productions, which tend to cap around 15 participants and require nearly as many performers. While some organizations understandably balk at the logistical hurdles to producing AIST, the historical exclusion of this audience creates an imperative for TYA companies to produce work which includes all members of the communities they serve. While further productions of these shows are ideal for the advancement of the art form, there are also many insights that practicing AIST companies can provide for producers of any type of children's programming, both artistic and educational. AIST companies should be looked to as innovators in the field of inclusion because of their profound acceptance of diverse aesthetic experience and expression.

Sensory Friendly performances are a valuable resource which should continue to be developed, but they will never serve the audience that needs the intimate setting and personal connection provided by AIST performances. Sensory Friendly performances are a strategy to make room in neurotypical space for people with autism where AIST performances create autistic space.

Notes

1. Members of Autism Network International, inspired by the Deaf community, prefer to be called "Autistic" rather than a person with autism, because they believe that person-first ordering implies that autism is a negative trait which must be separated from the Autistic person (Silberman, 2015, 441) I will use the term "Autistic" for individuals who have expressed a desire to be identified as such, and continue to use "person with autism" at other points in this paper to follow current standards.
2. Having a publicly-available generalized social story about coming to the theatre has quickly become best practice in the TYA community. Most AIST productions try to create a more specific social story for the production which includes pictures of the characters in costume, the performance space, transition areas, and sometimes major plot points.
3. The ethical inclusion of observers is a difficult and nuanced topic. I plan to write on this issue in an in-depth way in the future.

Disclosure statement

No potential conflict of interest was reported by the author.

References

Autscape. About autscape. http://www.autscape.org/. n.d. (Accessed 1 June 2018).
Berrick, S. (2018). *Sensorium theatre* application. (version 1.0.4). Sensorium Theatre.
Brown, M. (Ed.). (2012). *Oily cart: All sorts of theatre for all sorts of kids* Trentham Books.
Fletcher-Watson, B. (2015). Relaxed performance: Audiences with autism in mainstream theatre. *Scottish Journal of Performance*, *2*(2). https://doi.org/10.14439/sjop.2015.0202.04.
Grandin, T. (1996). *Thinking in pictures: And other reports from my life with autism*. New York: Vintage Books.

Performing Lines WA. (2015). *Sensorium's Oddysea to the Sydney opera house.* https://www.youtube.com/watch?v=3zYK3BW9Oow.

Schmidt-Chapman, J. (2018). *Creating theatre especially for audiences with autism. Presented at the big umbrella festival.* Lincoln Center Education. April 19.

Sensorium Theatre. (2018). *ODDYSEA (deconstructed and with commentary).* April 20.

Shaw, K. (2010). *The red kite project.* Rascals and Rogues with Chicago Children's Theatre.

Silberman, S. (2015). *Neurotribes: The legacy of autism and how to think smarter about people who think differently.* Allen & Unwin.

Sinclair, J. *History of ANI. Autism Network International.* http://www.autismnetworkinternational.org/History_of_ANI.html. (Accessed January 2005).

Sinclair, J. (2010). Being autistic together. *Disability Studies Quarterly, 30*(1). https://doi.org/10.18061/dsq.v30i1.1075.

Trusty Sidekick Video. (2015). *Creating UP AND AWAY at Lincoln Center.* https://www.youtube.com/watch?v=IJ1tEHgJ-_U.

Webb, T. (2018). *Creating theatre especially for audiences with autism. Presented at the big umbrella festival.* Lincoln Center Education. April 19.

Williams, D. (1994). *Somebody, somewhere.* Times Books.

CHAPTER THIRTEEN

The virtual dementia experience

Tanya Petrovich*
Dementia Australia, Parkville, VIC, Australia
*Corresponding author: e-mail address: Tanya.Petrovich@dementia.org.au

Contents

"If you have met one person with dementia, you have met one person with dementia."

Kitwood, 1997

1. Introduction

The experience of dementia is unique to each person living with dementia. With over 55 million people living with dementia worldwide (Alzheimer's Disease International) that makes for many unique experiences that we need to assimilate to create a common understanding of the disease. The World Health Organization defines dementia as a syndrome that leads

Developments in Neuroethics and Bioethics, Volume 5
ISSN 2589-2959
https://doi.org/10.1016/bs.dnb.2022.08.010

to deterioration in cognitive function (WHO, 2021). It affects memory, thinking, orientation, comprehension, calculation, learning capacity, language, and judgment. For each person living with dementia, the extent to which these cognitive functions are affected varies. For some people living with dementia, memory is not the main concern, but rather, difficulty with concentration and attention, extreme confusion and difficulty judging distances often resulting in falls.

To deliver services to support the individual with dementia and their care givers, a sound understanding of the impact of the disease on the individual is required. Through understanding the impact, it is possible then to build and develop products and services that address the concerns of the individual and their carers.

In this chapter, I will explore the experience of living with dementia, how dementia affects the senses, and the impact of the built environment on the person with dementia. I begin by providing some background on Dementia Australia, its general purpose, and its approach to education. I will then describe the rationale for the Virtual Dementia Experience, the processes that led to its development, and how the "experience" of dementia was created. The virtual experience recognizes and incorporates the role of the built environment and physical spaces in shaping peoples' experiences of dementia. A well-built dementia-friendly environment can be empowering for the person living with dementia supporting them to be engaged and live well with a good quality of life. An architect with a good understanding of dementia and the knowledge of what constitutes a dementia-friendly environment is essential to providing a better quality of life for those living with dementia.

2. Dementia Australia

Dementia Australia was founded by carers more than 35 years ago, and today it is the national peak body for people impacted by dementia in Australia. Its purpose is to support and empower the estimated half a million Australians living with dementia and almost 1.6 million people involved in their care. Dementia Australia strives to involve people impacted by dementia and their experiences in all its activities and decision-making to ensure it is representative of the diverse range of dementia experiences across Australia. The focus of Dementia Australia's work has been defined through extensive consultation with consumers and is reflected in three priority areas for strategic activities: 1. timely diagnosis and support; 2. improved quality of care; and 3. reduced discrimination. For all three of these priority activities' education is the key.

To provide timelier diagnosis and support, it is necessary to educate people at risk of developing dementia. Improving the quality of care in Aged Care requires education of workers who provide support in dementia care. Reducing discrimination of people living with dementia requires education of the general public to better understand dementia and its impact. 63% of people say they know very little about dementia and 41% of people say that they feel confronted when talking to someone with dementia (Dementia Australia—Strategic Plan, 2018).

For over 30 years Dementia Australia (formerly known as Alzheimer's Australia) has delivered education workshops for family carers and professional carers. These workshops are delivered in the usual manner with an experienced educator leading discussion using PowerPoint slides, encouraging a sharing of peoples' experience through discussion groups, videos, case studies and at times role play. These were always well received. Participants leave these sessions with a good understanding of dementia, the impact that dementia has on the person and what can be done to improve care and support. However, this model of education has not been impactful enough on dementia practice. Education that is impactful should motivate participants to change practice and improve dementia care, in so doing improve the lives of those living with dementia.

Because of this, Dementia Australia has sought to engage in more Experiential Learning, allowing the participants to learn by doing. By engaging carers in the experience of living with dementia, they are better able to connect the theory and knowledge of dementia to the practice of caring for the person living with dementia.

3. Experiential learning

Experiential Learning requires a shift in education delivery from being teacher-centered, where the educator provides the knowledge to the learner, to a more learner-centered approach. The education is less structured, and the participants must share and learn through the experience (Kolb & Kolb, 2009). Kolb describes experiential learning as a process whereby new learning is created through the transformation of experience. (Kolb, 1984).

How do we build a room to support this new style of education, a built environment that allows learners to experience how to provide better dementia care? At this time, I had my first experience of a "Human Room" at McCulloch House; a 16-bed palliative care unit at Monash Medical Centre, Clayton. Efterpi Soropos, the architect of Human

Rooms (Soropos, n.d.), develops the rooms through a codesign process. Human Rooms are immersive sensory rooms that use colored lighting, surround sound, purpose made film and music to create an environment that is soothing and relaxing for the user. Efterpi interviewed patients, families and staff and used questionnaires to develop video content based on common themes of place, space, and memories. Music was selected based on the most common preferences for music, instruments, and emotions. Color lighting was also introduced to provide choices in the environment that resonated for people. Spending time in the Human Room can have a very therapeutic experience on the user. The room can provide a sense of calm, of feeling nurtured and the feeling that time has slowed down. Human Rooms can transport the user, especially those in palliative care, to another place and provide a sense of calm and relaxation, such as the forest experience seen in Fig. 1. This process is akin to the act of mindfulness, to experience feelings and emotions which the user has actively created by engaging with the technology and being immersed in elements that could activate an embodied experience of memory.

At the same time, Alzheimer's Australia Vic collaborated with Swinburne University in Melbourne, Australia, to commence an innovative project in the Media and Communications Department, led by Ph.D. candidate Mandy Salomon. Mandy's thesis was titled *Finding Self in a Virtual World: Digital Engagement for People with Moderate to Severe Dementia* (Salomon, 2016), and it explores the use of Virtual Environments on touch screens and their use for people living with dementia. Initially, Mandy looked at the human interaction with the iPad: how do people living with dementia interact with virtual environments? can they swipe? Touch? pinch? What works for them? Do they relate to the virtual activities?

Mandy tested these questions by working with people living with dementia in a Jewish residential care facility, and by making use of existing games that were available online (as seen in Fig. 2).

Mandy focused on the home environment and built a virtual home with many activities within it. Music could be played in the lounge area, games played in the garden, birds could be interacted with, and there were food activities in the kitchen. Mandy also explored the use of virtual companions which was positively accepted by most people living with dementia.

A virtual companion called Julie, shown in Fig. 3, was designed to anchor the participant within the environment. Julie was developed through a co-design process. People living with dementia in a care facility were

Fig. 1 Photo by Efterpi Soropos "Forest" experience from Human Room at McCulloch House Inpatient Palliative Care Unit, Monash Health, 2010.

shown a wide range of images, and Julie Andrews was by far the most popular. The virtual Julie grew from there. Mandy Salomon conducted studies that showed the acceptance of 3D virtual worlds for people living with dementia and the psycho-social advantages of using them.

Through working with Mandy Salomon, I was introduced to Opaque Multimedia (now known as Opaque). Opaque consisted of a team of PhD students from Swinburne University, who were game developers and who were building virtual environments for Mandy Salomon's project. In our conversations we talked about the relationship between video games, engagement and education. If you observe children playing new video games, they do not read instructions on how to play the game. Rather, they learn through playing the game itself. Could aged care workers learn how to improve their dementia care skills through virtual reality game playing?

Fig. 2 Mandy Salomon, Swinburne University, PhD candidate working with a resident living with dementia to codesign the new application 2013.

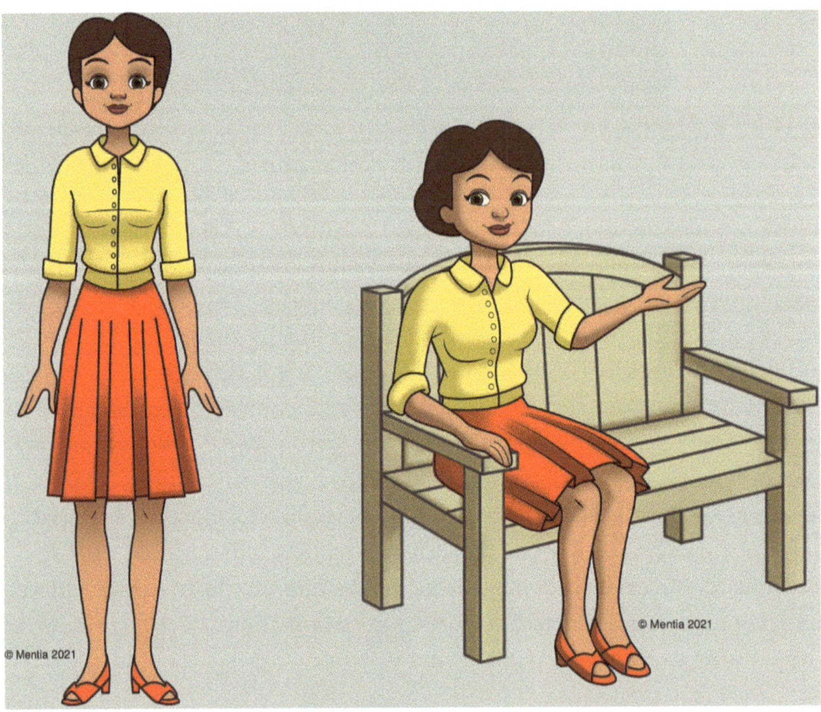

Fig. 3 A virtual companion called "Julie" co-designed by Mandy Salomon.

Experiential Learning was starting to take shape: an immersive room that tapped into the senses, virtual reality game playing, lots of physical interaction and less sitting and listening. Interestingly, when we teach carers how to better engage with people living with dementia, we talk to them about the importance of connecting to people with dementia through the senses. We encourage aged care workers to not rely on words alone: use hand gestures. Telling someone with dementia that dinner is ready may not be enough information. The smell of dinner cooking and the words together may be needed for a person living with dementia to make sense of what is to happen next. Likewise, our new training room needed to tap into the senses to give the learner a more complete understanding of the impact of dementia.

The training room was designed in 2010 with FPPV Architects and Opaque Multimedia, and was named the "Perc Walkley Dementia Learning Centre." The room needed surround-sound, a colored lighting system, large touch screens, and a large interactive media wall for projection of video games. A huge screen was designed that is 10 m long and 2.5 m high, with 2 large projectors, and a Kinect sensor to allow for interaction with the screen. The room would be controlled by remote to create different settings. For example, a red hostile room is created with red lighting, a busy street scene on the screen and the loud sound of chaotic street scene with traffic and people. Asking learners to test their memory in this room setting causes poor memory recall. The experience quickly demonstrates to participants the impact their environment has on their performance. From this experience the participant then reflects "If I perform poorly in a busy and chaotic environment, how then does a hostile chaotic environment impact a person living with dementia?"

The room can be changed to a green, peaceful, and serene environment using lighting, a tranquil garden scene on the screen and the surround sound of soft mindful music, as shown in Fig. 4.

Under these more tranquil conditions participants are again asked to test their memory. Their recall is much improved, and it becomes clear to them that a calm environment improves their performance. The participant (family carer or professional carer) then immediately reflects on their own home environment. Is their home peaceful and calm, or is there a lot of activity? Does the person living with dementia, find the home over stimulating, confusing, anxiety producing and upsetting? Or is the home calm and inviting and induces a sense of wellbeing just by being in that setting? These are questions participants immediately reflect upon. Often, they stop and reflect on these questions without prompting from the facilitator. The experience brings the two environments to life side-by-side and the participant can

Fig. 4 The Perc Walkley Dementia Learning Centre set up to demonstrate a serene environment that is calm and more conducive to better cognitive performance.

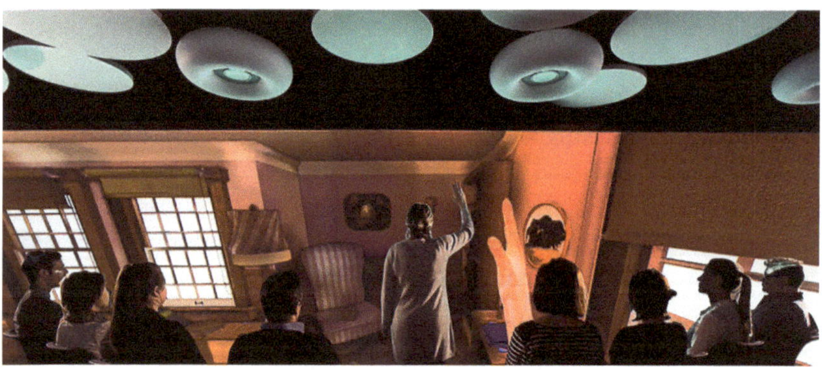

Fig. 5 The Virtual Dementia Experience game is projected on to the wall of the Perc Walkley Dementia Learning Center.

see how improved quality of life can be for the person living with dementia, simply by creating a quieter and less busy environment.

The training room is unlike anything used previously for education delivery. The room becomes a stage. On to this stage the virtual experience of living with dementia was introduced. The Virtual Dementia Experience (VDE) shown in Fig. 5, used game technology (Unreal engine developed by Epic Games) to create a virtual world that could be experienced by the learner. The game developers at Opaque Multimedia worked with the Dementia Australia team to create a scenario that would bring the experience of living with dementia to life. While the virtual experience may take only 5 min to go through, which is of course only a glimpse of the reality for the person living with dementia.

4. Co-Designing the virtual experience

How do you represent the experience of living with dementia? Medically, dementia is described as a set of symptoms that includes a decline in the ability to think, it affects memory, behavior, and mood, and ultimately affecting the person's ability to perform everyday activities. The actual way in which these symptoms manifest in each individual may differ, and so the experience of dementia may vary from individual to individual. In addition, there are many different types of dementia and some affect memory while others may cause issues with visual perception.

The signs and symptoms of dementia can vary but may include the following: Experiencing memory loss, poor judgment, and confusion; Difficulty speaking, understanding, and expressing thoughts, or reading and writing; Wandering and getting lost in a familiar neighborhood; Trouble handling money responsibly and paying bills; Repeating questions; Using unusual words to refer to familiar objects; Taking longer to complete normal daily tasks; Losing interest in normal daily activities or events; Hallucinating or experiencing delusions or paranoia; Acting impulsively; Not caring about other people's feelings; and Losing balance and problems with movement (National Institute on Aging, 2022).

With these widely documented symptoms as a starting point, we set out to gain greater insight into the lived experience from those living with dementia and their carers. We met with family carers and people living with dementia, visited private homes, and spent time in a residential care facility. The purpose was to collate the lived experiences of people with dementia and to co-create a lived virtual experience. One of the facilities involved in this evidence gathering exercise was a Jewish facility located in Melbourne. The professional carers shared stories of the difficulty in showering some residents and the reluctance they had in entering a shower. The carers did not know why the person living with dementia was unwilling to have a shower, but they could see that some were fearful of the shower. A care worker did note, however, that some of the residents in the facility were Polish Jews that had survived the holocaust. Undoubtedly these residents had experienced significant trauma in their younger lives, but the details were not always known by the carers.

In addition to our interviews, we reviewed online content and books in which people living with dementia have shared their lived experiences. Kate Swaffer and Christine Bryden have shared their experiences widely, in personal presentations, online and in books. Reading and listening to these provided additional experiences that gave insight into the lived experience and reiterated the stories gathered. Many experiences spoke to the impact that loss of memory had on the ability to perform everyday tasks, to hold conversations, to participate in social activity. There were also examples of behavioral changes and the impact of these on carers. We focused our investigations on the changes and impact on the senses.

5. Dementia and sensory changes

Dementia is often associated with a loss of ability to recognize and understand information gathered from the senses of vision, hearing, touch, taste, and smell. In our research many examples were provided where the person with dementia misinterpreted sensory information. This has more recently been verified by publications written by people living with dementia (Houston, 2015).

Dementia may affect a person's visuo-perception and spatial awareness. This can mean that a simple task like taking a seat in a chair could become quite challenging. As healthy individuals we take for granted the simple act of seating ourselves in the car or in a chair. We have a sense of our surroundings and our body positioning in them, also known as proprioception. With dementia these common everyday activities begin to challenge the individual. For a person with dementia, the act of taking a seat may require some thought and planning, touching the chair to aid vision and getting a sense of its location; and slowly edging into the seat. In addition, if the seat has no contrast with the area surrounding it, then depth perception can be more difficult and so taking a seat is more challenging. Often locating a toilet and taking a seat is challenging because the toilet is white and the surrounding area is also white, as illustrated in Fig. 6. This color scheme is most common in healthcare, aged care settings and private homes and can lead to continence issues for the person living with dementia.

Carers gave examples of visual hallucinations experienced where the person perceives something or someone that is not real. Hallucinations are more commonly experienced with people diagnosed with Lewy body dementia (Dementia Australia, 2018a, 2018b). An example shared with us was of person who mistook the zigzag patterns of a bed cover for slithering snakes.

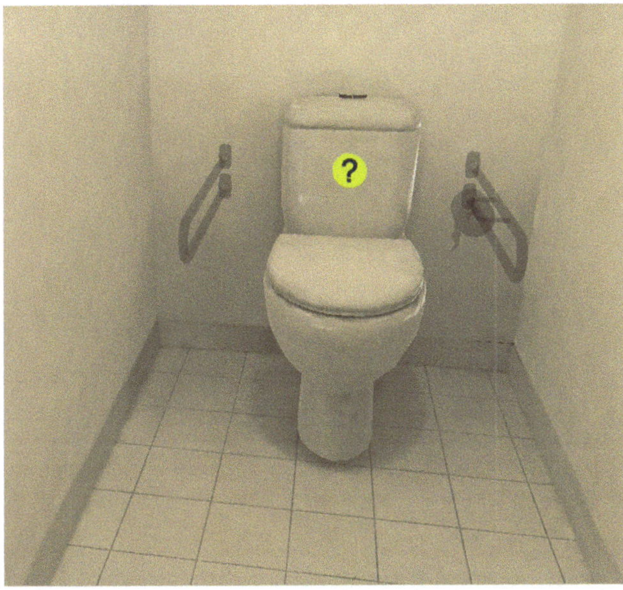

Fig. 6 A monochrome bathroom is challenging for a person living with dementia. Finding the toilet and taking a seat may prove difficult. Photo taken from the Dementia-Friendly Home application (2017).

Naturally, the person with dementia refused to get into the bed. The person may understand the hallucination is not real and may describe what they are seeing. The appearance of children, snakes, spiders are all common themes in hallucinations (Figs. 7 and 8).

Environments that are visually busy can cause distraction and imped a person's focus on the activity at hand, or it can be disorienting, leading to unsteadiness. For example, doorway thresholds where two different floor coverings intersect may cause difficulties and confusion. If floor coverings of very different colors intersect (e.g., where a carpet with a busy patterns meets a tiled floor) this can be disorienting for the person with dementia. Patterned carpets may even seem like leaves on the ground. Leaves that need to be picked up and removed, and can, then, very disorienting for the person living with dementia and also a risk for potential falls (Fig. 9).

It is, of course, important to consider the challenges that arise in addition to those associated with dementia. The aging eye, for example, can be affected in many ways: the lens becoming less flexible and may "yellow" in color with age. This can impede the clarity of vision and requires

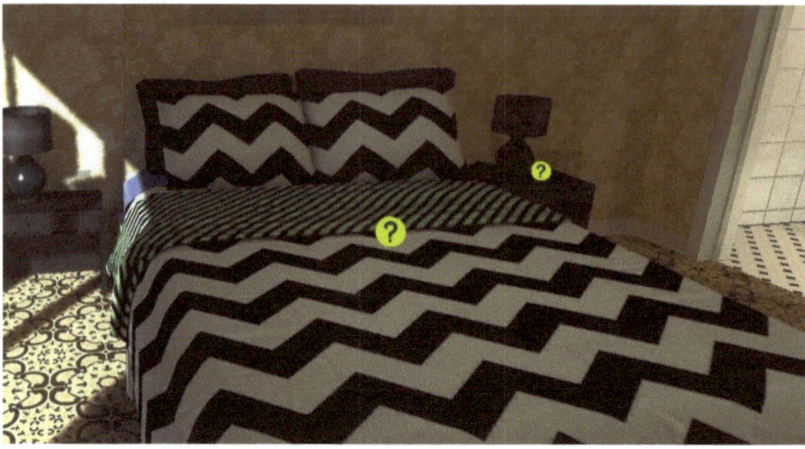

Fig. 7 Stripes and ZigZag shapes can appear to be moving for the Person living with dementia. Image taken from the Dementia-Friendly Home (2017).

Fig 8 Carpets with patterns can be interpreted as covered in crawling insects for the Person living with dementia. Image taken form the Virtual Dementia Experience, courtesy Opaque (2014).

Fig. 9 A doorway threshold where patterned carpet meets tiled floor may cause disorientation. *Photo taken from Dementia-Friendly Home (2017).*

brighter light. Indeed natural light is recommended as the best light source for aging eyes, followed by incandescent light. Poor lighting adds to the difficulties of visuospatial interpretation for the person living with dementia (Fig. 10).

While hearing functionality may be normal, a person with dementia may be hypersensitive to loud noise, and this may lead to a sensory overload experience if too much information is being given at the same time. Watching television and having a conversation at the same time may become impossible. Auditory agnosia is a condition where the person may hear normally but is unable to associate the spoken word with the meaning, or is unable to identify the sound from the object creating it. Many examples were seen and described where the person living with dementia may become more agitated and disoriented in a lounge area with lots of competing noise. Televisions being on while no one is actively watching, can be distracting and at times can cause distress. Similarly, radios can cause distraction and agitation. Air conditioners can create a lot of noise in large open spaced living areas. In care facilities it is not unusual to see televisions on, not being

(A)
(B)

Fig. 10 A and B: The Home- as viewed through the aging eye leads to less definition in the environment. (A) Shows the Home with normal vision. (B) Shows the Home with aged vision. Images taken from The Virtual Dementia Experience, courtesy Opaque (2014).

watched, while vacuuming is also in progress. Care staff then try to administer medication or engage the person living with dementia and this becomes almost impossible.

Dementia causes the brain to dysfunction and no longer interpret the different sensations of touch. The touch receptors in our skin respond to four sensations, including Pressure, Hot, Cold and Pain. Dementia may lead to a hyper- or hypo- sensitivity to touch. The person may not be able to differentiate between hot and cold for example, which could have very severe consequences.

Carers shared stories of their family members with dementia not wanting to have their hair brushed and complaining that it hurt. In some cases, even being touched would elicit a painful response. Showering could feel like pins and needles on the skin. The sensation of water falling over the head and face could be overwhelming and upsetting.

The sense of smell may change with dementia. Often it decreases. The person may experience a lack of association between object or food and its smell. Some individuals may have olfactory hallucinations often with unpleasant smells that are not there, like a fire burning, when nothing is evident.

6. Bringing the experience of dementia to virtual reality

Having gathered these experiences, we set about creating the virtual scenario. Working with Opaque Multimedia company, we focused on a scenario that would work well for carers to understand the lived experience.

Firstly, the setting was chosen as a private home as an appropriate and relatable setting for all carers be they professional carers or family carers. The virtual home would be the setting for an immersive multi-sensory experience, engaging the senses of seeing, hearing, and touching. Smell and taste would not be a part of the experience.

The home would need to demonstrate the following sensory issues:

- Memory loss issues
- Visuo-spatial issues
- Hearing issues
- Language and communication issues
- Thinking problems- Agnosia
- Hallucinations
- Mobility issues

The virtual experience needed to focus on an activity of daily living to make it more meaningful and applicable for the learner. We explored many activities of daily living that were impacted by symptoms of dementia. These included:

- Getting out of bed and getting dressed
- Using the bathroom
- Making a cup of tea
- Going shopping
- Using public transport

It was decided that the focus of this work should be on the bathing experience.

Showering and bathing are known to be activities that are often challenging for people living with dementia and their carers. It is an activity that

Dementia Australia often is asked for "strategies" to improve the caring support for the person living with dementia.

Showering and bathing are intimate activities that may cause discomfort. There may also be other issues such as depth perception, making it frightening to step into water. The person living with dementia may not perceive a need to bathe, as they believe they had done so recently, or they find it a cold, and uncomfortable experience.

7. Designing the virtual experience of living with dementia

The virtual experience focuses on the daily activity of waking in the morning to then find the bathroom in preparation for the day. The person needs to wash their teeth and then have a shower, all in preparation for a visit to the doctor.

The experience is described in 6 tasks. Each task breaks down the experience into the hurdles for the learner to overcome before moving on to the next part of the activity.

Dementia Australia worked with Opaque Multimedia to design the experience. We described the scenario in text and the game developers then sketched the scenario. Through this process, we built this virtual world unfolding the daily activity and the issues that the person living with dementia may confront (Fig. 11).

7.1 Task 1: Waking up in the bedroom

In this task, the experience focuses on getting out of bed, and the learner must procure eyeglasses and dentures. The action of getting out of bed could be difficult in a room with cream-colored walls and floors, cream bed linen, no or little contrast is provided in the bedroom. The colors reminiscent of the trend in home and residential care settings with furnishings of neutral pallets.

If the learner looks toward the floor, visuo-spatial difficulty makes the floor appear far away, much further than one would expect. In addition, there is the TV noise in the background and someone calling out "It's time to get up." The instructions are loud and disorientating. It is not clear from which direction the voice is coming from.

At some point the user puts on their eyeglasses, the difference between aided and unaided vision is demonstrated, as is the difficulty of finding objects with poor eyesight, especially when habit and memory cannot be relied upon. In addition, even with glasses, visual quality is far from

Fig. 11 Waking up in the bedroom a feeling of disorientation, searching for eyeglasses, sketch courtesy Opaque 2013.

normal, there is still some unsteadiness to the vision (which can be mimicked through camera movement) as is common with Alzheimer's and Lewy body disease.

Walking towards the bathroom door, the wallpapered walls with striped patterns appear to be moving. This would exaggerate the feeling of unsteadiness for the learner. Movement of the screen would create a feeling of unsteady gait for the learner (Fig. 12).

7.2 Task 2: Finding the bathroom door

We take for granted that we know where the bathroom is in our own home. But when the doors in a hallway are all the same and the person must remember which door is the bathroom door this becomes a challenge. This experience is common for people living with dementia. In residential care we were often told that residents would walk into the wrong room, having difficulty finding their room amongst a long corridor of identical doors.

A bathroom that is thermally cold and aesthetically clinical and uninviting also added to the apprehension of older people having a shower, with or without dementia. Cold rooms and not warm enough water temperature could add to the pain and discomfort of the experience, especially for those with comorbidities such as arthritis (Fig. 13).

Fig. 12 The striped wallpaper in the Virtual Dementia Experience appears to be moving as the learner walks through the hallway. Image from the Virtual Dementia Experience (2014).

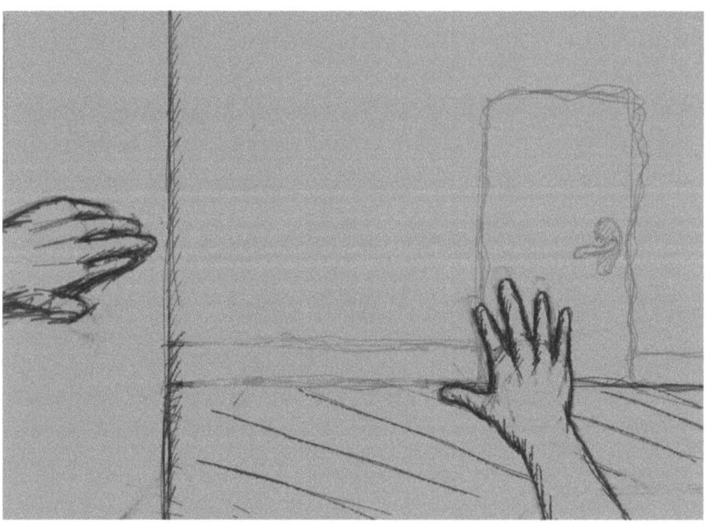

Fig. 13 Trying to find the bathroom door, sketch courtesy Opaque Media (2013).

Upon entering the room, the learner notices that the floor between the hallway and the bathroom is strongly contrasting, as hallway being carpet and the bathroom being tiled. The bathroom tiles have a shimmer to them making them appear to be wet. A dark mat on the bathroom floor is black in color and as you move closer it falls away into a hole in the floor.

Observing the room, the toilet and hand basin are a similar shape and size. Both are white, as are the surrounding tiles.

7.2.1 Task 3: Locating the wash basin

Once in the room, the learner must find the wash basin. The room is very white with lots of glare, making objects difficult to distinguish. Low contrast environments with bright lights and reflective surfaces can be both visually and spatially confusing to persons living with dementia. If this is combined with bathroom fittings of similar shape, persons living with dementia can struggle with routine parts of daily life.

In this task, the learner must distinguish the toilet from the wash basin or laundry basket. We were told by many carers in health care that people living with dementia would often mistake a basin as a toilet and relieve themselves in the basin. The learner must decide which white object is the toilet, and which is the hand basin (Fig. 14).

7.3 Task 4: At the wash basin Brushing Teeth

Once the wash basin is found, lots of clutter appears around the basin and the learner is asked to wash their teeth (Fig. 15).

Poor eyesight, confusion about the intended use of an object or inability to remember the detailed shape or orientation of objects can lead to the misidentification and misapplication of similar objects. The toothbrush and toothpaste are stored in a bathroom cabinet, alongside a razor, soap and shaving and medical creams. These other items are deliberately chosen for their

Fig. 14 A linen basket and the toilet bowl appear a similar shape and size through the eyes of the person living with dementia. Image from the Virtual Dementia Experience (2014).

Fig. 15 A cluttered bathroom counter makes finding a toothbrush confusing for the person living with dementia. Image taken from the Virtual Dementia Experience (2014).

Fig. 16 Cluttered environments may cause confusion. Sketch courtesy Opaque (2013).

resemblance to the toothbrush & toothpaste and their potentially deleterious effects if they are applied instead of a toothbrush & toothpaste (Fig. 16).

The learner proceeds to select the toothbrush from the many articles surrounding the basin, the learner grabs what looks like a toothbrush but on bringing it closer becomes a razor. When items are placed down on what appears to be a ledge on the edge of the sink, they fall into the sink. Hot and cold-water taps are not distinguishable from each other.

Fig. 17 The simple task of brushing teeth becomes nightmarish when mistaking the razor for a toothbrush. Sketch courtesy Opaque (2013).

The learner is unable to distinguish between the implements in the cupboard, and randomly selecting will always result in picking up the wrong implement (Fig. 17).

7.4 Task 5: Entering the shower

Many carers—both family and professional—commented on the fact that bathing and showering were activities that led to a lot of frustration and angst for the carer and the person living with dementia. The reasons for this varied greatly from one person to another and in some cases the reason was not clear at all. For some people showering is an intimate experience and being asked to undress in front of a carer is too much to ask. In addition, if they are being asked to undress and shower in an environment that is not recognizable, not homelike, or friendly, but rather clinical and thermally cold, the anxiety is increased (Fig. 18).

Bathmats at the shower entrance is a common occurrence. If the bathmat is a dark color and contrasts with the light-colored tiles or flooring surrounding, it may be a hinderance to people living with dementia from entering the shower. There are many examples where the person living with dementia refuses to enter a shower or a room where a dark-colored mat lies at the threshold. The person with dementia may perceive these mats as holes, and in asking them to enter the shower or the room, we are, as far as they perceive it, asking them to step into a hole. The person with dementia may be able to explain they see a hole, but not always (Fig. 19).

Fig. 18 Showering can be a frightening experience for people living with dementia. The shower may appear cold and unwelcoming. Sketch courtesy Opaque (2013).

Fig. 19 Black mats on light colored flooring can appear to be a hole for the person living with dementia. Image taken from the Virtual Dementia Experience (2014).

7.5 Task 6: Having a shower

To bring this experience to life we worked on a showering experience that would be hostile and frightening to anyone who experienced it. The environment needed to be cold and uninviting. The noise and pressure of the water need to be unexpected and intimidating to the learner. The volume of the shower is loud and appears hot and steamy. The pressure unusually abrupt and loud. Surround sound was used to create a loud base noise that almost shook and vibrated through the room.

The sound and feel of the shower needed to instil fear in the learner and increase the perception of threat. The learner would be unable to perceive details in the shower once the water and steam were on and a lack of fine motor control of the shower taps resulting in a frightening or even harmful event. A sense of unsteadiness could be created when walking on the slippery shower surface, with a slight movement of the projection.

Some carers and people living with dementia shared that the experience of showering itself may be painful or frightening. The sensation of water flowing onto the head could be overwhelming for some people. Others likened the sensation of the water on the skin as being like pins and needles falling on to their body. No wonder showers become frightening and even painful experiences (Fig. 20).

Fig. 20 The sensation of water on the skin or body can cause pain or be frightening. Sketch courtesy Opaque (2013).

8. The virtual dementia experience launch

The Virtual Dementia Experience (VDE) was launched October 2013 at the specially designed Perc Walkely Dementia Learning Centre in Parkville, Victoria. It has been experienced by more than 14,000 people to date including aged care professionals and family carers, healthcare professionals, high profile business and community leaders, and Victorian State and Federal politicians.

The VDE is delivered in a 2-h workshop that takes the participant through a multi-sensory simulation using light, sound, color, and visual content while incorporating "serious games" technology to create a virtual reality experience. The intention is to lead our course participants into thinking differently to their approach to dementia care and to develop empathy for the person living with dementia. Through the virtual simulation participants develop an empathic understanding of living with dementia. They gain a deeper understanding of the impact dementia has on the person's life. The experience is impactful and has led to participants reflecting on their practice and inducing practice change.

The author Melanie Joosten visited the Perc Walkley Dementia Learning Centre in 2015 and participated in the VDE workshop. Below is a walk through the virtual experience through Melanie's eyes. Melanie described the virtual experience in her book, *A long time coming; Reading into dementia* (Joosten, 2016) (Fig. 21):

> *"Hurry up and jump in the shower, Mum. We're in a rush."*
> *Myrtle looked at her hands as though surprised to see them there on the ends of her arms. She sighed and stepped forward. The black and white walls of the hallway rippled and bulged in grotesque* Beetlejuice *fashion, billowing as she walked past. At the end of the corridor she paused at a door, hand outstretched.*
> *"Get yourself in the shower, Mum. We haven't got all day. Yell out if you need help."*
> *'Oh dear.' Myrtle's voice was faint. She took a deep breath and fumbled with the door handle. Everything was blurry; she couldn't quite make out the edges of things. She was about to step through the doorway when she saw that it wasn't a room the door opened to, but a cupboard. Where was she going?*
> *She looked down at her feet, but the bugs were back, crawling about the carpet. Her house never used to be dirty—she always kept it well. She tried another door and it swung open, the roar of the exhaust fan leaping out at her. The bathroom. She squinted in the glare of the light bouncing off the tiles and took a tiny step forward. But there was a mammoth gaping hole in front of her.*
> *'Mum! You're not even undressed yet! Don't you remember we have to go to your doctor's appointment?'*

Fig. 21 The Virtual Dementia Experience in session. Image taken from the Virtual Dementia Experience (2014).

Be brave, Myrtle told herself, and stepped forward, waiting for the fall. Luckily, the floor rose up to meet her, like a flying carpet, and she shuffled onward. It was the bathroom, she was sure of it. But she couldn't remember what she was supposed to do there. She walked towards a white blob. What was it? And then she saw her mother's face, staring at her in the mirror. Was she supposed to be taking her somewhere?

Myrtle's breath was coming quickly. Her teeth needed to be brushed—that's what she must do. But when she took up the toothbrush, it didn't feel right. It was something else she was holding in her hand; she couldn't put that in her mouth. She dropped the razor in the sink and turned towards the shower. It was grimy inside, the small space dark. There were stains on the walls and a rusted shower rose directly above. Her heart started beating rapidly. It reminded her of something, a terrible memory—no, not a memory. A terrible thing was happening. Her hand reached for the tap, and when she turned it, hot water pummeled at her body, the pipes squealed, and the water roared. It was too much, it was too noisy. She just wanted it to stop.

9. Impact of the virtual dementia experience

The innovative nature of the VDE has allowed Dementia Australia to increase awareness of the disease and engage, inform and inspire a wide political audience and helped transform this audience's thinking about dementia in all its forms. The health services space in which Dementia Australia operates is highly competitive, but the VDE's marketability has attracted the attention of politicians and has precipitated a deeper understanding of dementia and strengthened these vital relationships. Politicians who have been engaged with the VDE over the past 8 years include Victorian State Minister for Health and Aging, David Davis MP; the Senate Standing Committee on Community Affairs; Tasmanian Senator Helen Polley; and Member for Wills, Kelvin Thomson MP, and a previous Federal opposition leader, Bill Shorten.

In March 2014, the Senate Standing Committee on Community Affairs acknowledged the ground-breaking work of the VDE and recommended that all states develop installations similar to that at the Perc Walkley Dementia Learning Centre. (Senate Standing Committee Community Affairs, 2014).

Consistent feedback from our clients has been that the VDE is a powerful experience which has helped them to empathize with the person living with dementia and reflect on their practice to incorporate practice change in their daily activity.

Feedback included statements such as: *"I am so grateful to have been given the opportunity to take part in the virtual experience. I had an 8 hour shift in the facility the next day (up at 4:30am!) and I was able to use the experience several times, in different ways, throughout the day. It was such an incredible feeling!"* A facility manager remarked that *"I will be encouraging everyone I come into contact with to go and participate in a VDE session. I actually believe so much in the depth of insight the VDE gives you that I have offered to pay for our Assistant Manager to attend a session."*

10. Dementia-friendly design

The virtual experience was used to highlight the importance of dementia-friendly environments also known as dementia enabling environments, based on the research of Fleming and Bennett (2017). In Fig. 22 the virtual home is modified with contrasting colors on walls and floors to help with visuo-spatial difficulties, clutter has been reduced in the bathroom and labels have been used to improve wayfinding. Using the game technology, we are able to modify the home to be more enabling and illustrate the 10 principals of dementia-friendly design (Fig. 22).

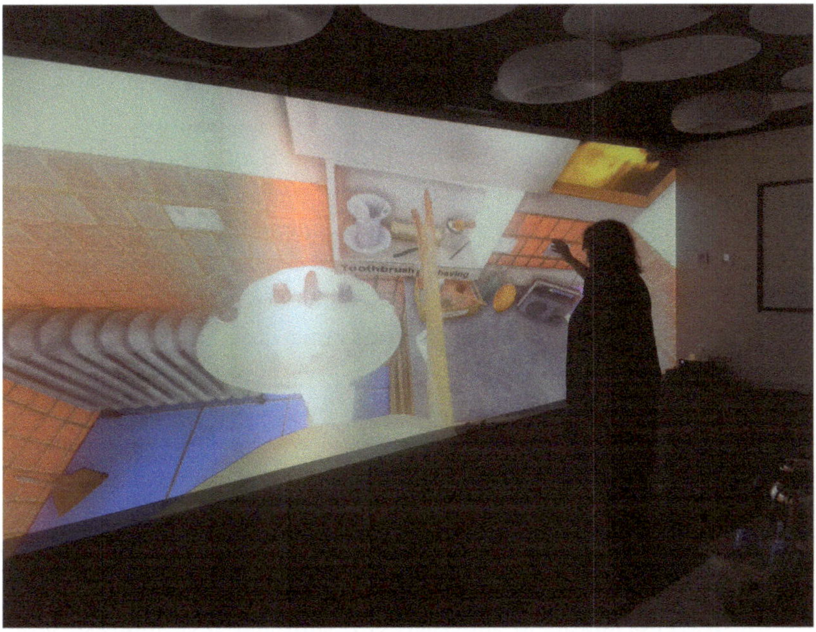

Fig. 22 The virtual bathroom can be modified to show how the built environment can be changed to make the environment more dementia friendly. Image taken from the Virtual Dementia Experience (2014).

The principles developed by R. Fleming and K. Bennett are described in the application Dementia-Friendly Home, which is available for download from the Google and Apple stores as:

1. Unobtrusively reduce risk- make the environment safe and secure by removing hazards in a way that does not detract from quality of life.
2. Provide a human scale—a person should not be intimidated by the size of the surroundings or confronted with a multitude of interactions and choices. Rather, the scale should help the person feel in control.
3. Allow people to see and be seen- The environment should help people with dementia recognize where they are, where they have come from, and what they will find if they head in a certain direction.
4. Reduce unhelpful stimulation—Minimize exposure to stimuli that are not helpful. The full range of senses must be considered: Sight, smell, hearing, touch, and taste.
5. Optimize Helpful stimulation—Enable the person with dementia to see hear and smell things that give them cues about where they are and what they can do. This can help to minimize their confusion and uncertainty.

6. Support movement and engagement- Aimless wandering can be minimized by providing a well-defined pathway, free of obstacles and complex decision points, guiding people past points of interest and opportunities to engage in activities or social interaction.
7. Create a familiar space- the person with dementia is more able to use and enjoy spaces and objects that were familiar to them in their early life. The involvement of the person with dementia in selecting familiar objects should be encouraged.
8. Provide opportunities to be alone or with others- People with dementia should be able to choose to spend time alone or with others. This requires the provision of a variety of spaces, some for quiet conversation, some for larger groups, and some where they can be themselves.
9. Provide links to the community—The environment should include spaces that encourage visitors to come and spend time, as well as opportunities to interact with the wider community.
10. Respond to a vision for way of life—The environment should support the person with dementia to lead a life that has meaning and value to them. Opportunities to undertake enriching activities should be clearly stated in the design of the home.

11. Awards and success

From 2014 the Virtual Dementia Experience has been recognized through multiple awards as an innovative tool in providing education to aged care workers. The list of awards is diverse and includes: Winner 2014, AIIA Victorian iAwards eLearning, Winner 2014, National iAwards Education Category, Winner 2014 Asia Pacific ICT Alliance Awards APICTA for eLearning, and Finalist for Melbourne Awards Contribution to Profile.

In 2015 it was awarded Gold Winner Victorian Public Healthcare Award Excellence in person, family and community-centred care, Winner with Opaque Multimedia Virtual Dementia Experience™, Microsoft Imagine Cup World Citizenship Award and Finalist HESTA Community Sector Awards Outstanding Organization.

In 2017 the Virtual Dementia Experience was part of the submission that won Small Training Provider of the Year Award Victorian Training Awards, presented by the Victorian Department of Education and Training and Winner Best Aged Care Software Development and Deployment Award, Information Technology in Aged Care Awards (ITAC).

The Virtual Dementia Experience has become a powerful advocacy tool to engage a wide audience in increasing the awareness of dementia and the impact it had on individuals.

12. Conclusion

Dementia is a complex syndrome that has no cure. The person living with dementia may experience many disabilities in performing everyday tasks that progressively worsens over time. What we can do is focus on improving the quality of life of the person, by modifying the built environment. To do this we need to have a deep understanding of the dementia experience and a true compassion for the person. Technology has provided the opportunity to have virtual experiences that can profoundly affect our understanding, empathy, and compassion Virtual Reality has allowed us to bring to life the experience of living with dementia, and through that, a deep understanding of how the environment and the built environment can affect our quality of life. Using this understanding and through co-design, care facilities can be built that better support the person and provide an enabling and engaging environment that empowers the person to live the best possible quality of life. Dementia Australia, as the peak body advocating on behalf of people living with dementia, is well placed to connect with and bring together the voices of people living with dementia, their carers and the providers servicing the aged care industry. So often we see aged care facilities built to meet the need of function for staff and without consideration for the people living there. The starting point for all new builds in aged care must be the person in residence. In residential aged care more than 70% of people are living with dementia. In the words of our consumers: "If you can get it right for dementia, you can get it right for everyone else" Person living with dementia (Dementia Australia, 2019).

References

Dementia Australia. (2018a). *LewyBodyDisease01- LewyBodyDisease [Help sheet]*.

Dementia Australia. (2018b). *LewyBodyDisease04- MotorChanges [Help sheet]*.

Dementia Australia. (2019). *Our Solution: Quality Care for people living with dementia*.

Dementia Australia Strategic Plan. (2018). https://www.dementia.org.au/sites/default/files/documents/DA-2018-Strategic-Plan.pdf.

Dementia-Friendly Home. (2017). https://www.dementia.org.au/information/resources/technology/dementia-friendly-home-app.

Fleming, R., & Bennett, K. A. (2017). *Environmental design resources*. Dementia Training Australia.

Houston, A. (2015). *Dementia & sensory challenges. Dementia can be more than memory*. Deal With Dementia.
Joosten, M. (2016). Reading into dementia. *A long time coming* (pp. 225–226).
Kitwood, T. (1997). *Dementia reconsidered: The person comes first*. Buckingham: Open University Press.
Kolb, D. (1984). *Experiential learning: Experience as the source of learning and development*. Englewood Cliffs, NJ: Prentice Hall.
Kolb, A., & Kolb, D. (2009). The learning way: Meta-cognitive aspects of experiential learning. *Simulation Gaming*., *40*(3), 297–337.
National Institute on Ageing. (2022). https://www.nia.nih.gov/health/what-is-dementia.
Salomon, M. (2016). *Finding Self in a Virtual World: Digital engagement for people with moderate to severe dementia*. (PhD Thesis).
Senate Standing Committee Community Affairs. (2014). http://www.aph.gov.au/Parliamentary_Business/Committees/Senate/Community_Affairs/Dementia.
Soropos, E. (2017). *Human Rooms, CSPA Quarterly, no 16 on expanded Scenography–Spring 2017* (pp. 16–21). Center for Sustainable Practices in the Arts.
WHO (2021) https://www.who.int/news-room/fact-sheets/detail/dementia (2021).

Further reading

ADI. (2020). https://www.alzint.org/about/dementia-facts-figures/dementia-statistics/.
Gilmartin-Thomas, J.F.M., McNeil, J., Powell, A., Malone, D.T., Larson, I.C., O'Reilly, C.L., Kirkpatrick, C.M., Kipen, E., Petrovich, T., Ryan-Atwood, T.E., and Bell, S.J. (2018a). Qualitative evaluation of how a virtual dementia experience impacts medical and pharmacy student's self-reported knowledge and attitudes towards people with dementia. Dementia 19, 205–220.
Gilmartin-Thomas, J. F. M., McNeil, J., Powell, A., Malone, D. T., Wolfe, R., Larson, I. C., et al. (2018b). Impact of a virtual dementia experience on medical and pharmacy student's knowledge and attitudes toward people with dementia: A controlled study. *Journal of Alzheimer's Disease*, *62*, 867–876.
Royal Commission Final report-list of recommendations March. (2021). https://agedcare.royalcommission.gov.au/publications/final-report-list-recommendations.
Salomon, M. (2014). Mandy Salomon tracks digital trends in dementia care. *Australian Journal of Dementia Care*, *3*(5), 17–19.
Shorten, B. (2017). https://www.billshorten.com.au/tackling_dementia_our_generation_s_duty_sydney_tuesday_21_november_2017.
Stargatt, J., Bhar, S., Petrovich, T., Bhowmik, J., Sykes, D., & Burns, K. (2021). The effects of virtual reality-based education on empathy and understanding of the physical environment for dementia care workers in Australia: A controlled study. *Journal of Alzheimer's Disease*, *84*, 1247–1257.
Stargatt, J., Doube, W., Bhar, S., Petrovich, T., McGuire, L., & Willison, A. (2021). Increasing understanding of environmental modifications using the virtual dementia experience for professional carers of people living with dementia. *Gerontology &Geriatrics Education*. https://doi.org/10.1080/02701960.2021.1967152.

CHAPTER FOURTEEN

Designing with neurodiverse children and adults: learning a different lesson with every engagement

Sarah Wigglesworth* and Clare Bond
Sarah Wigglesworth Architects, London, United Kingdom
*Corresponding author: e-mail address: sarah@swarch.co.uk

Contents

1. Introduction

Sarah Wigglesworth Architects (SWA) takes pride in its record of 'learning through doing,' and we like to build our knowledge base partly through undertaking project work with a clear research agenda. Often this means we are learning about new aspects of design by collaborating with clients who are able to inform us about their world from the inside.

Developments in Neuroethics and Bioethics, Volume 5
ISSN 2589-2959
https://doi.org/10.1016/bs.dnb.2022.05.002

Moving into a new area of expertise or building type helps challenge our prior knowledge but also permits insights by encouraging different perspectives on common problems.

Since committing in 1991 to addressing climate change, with its attendant issues of social equity, SWA has established a dedicated Equality, Diversity and Inclusion (EDI) team aiming to raise the discourse and inform ourselves on aspects of design that affect our work and could be blind spots. This could encompass issues that impact clients & users, our own workforce or our approach to working and design processes. We explicitly gravitate toward work that carries positive social value to our team and to those affected by our work, whether this be occupants or collaborators. When targeting or undertaking projects we look for opportunities to leverage greater knowledge or value by building on what we already know. Sometimes this is with the blessing of, or in collaboration with, our clients and users and sometimes it remains a private ambition or additional service within the SWA team.

We have also established a modest research fund, with time off to carry it out, paid to an individual with the passion and the curiosity to explore a topic of great interest to them. We capture the learning for future reference and to ensure the whole office can benefit from the knowledge gained. By sharing this work and regularly hosting other EDI-focused reviews and project shares with the office, we hope to spread knowledge, forge a collective culture and create a learning loop that feeds back into actions within our project work.

We have been fortunate to work on several projects that address designing for neurodiverse and impaired users. This chapter will explore our approach to two different projects, one a completed building and one a research project that aims to conduct primary research into the relationship that people with impairments have with the built environment, exploring methods of engagement and what makes an accessible, inclusive space. These projects build on research relating to other types of specialist groups such as older people and children.

2. Designing for neurodiverse children: Mossbrook school, classroom for the future

The Classroom of the Future (2001 – 2) was a project funded by the DfEE aimed at exploring prototypes for the future of classroom design (Fig. 1). Local authorities were invited to bid for funding exemplary projects aiming to explore what form and character a Future Classroom might take,

Fig. 1 Front Cover of the Publication Classrooms of the Future by DFEE. *Images are owned by Sarah Wigglesworth Architects. Courtesy UK Government, Department for Education.*

with the express remit promoted by the DfEE to take on board the theme of technology in the 21st century classroom. We understood the DFEE to be interested in how information and communication technology (ICT) could enable an enhanced learning experience for children growing up and naturalized to ICT. Yet this agenda provoked us to consider how the school client could best use ICT as a tool to engage their pupils. We had no idea how to address this at first, which was challenging, but as the project progressed ideas began to form and, in their final manifestation the proposals to incorporate ICT took on an imaginative and exciting aspect.

The Classroom for the Future at Mossbrook Special Primary School was one of four projects that won funding in the Sheffield area. Our client was a school teaching mainly autistic children, but the new classroom was also intended as a shared resource for pupils from all schools throughout the Sheffield area. The pupils at Mossbrook had been assessed as having learning difficulties ranging across the autism spectrum as well as physical restrictions, so the classroom had to address a wide range of needs. This dual brief is of interest, because it demonstrates that, by looking at design challenges from the viewpoint of the (generally excluded) 'other,' we can make a world that embraces the needs of and enhances experiences for a more diverse range of abilities.

2.1 Teaching science to children with autism

The school teaches the National Curriculum. The project brief, developed by school staff, requested a science teaching space designed primarily for learning about the natural environment. For children with learning difficulties, science at Mossbrook is taught in a demonstrative and experiential way, through direct interaction with the qualities and physical characteristics of the world around them. We felt this offered the potential to work with the traditional principle of science, which is the direct observation, recording and interpretation of the natural world. This idea can encompass everything from experiential encounters through visual exploration to data recording and technical measurement. In acknowledging climate change, there was an additional drive to link pupils with the broader ecologies around them.

In the Future Classroom, children learn about sustainability by discovering how they form part of a large, complex set of related interactions with their environment. In this context we felt that technological gadgetry and tactile and sensory experience could be brought together as twin and complementary ways of experiencing and interpreting the world. We saw the

Classroom of the Future as a learning apparatus in its own right. This meant thinking about the space within the building as a vessel for capturing and manifesting phenomena—using light, space, texture and materials as primary stimulation, in addition to the imaginative incorporation of virtual technology. As regards the latter, we sought to go beyond the expectations of the DfEE and think critically and broadly about what kind of technology could meaningfully engage children at Mossbrook in learning science. This is discussed later in the chapter.

2.2 Engagement with the school

Engagement was undertaken by postgraduate students from the University of Sheffield Department of Architecture who consulted widely with staff and pupils as part of the process of brief writing. Students worked with small groups of pupils and used props such as image cards to establish what they would like to see in their new classroom, and the kind of materials, atmosphere and feel it should have. Pupils were also invited to draw their ideal school. As expected, many focused on the outdoors. Some of their proposals were particularly imaginative, mentioning towers, bridges, space travel and rockets. Others wanted to see a McDonalds on site! SWA also carried out extensive discussions with key staff members to understand how teaching worked in the school, how pupils are managed, how dining took place and how classes were organized and run. We spent time at the school as observers of all these aspects.

These two elements of the engagement indicated the breadth of issues the design had to address, and while the big imagination of the pupils could not always be delivered, the pupils' ideas were formative in helping us remain ambitious for the classroom despite the need to address practical matters and safeguarding.

Knowing nothing about designing specifically for autism, we learned a good deal about the challenges through our many conversations with experienced and committed staff, in particular the Head Teacher Maggie Brough and her Deputy and Science Champion, Heather Wood. Having made a bid to the Local Authority to win the funding, they were clear about their intentions and more than generous in helping us understand how we could address the needs of the children in their care. Working so directly with knowledgeable and enthusiastic staff gave us confidence in what we were doing, allowing us to create a nuanced proposal that answered to the school's ambitions.

2.3 The opportunities of location

Mossbrook Special School is located on the very outskirts of the city, in Sheffield's green belt. Close to fields and farms, it has a rural character that offers great benefits for experiencing and being in nature. A swimming pool within the school structure is a controlled space for assisted exercise, however, for all children (some have restricted mobility, some are poorly sighted, some are hearing impaired), being able to access the outdoors is critical for their enjoyment, stimulation, health and wellbeing. The potential the setting presented for a relationship with the natural world gave us the opportunity to facilitate engagement, sensory stimulation, immersion and observation with the natural habitats of plants and animals all around the Future Classroom. (Figs. 2–4).

Over a number of years the school had been developing a pond as a valuable teaching resource for the children, but which was difficult to access due to steep slopes all around it. Following discussion with the school community it was decided to locate the building on the edge of the playground, close to the pond. Due to planning restrictions on green belt development, we pinpointed a site on the footprint of a temporary structure housing an indoor soft play area or 'Ball Pool.' This allowed the new building to form a gateway into the Conservation Area so that access could be controlled safely. The main teaching space looks directly over the pond and to the north, the Future Classroom adjoins the school's Sensory Garden. As a result of the Ball Pool's displacement, our brief expanded to include a new space for indoor exercise in addition to the classroom itself. This deep connection to nature is a celebrated aspect of the final project, however, if planning constraints had been less restrictive, SWA would have liked to explore a location which was better connected to the main school as a means of further improving the accessibility and usability of the Classroom and Ball Pool all year round.

2.4 Design for autism

Understanding that children with autism often respond better to simply-delineated, clearly-contained spaces (Fig. 5,) the classroom itself is a large, intelligible room whose volume is created by a series of thick portal frames made of plywood and softwood. We employed a variety of materials in its fabric as primary stimulation. Our idea was to incorporate different substances in its construction so the children could learn about them

Fig. 2 Site plan showing classroom adjacent to the pond and school. *Images are owned by Sarah Wigglesworth Architects and courtesy: Peter Lathey, Susan Collins, Sue Barr and Sarah Wigglesworth.*

Fig. 3 View of pond 1. *Images are owned by Sarah Wigglesworth Architects and courtesy: Peter Lathey, Susan Collins, Sue Barr and Sarah Wigglesworth.*

through close scrutiny and interaction—another example of the building itself being used as a tool for education. The classroom's geometry helps sound attenuation and its ceiling is lined in acoustic board to promote optimal conditions for comprehension and reduce distracting noise. Externally the Future Classroom is clad in a three successive layers—corrugated steel, oak and polycarbonate sheeting—which progressively peel away to reveal

Fig. 4 View of pond 2. *Images are owned by Sarah Wigglesworth Architects and courtesy: Peter Lathey, Susan Collins, Sue Barr and Sarah Wigglesworth.*

different glimpses of the inside, while simultaneously framing different views of the exterior world (Fig. 6).

The classroom has good daylighting from three sides. Lit by high level glazing to the north (Fig. 7), it has three child-height south-facing windows overlooking a meadow and full height glazed doors giving onto the pond area to the west. The south-facing windows are set deep within the thickness

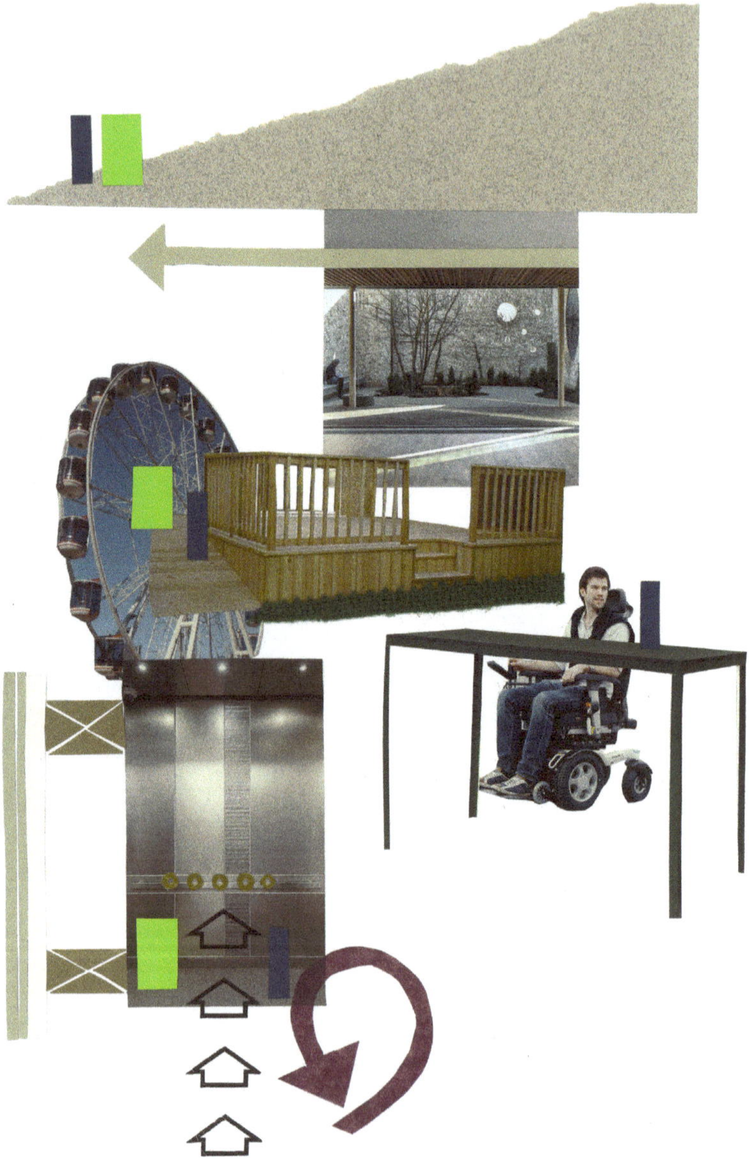

Fig. 5 Labeled ground floor plan. *Images are owned by Sarah Wigglesworth Architects and courtesy: Peter Lathey, Susan Collins, Sue Barr and Sarah Wigglesworth.*

of the portal frames which creates an overhang to avoid overheating and glare, though they are also fitted with blinds so daylight can be moderated (Fig. 8). Three top-lit resource rooms along the north wall of the building store props for teaching. These are categorized by Material, Physical and

Fig. 6 View of overhang with three cladding layers. *Images are owned by Sarah Wigglesworth Architects and courtesy: Peter Lathey, Susan Collins, Sue Barr and Sarah Wigglesworth.*

Biological themes and their doors are colored accordingly for easy recognition. To create a different atmosphere in each, one was lined in felt, one in metal and the third in wood. Equipped with shelves and seating, these rooms also allow small numbers of children to gather and focus on specific demonstrations or to study quietly away from the remainder of the class (Fig. 9). They provide a breakout room for one-to-one teaching and can

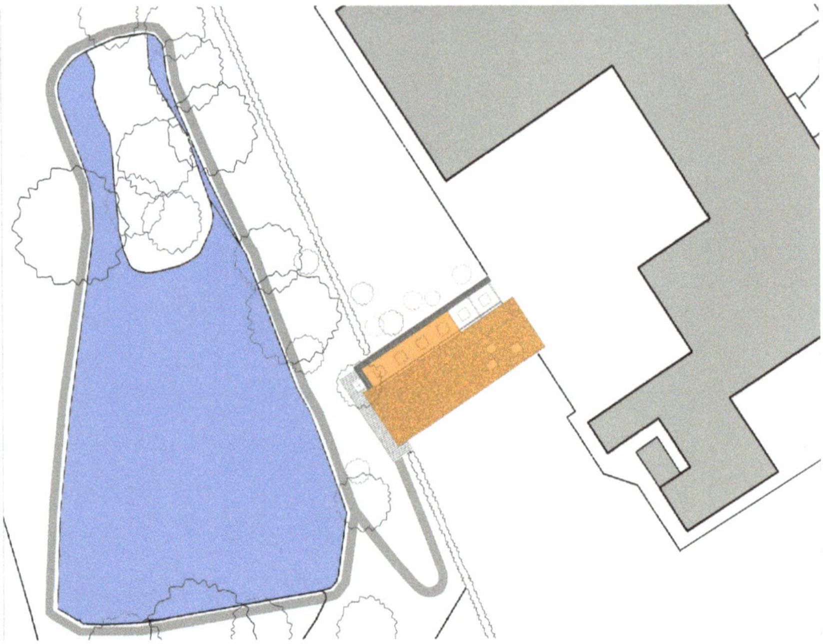

Fig. 7 Collage of building showing clerestorey glazing to north. *Images are owned by Sarah Wigglesworth Architects and courtesy: Peter Lathey, Susan Collins, Sue Barr and Sarah Wigglesworth.*

Fig. 8 Collage view from south lawn. *Images are owned by Sarah Wigglesworth Architects and courtesy: Peter Lathey, Susan Collins, Sue Barr and Sarah Wigglesworth.*

Fig. 9 One of the small resource rooms being used for small group teaching. *Images are owned by Sarah Wigglesworth Architects and courtesy: Peter Lathey, Susan Collins, Sue Barr and Sarah Wigglesworth.*

offer a respite space with a calm & contained atmosphere when a pupil feels over-stimulated. This is an example of where different needs can be addressed through the provision of specific, yet flexible, spaces and where each space can have several purposes besides the one initially envisaged.

In and around the site, evidence of bats, badgers, foxes and other wildlife showed that the building should act as host (not deterrent) to their patterns of occupation. (Fig. 10) We set a glass window into the floor allowing a view into the void under the building for surveying the movement of living things in the gradient below. Another 'window on the world' was located in the lobby floor, this time, a virtual window housing a plasma screen. We hoped to encourage children to closely observe physical and natural phenomena under safe, supervised conditions. The sensory garden meets the edge of the building by turning vertically up the living crib wall while providing a protective, insulated face to its exposed north edge (Fig. 11). Pupils that have difficulty bending down are able to garden at heights to suit them, making use of pockets of soil housed within it. A ramp zig-zagging down the slope to the pond encourages wheelchair users to enjoy the waterside alongside the other pupils (Fig. 12).

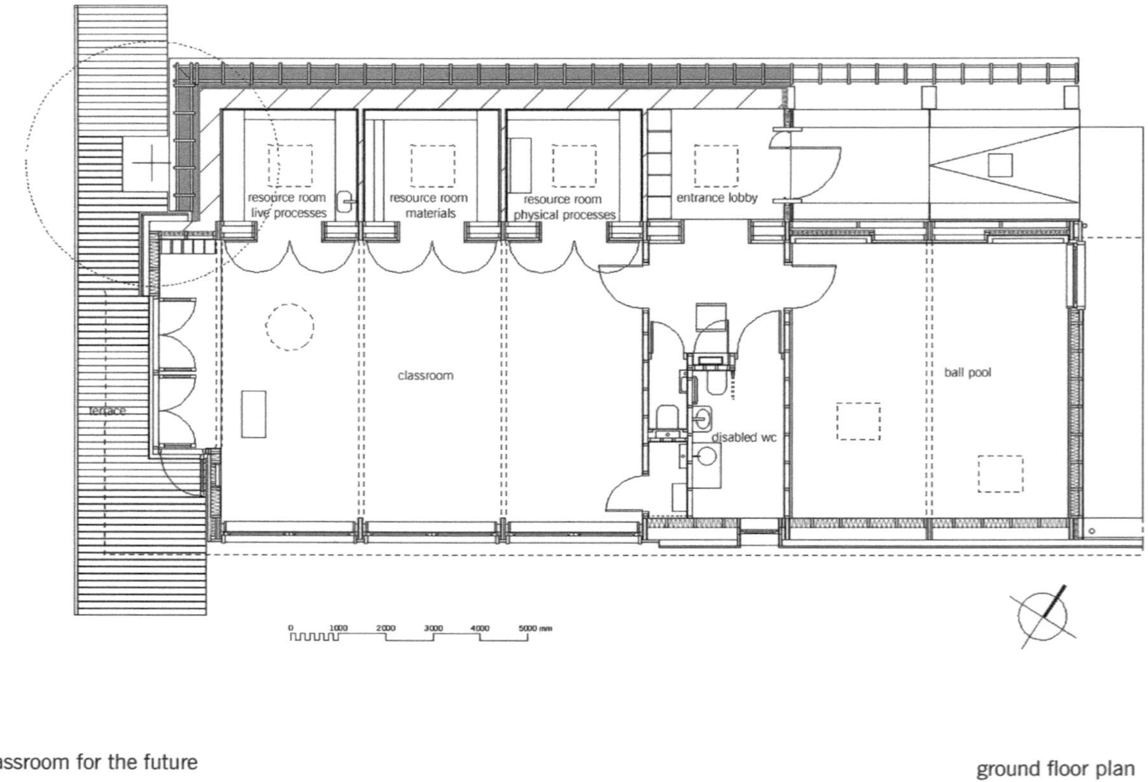

Fig. 10 Collage showing co-location of the classroom with nature. *Images are owned by Sarah Wigglesworth Architects and courtesy: Peter Lathey, Susan Collins, Sue Barr and Sarah Wigglesworth.*

Fig. 11 View from playground with crib growing wall. *Images are owned by Sarah Wigglesworth Architects and courtesy: Peter Lathey, Susan Collins, Sue Barr and Sarah Wigglesworth.*

2.5 Enhancing the learning experience

The brief challenged us to find relevant and appropriate solutions to the use of new technology that would augment the learning experience of these children. Given the mission pursued by Mossbrook for haptic experience of phenomena as a valid way of learning science, we were looking for a way to use digital technologies to augment the physical. A grant funding call from the Royal Society of Arts' 'Art for Architecture' programme drew our

Fig. 12 EXTERIOR with ramp and web cam. *Images are owned by Sarah Wigglesworth Architects and courtesy: Peter Lathey, Susan Collins, Sue Barr and Sarah Wigglesworth.*

attention to the potential of collaborating with an artist at the forefront of experiments in the use of digital capability. We were privileged to work with artist Susan Collins to develop, with the children, an art project that addresses the aims and ethos of the Classroom of the Future. Collins works with virtual and electronic media, exploring their relationship with architectural spaces and her involvement in this project added another dimension to the incorporation and interpretation of technology in the context of the teaching of science and environmental issues.

Collins and ourselves developed four projects relevant to the Future Classroom (Fig. 13). In one, web cams located on and under the building and overlooking the Conservation area are triggered by body heat; these record animal movement and direct images into the Classroom through the plasma screens located) in its floors and walls (Fig. 14). In another, a boat on the pond was fitted with an underwater camera and children were able to remotely control its movement to observe underwater pondlife (Fig. 15.) In one of the resource rooms its rooflight was replaced by a Camera Obscura projecting over the rooftop. (Figs. 16 and 17)

Fig. 13 Sarah W and Susan Collins collaborating on the Art for Architecture. *Images are owned by Sarah Wigglesworth Architects and courtesy: Peter Lathey, Susan Collins, Sue Barr and Sarah Wigglesworth.*

Using a car jack to focus the image on a small round table in the room, children are able to watch real-time images of the exterior from the comfort of their classroom.

The fourth project involved the new Ball Pool. This simple, square room full of soft play equipment was a favorite with children, allowing them to take playful exercise with the supervision of teachers. Here, Collins worked with the children to develop musical sequences that could be played back into the room through tiny speakers located in the wall mattresses. When hidden LED beams are broken through the children's movement in the soft play, the music plays. The room quite literally sings, as the bodies of the children trigger it in sequence (Fig. 18).

Other ideas incorporated in our design involved the development of a clear Perspex sink, located within the classroom, for demonstrating the characteristics of water—floating/sinking, freezing/thawing, wet/dry, wave motion, etc. (Fig. 19) In the lobby, outside the Changing Places WC, we put a transparent plastic cistern behind a window. This was a fun way to

Fig. 14 Interior of classroom with plasma screen. *Images are owned by Sarah Wigglesworth Architects and courtesy: Peter Lathey, Susan Collins, Sue Barr and Sarah Wigglesworth.*

demonstrate the cause-and-effect of the handle on the water, which is usually hidden from view! (Fig. 20).

This project has generous space standards, including door widths and lobby areas, liberal storage (including boot/wellie cubbyholes), excellent

Fig. 15 Launching boat w underwater camera. *Images are owned by Sarah Wigglesworth Architects and courtesy: Peter Lathey, Susan Collins, Sue Barr and Sarah Wigglesworth.*

Fig. 16 Camera obscura on roof. *Images are owned by Sarah Wigglesworth Architects and courtesy: Peter Lathey, Susan Collins, Sue Barr and Sarah Wigglesworth.*

Fig. 17 Children playing with the camera obscura.

Fig. 18 Ball pool play space.

Fig. 19 Children playing with water in the bespoke sink.

Fig. 20 wc cistern. *Images are owned by Sarah Wigglesworth Architects and courtesy: Peter Lathey, Susan Collins, Sue Barr and Sarah Wigglesworth.*

daylight and an expanded brief that aimed at creating a memorable indoor and outdoor environment for learning science. Nonetheless, it was the product of a dedicated and knowledgeable staff that had high ambitions for their pupils and we are grateful for the time and energy they offered to help us develop this project. Working collaboratively with children, staff and artists showed how important it is to address directly the needs of users, and tap into years of educational experience. We were privileged to have had this opportunity.

3. Designing with neurodiverse adults: What does an accessible building look like?

Mossbrook School, Classroom for The Future is a built project with extended briefing and learning built into the process. It was conceived and delivered with children in mind, focusing on their specific, yet diverse set of requirements for a learning environment. That said, we of course design places for both children and adults and opportunities to improve our knowledge are not always in the form of buildings. With all built projects, there are wide-ranging constraints that can affect the extent that early ambitions for a project are realized. This can be a frustration; however, self-initiated research projects form a platform for us to continue to test unrealized ideas and equip ourselves with the relevant experience or knowledge to help demonstrate ways that these ambitions can be realized in future, 'real' projects.

An openness to other methods of learning is a critical part of SWA's culture and often requires a creative approach to practice. This second project is self-made and is an example of how SWA 'layers' cultural capital across time and space. This time we focus on adults and explore the briefing process itself, aiming to demonstrate how engagement with real lives can begin to undo the potential negative stereotypes that can flow from too narrow a concept of 'user.'

3.1 Spotting opportunities to learn

Sarah Wigglesworth Architects strongly believes that good architecture is supported by knowledge and passion and that making time for research and reflection is central to enabling this. We are a process driven practice who engage with architecture and all of its mess and complexity as a continuous learning loop. Both as individuals and collectively, we are at our best when we are learning something new about what we feel passionately about—caring for people and our planet.

Led by one of our Architects (and Diversity Champion) during summer 2020, the team at SWA collaborated on an essay competition which asked entrants to consider how the design and construction of homes and neighborhoods could develop over the next 50 years in order to be inclusive regardless of impairment. The essay was one of 10 shortlisted and went on the be awarded a runner up prize. The winning entry was from another Architect, whose essay calls for a re-evaluation of town planning and national policy to ensure accessible housing is delivered in all new developments in a top-down approach.

We wanted to write this essay for a number of reasons: our Equality, Diversity and Inclusion agenda, continued interest in inclusive design and the opportunity to conduct research which could meaningfully impact the work we do as designers. Rather than simply pointing out all that's wrong in our industry, entering this competition was an opportunity to propose a framework for ethical change which could help to inform actions, as well as discussion. Internal workshopping, sharing knowledge and experiences and conducting desk-based research all formed part of SWA's process when writing the essay.

The essay competition was organized by Habinteg Housing Association to celebrate their 50th birthday. At the heart of Habinteg's ethos is a belief that having an accessible home in an inclusive setting can transform the lives of disabled people and those around them. They are a leading social housing provider, delivering and promoting accessible homes and communities across England and Wales. SWA share Habinteg's vision for communities which include places to live that meet a diverse range of needs and provide the highest levels of independence, choice and control over daily life.

The essay argued that to have to best chance at creating inclusive neighborhoods, we need a diverse set of designers. It proposed a simple, three step framework that designers could use as a guide to implement changes in their design thinking and processes. Step one increased diverse community voices through engagement and co-design, step two was increasing the diversity of the profession (partially facilitated through engaging hard to reach young people in architecture during step one) and step three introduces a continuous feedback loop of diverse voices and inclusive processes (Fig. 21).

3.2 SWA research grant and a piggy-backing approach

In order to embed research more firmly in SWA's culture and support all staff in developing their own capabilities, interests and careers, SWA offers a bursary once a year to enable a member of staff to take some time out for such a pursuit. The project proposal can cover anything, but should enhance interests, provide stimulus and have a clear output that will feed back into the work of the practice, helping to build up intellectual capital. This could include a short course, a visit to study a building or place of interest, making something explorative or learning a new skill. A key aspect of the research has how it is communicated back to others both in the practice and beyond. The SWA research grant began in 2018 and projects have included

Fig. 21 Diagrams illustrating the existing scenario and three-step framework to increasing diverse voices within the design and delivery of our homes and neighborhoods. *Images are owned by Sarah Wigglesworth Architects and Clare Bond.*

exploration of traditional timber joinery through furniture making and a photographic study of SWA's dance studio project, exploring form and materiality in relation to function and resulting in the publication of a zine.

In Summer 2021, SWA's research grant was awarded to conduct primary research into the relationship that people with impairments have with the built environment. This research project piggy-backs onto the Habinteg essay which was completed a year earlier. Step one of the essay framework was meaningful engagement with people with disabilities and the research grant puts the first step into practice by exploring a method to find, listen to, understand, record and share diverse voices and experiences.

3.3 Opening a conversation, listening to diverse voices

In collaboration with Habinteg Housing Association residents and insight groups, we have been seeking to update and deepen our understanding of how adults as well as children with neurodiversities and other physical impairments navigate the built environment within everyday life.

Four people living with a variety of abilities took part in the research, which consisted of individual interviews focused on their unique lived experience. The questions focused on everyday life at home, but also addressed people's ability to participate in life outside of home, at workplaces, entertainment spaces, parks and transport.

Neurodiversities are wide-ranging, but often form part of a individuals complex mental and physical needs. Three of the interviewees had lived experience of one or more mobility and dexterity impairment, including paraplegia, immune system sensitivity, multiple severe chronic illnesses (including Mast Cell Activation Syndrome, Ehlers-Danlos Syndrome and Crohn's), hearing impairment, neurodiversities including learning disabilities and mental health issues, including depression. One of the interviewees was, for 12 years, the sole career for her severely disabled late husband and generously shared her experience of navigating the built environment alongside him.

The interviews were virtual due to Covid-19 and although it was disappointing to not meet the interviewees face to face, the online nature of the research removed any physical barriers which could have made taking part in the project inaccessible. Although physical barriers were removed, it was important to acknowledge neurodiversites when preparing for this engagement. Many people (regardless of their neurology) don't like the feeling of being unprepared, therefore, interview questions were sent out in advance to allow the interviewee time to prepare and reflect on their answers, should they wish. All interviewees were offered guidance on using MS Teams ahead of the interview taking place, alongside ongoing, pre-interview correspondence to help build a trusted relationship ahead of the main engagement. Each interviewee was asked the same six questions and additional time for more detailed, personal conversations was allowed for in the meeting invite. It was important that videos were on for the interviews to help with social cues. and communication, however, advance warning of this requirement was given to allow interviewees to raise any concerns prior to the interview and avoid any surprise or distress during the call. The interviews were audio-recorded, but it was agreed in advance that the recordings would not be shared externally.

3.4 Evidence gathered from the recordings

The experiences shared during the interviews were highly personalized. That said, common ground was evident in some of the challenges faced by each person. Unsurprisingly, access was a shared frustration, as well as a lack of correct knowledge or understanding from able bodied people.

The discussions included conversations about what makes a space joyful and examples of accessible and inclusive public places. We concluded that accessibility relates to what is physically built. Inclusivity is a mindset, a willingness and an empathy by the designer for those who work in or occupy physical space. Consensus emerged around specific high-profile places that were deemed by these users to be successful. These included the Royal Festival Hall, the British Film Institute and the British Museum in Central London. A visit to the Ashton Gate Bristol City Football stadium was also referenced as a seamless experience for a wheelchair user. It was also noted that many of these examples are historic buildings with heritage value, demonstrating that this is not a barrier to accessibility or inclusivity, despite often being used as an excuse for poor user experience.

At Mossbrook School, sensory experiences were created for neurodiverse children, including music activated by the body running, jumping and playing. During the Habinteg interviews, one interviewee spoke enthusiastically about the singing lift at the Royal Festival Hall, Southbank Centre in London. It is the only lift which accesses all six floors of the building and is used by those with specific access needs. As visitors travel up and down in the lift they are accompanied by scales in four-part harmony which correspond with the progression of the lift—a wonderful example of how responding to physical needs can also celebrate neurodiversities in a joyful and engaging way. When talking more generally about lifts and the value of independence, one interviewee who also uses a wheelchair noted that they are unable to turn around once they have entered the lift. If the buttons to operate the lift are only found next to the door and they didn't reverse in, they are unable to use the lift on their own.

Management of spaces and social expectations were also discussed. One interviewee had very specific, medical dietary requirements—inclusive, indoor dining opportunities alongside cultural and civic buildings, where they could bring their own food while allowing friends to purchase something from a vendor, were very hard to come by. One interviewee asked 'where do guide dogs drink water or go to the toilet when their owner visits concert halls or cultural buildings?'

During Covid-19, people were told to 'stay local.' However, it became evident through the discussion that many smaller, local, independent companies were unable to accommodate those with impairments. This was sometimes due to a lack of accessibility, but sometimes a lack of inclusivity and wrong information given prior to a visit. One interviewee did speak highly of the Picture House in Crouch End, North London. A positive of the Covid-19 rules meant that there was more space available in these venues. Also, that table service became much more commonplace meaning those who were unable to order at the often crowded (and high!) bar did not feel excluded, for example.

A heart-warming conversation with one interviewee focused on the needs of family and friends as much as their own—an elderly mother who walks with a stick and a partner with a broken toe. This interviewee uses a mobility scooter and flagged that they didn't experience the same pain or fatigue in scenarios that required lots of walking. A small move such as providing sufficient space next to a bench for a mobility scooter or wheelchair to pause without blocking a thoroughfare allows for a more inclusive experience for all users. All of the interviewees shared the view that a truly joyful space is one where you don't notice it's been designed for accessibility—it is inclusive of all of its visitors' diverse needs.

3.5 Recording and sharing knowledge

The interviews provided a valuable user insight, but the challenge was to explore how to take the information learnt from the interviews so it could be shared in a meaningful and engaging way.

The first option considered was to write up the interviews verbatim (as a good researcher would do); this felt like it would be the most accurate representation of each individual and their experience. But we also feared that few people were likely to read (and genuinely process) that quantity of text. Text that included the 'ums,' 'erms' and general 'chit-chat' which was an inevitable (and enjoyable) part of the interview process but might not make for the most concise, accessible or educational documentation.

We decided that an edited write-up, although now incorporating elements of the recorder's biases, would be more to the point and arguably a more useful resource for the recorder, SWA colleagues and broader peers. But this common way of recording interviews favors those whose strength is reading. This research wanted to explore different methods of communicating engagement to designers and other stakeholders, celebrating variety and diversity within both processes and people; a method that could be of increased benefit to those communicating the information, as much as those receiving it. Regardless of neurology, people learn differently. Some can learn in a variety of formats, others, often autistic people, generally favor one learning style and dyslexic people often find words more challenging, for example.

As Architects and creative, visual learners, we felt we would benefit from something visual, graphic and colorful. As a kinaesthetic learner, it was acknowledged that physically making something would further strengthen the connection to each person's story. The collages that have been created are as much a reflection of how the maker learns and communicates information as they are a representation of the interviews themselves. So a range of collages were the result, and are illustrated in Figs. 22–25

Accessibility takes many forms for many different people and we hope that these collages go some way in demonstrating the individualities and complexities of each person's experience and the importance of meaningful engagement to increase diversity of thought within every design process.

Fig. 22 Collage created based on an interview with Jai. *Images are owned by Sarah Wigglesworth Architects and Clare Bond.*

Fig. 23 Collage created based on an interview with Amy. *Images are owned by Sarah Wigglesworth Architects and Clare Bond.*

By continuing to be curious and investigate how others use, understand and navigate the built environment we hope to assimilate new insights that can help designers such as us improve the experiences of a wider range of people.

3.6 Acknowledging biases and celebrating experience

As authors of design, it is critical that we acknowledge that we all have biases based on our own lived experience, therefore, it must be said that these collages are the maker's interpretation of each interview and it is inevitable that they will have been influenced by their own biases.

Fig. 24 Collage created based on an interview with Andrew. *Images are owned by Sarah Wigglesworth Architects and Clare Bond.*

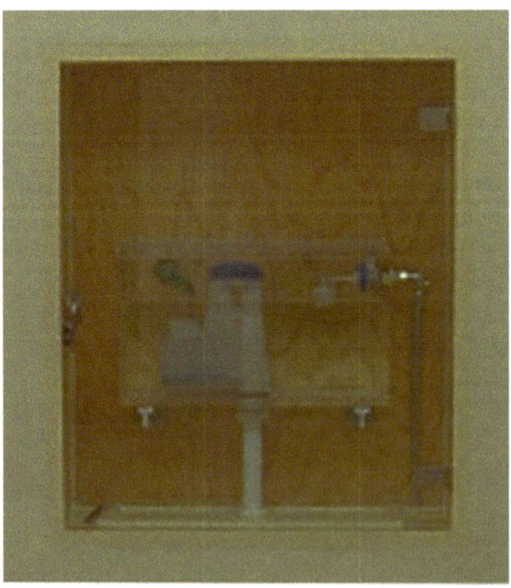

Fig. 25 Collage created based on an interview with Angela. *Images are owned by Sarah Wigglesworth Architects and Clare Bond.*

One might conclude that architecture will never be free of the bias of the designer, documenter or decision maker. Yet the point of seeking different viewpoints is to acknowledge this very fact: that there is no unbiassed truth, only knowledge interpreted through experience. This is the reason it is important to include a variety of voices, so that the designer can acknowledge different outlooks and assimilate them within their creative process. Recognizing that we each have a particular perspective granted us through our personal experience means we need to be alert to the perspectives of others, and make this a consideration in our design work. While it is unlikely that we will always satisfy everyone's needs and desires, it is the designer's role to inform themselves of different views and assimilate the full breadth of knowledge of how their work will be understood and received.

Prevailing processes, theories, regulations and policies contain implicit bias and favor specific groups of people. Yet we rarely see this because it is so normalized in everything we do through our education, training and practice. Inclusionary tactics aim to question this, bringing insights that expose this bias and seek to bring other voices into the equation. Diverse voices will enrich and broaden the relevance and appeal of our work and bring important insights for design. It is hoped that this research and these collages go some way to demonstrate the individualities and complexities of each person's experience and the importance of meaningful engagement to increase diversity of thought within the design process.

4. Conclusion

The two examples showcased here take different trajectories: one is a conventional building project with extended briefing and learning built into the process. The second explores the briefing process itself, aiming to demonstrate how engagement with real lives and experiences can begin to undo the potential negatives (and broad stereotypes) that can flow from a too narrow concept of 'the user.' Both projects illustrate the complications in society's acceptance of the 'universal' human, so often the subject of architectural guides and source books. When considered together, they illustrate how our practice's research works to open our minds to how people other than ourselves experience and understand our work. We think of our practice knowledge as a set of sedimentary layers, one overlaid on another over time which we can access by excavating through the strata. Thus, while these two projects are from very different eras they can coexist and be accessed in a conversation across time and space, allowing us to transfer knowledge in one area (and era) to another.

Critically, we hold that it is principally processes and systems that give rise to the built environment we make, and aesthetics are only part of those systems. By placing importance on thinking about our working method and how we communicate with our clients and users, we ask difficult questions about our own bias—how can this be mitigated and the experience become enriched for all involved? It is the context, who are the people included and what are the questions asked, that will offer the clues to the resulting design? Accordingly, we approach design, as we did at Mossbrook, with a recognition of our lack of knowledge as much as we acknowledge our skillset, and we aim to expand this through engaging with the experiences of those using our work. Documenting our work, be this questioning a received brief in collaboration with users, or through monitoring users' reactions and experiences of our work, or learning how other buildings have played out in diverse lives, helps develop greater empathy and, we believe, ultimately produces more relevant and enjoyable design work.

In considering the particular experiences of individuals engaging with the built environment we acknowledge the specificity of their understanding. Universal Design seeks to embrace all experience and find an accommodation for the whole of humanity. We take a slightly different view. While knowing that all human experience is in some sense specific to the individual and that we will never be able in a single design to address the wide breadth of human experience and engagement with the built environment, we nevertheless believe there is great value in examining unique perspectives in order to challenge mainstream assumptions. We do not strive for the perfect solution implied by Universal Design. Rather, we see these experiments in marginal experience as expanding a horizon in order to question received wisdom. And, in doing so, we hope to address the needs of specific users too.

4.1 '.....but what does an accessible building look like?"

The potential changes that could come about in our built environment by thinking in an inclusive way are wide-ranging and could be manifest in many ways: through organization, dimensions, colors, textures and contrasts, materiality, gradients, visual language, lighting and many, many more, encompassing the design of the whole public realm. These are the tools of design but the product depends on the designer's skill. Unless we get the processes, conversations and policy in place, we will not make changes that could normalize inclusive design. Just as in our response to the climate crisis, the ultimate aim would be for accessible architecture to be so commonplace that it is not even recognized. The question of what an accessible building or environment would look like misses the point: for we are still in the age of learning what this could be, experimenting within the confines of the current context. Huge possibilities could be heralded if we were to think more inclusively and ethically within the framework of inclusivity. The studies illustrate our small contribution to this experiment. It is by no means the end of the story, but just the beginning.

CHAPTER FIFTEEN

Relax and resist: Reflections on the Touretteshero Relaxed Venue methodology

William Renel* and Jessica Thom
Touretteshero CIC, London, United Kingdom
*Corresponding author: e-mail address: will@touretteshero.com

Contents

1. Introduction

Battersea Arts Centre (BAC) describes itself as a home for radical artistic ideas and believes in the power of performance and collective imagination to spark positive change (bac.org.uk). Based in an iconic Victorian building with a radical history in Southwest London, BAC welcome over

Developments in Neuroethics and Bioethics, Volume 5
ISSN 2589-2959
https://doi.org/10.1016/bs.dnb.2022.08.011

160,000 people each year for a rich and ambitious program of live performance, heritage, events and creative and community projects. Supporting people to take creative risks that inspire social change is at the core of what BAC does. In February 2020, BAC launched as the world's first Relaxed Venue, aiming to radically embed new forms of access and inclusivity across all of their activities by utilizing a new "Relaxed Venue" methodology. This methodology was developed by Touretteshero, a disabled and neurodivergent-led Community Interest Company, in partnership with BAC. Using Tourettes Syndrome as a catalyst, Touretteshero's mission is to create an inclusive and socially just world for disabled and non-disabled people through our cultural practice. The authors of this chapter are the Co-Artistic Director and Director of Research at Touretteshero. The authors identify as disabled and neurodivergent. We write with love and solidarity for disabled and neurodivergent people who continue to navigate the inherent ableism engrained in architectural and built environment practices. The chapter is richly informed by disability activism and disability justice as frameworks that center historically marginalized people with lived experiences at the intersecting junctions of oppression. The chapter is autoethnographic (Denshire, 2013) as we share and reflect on our social, political and cultural experiences of developing the Relaxed Venue methodology. The author Jess Thom's name is used within the text to highlight when direct quotes relating to personal lived experiences are shared.

In detailing the process of developing the Relaxed Venue methodology, including both the key moments of change whilst working at BAC and our personal reflections, our intention is not to create a new guide to accessible design or to suggest that taking a relaxed approach is the answer to unpacking the multitude of ways that ableism operates within the context of the built environment. Rather we are interested in new ways of thinking, being and doing within design practices that understand disabled and neurodivergent people's expertise and experiences as currencies of resistance. We believe that these currencies can shake up the normativity still entrenched in how the contemporary built environment is imagined, designed and maintained. New critical paradigms in Neurodiversity Studies and the growing understanding of "cognitive othering" (Rosqvist, Chown, & Stenning, 2020) are important frameworks for understanding the similarities and differences between disabled and neurodivergent people's experiences. Such issues cannot be disconnected from the wider intersections of race, gender, sexuality, age and class all of which affect experiences of marginalization and discrimination and inform acts of creative resistance. Building on the histories and legacies of our disabled and neurodivergent ancestors we propose the Relaxed Venue methodology as new form of anti-ableist resistance. A resistance through which we can forge new understandings of what it means socially, politically and culturally for a diversity of bodies and minds to be supported by designed worlds that value collective acts of liberation.

2. Time to relax

The Relaxed Venue methodology aims to identify and dismantle the barriers faced by disabled and neurodivergent people in accessing the built environment. The methodology builds on the social (Oliver, 1990, 2013) and strong social (Crow, 1996) modes models of disability[1] and the principles of "Relaxed Performances." Relaxed Performances are understood as those which adopt a relaxed approach to noise or movement from the audience, offer a warm welcome to people who may find it hard to follow the conventions of traditional theater etiquette, and that encourage everyone to respond without inhibition (Thom, 2016). In the UK, Relaxed Performances grew out of sensory-friendly film screenings in the 1990s and were formalized by The Relaxed Performance Project in 2012–2013 where eight high-profile venues hosted their first relaxed events (Sedgman, 2018). Taking a relaxed approach—in which normative assumptions about "the right way to act" in public environments are challenged—builds on a legacy of inclusive theater etiquette and practices dating back hundreds of years McNabb, 2016; Sedgman, 2018; Simpson, 2018). The Relaxed Venue methodology takes the principles of a relaxed performance and applies them to all of an organization's systems, spaces, structures and services. The concepts of the normate (Garland-Thomson, 1997), normate template (Hamraie, 2016) and disembodied environment (Imrie, 2010) are all useful in developing a theoretical framework to conceptualize the idea of a Relaxed Venue. Collectively these terms provide useful abbreviations "for the complex, critical notion that the world was designed with normative inhabitants in mind" (Hamraie, 2017: 20). Practically the methodology builds on work concerning the social (re) production of architecture (Petrescu & Trogal, 2017) which aims to question

[1] The social model of disability states that people are not disabled by their impairments but by a societal failure to consider difference in the way that environments, systems and services are designed and maintained. The strong social model builds on the foundations of the social model and considers a 'more complete understanding of disability and impairment as social concepts; and a recognition of an individual's experiences of their body over time and in variable circumstances' (Crow, 1996: 218).

the professional agency of build environment practitioners through radical politics and "offer an alternative trajectory for architecture's production of the social" (Hyde, 2017: 308).

3. Time to resist

There are a series of real-world events and experiences through which the concept of a Relaxed Venue emerged. These took place prior to 2017, when the formal process of developing the methodology began. The first event is a negative lived experience that Jess Thom had when she was 15. Jess has Tourettes, a neurological condition that means she makes involuntary movements and noises called tics. The experience happened on a school trip to the Old Vic theater in London:

> *"It was one of the first places where my body had been placed in an environment in front of my peers where I wasn't able to follow a set of rules about how to act. My tics weren't an issue in the classroom or other familiar environments but at that theatre it was the first-time other people picked on me for rocking, shaking and moving in an unpredictable way. It showed me the intersections of architecture and the medical model of disability. So often architecture seems to be made with a medical, individualized understanding of disability—the 'be grateful and make do' model."*
>
> ***Jess Thom***

Another negative lived experience that informed the concept of the Relaxed Venue methodology occurred when Jess, as an adult, attended a performance of Extreme Rambling by Mark Thomas:

> *"During the interval I was asked if I'd move and sit in the production booth at the side of the stage because some members of the audience couldn't tolerate the noise of my tics.*
> *I'd made sure the theatre and Mark knew I had Tourettes before I arrived. The staff were welcoming and Mark came to meet me before the show. With my permission, he'd explained to the audience at the beginning that I would be making some unusual noises.*
> *When the theatre manager asked me if I'd move, he made it clear that I didn't have to. But inevitably, when I heard people around me had complained and didn't want me to be there, I felt extremely uncomfortable and I agreed to move.*
> *Once in the booth I was hit by a wave of humiliation and sadness. I started to cry. Part of me wanted to leave and never go to the theater again."*
>
> ***Thom, 2011***

In the examples above the built environment becomes exclusionary not because of physical barriers but by normative rules held within the space that prioritize non-disabled and neurotypical ways of being. This highlights the importance of social factors when designing built environments with neurodivergent inhabitants in mind. As Hamraie (2013: 06) notes, "buildings are not simply static structures in which human interaction occurs; on the contrary, buildings actually produce lived and embodied experiences for spatial inhabitants and. .. [therefore] produce and reinforce structural inequalities." Within the contemporary trajectory of Universal and Inclusive Design there are a number of designers who are challenging a depoliticised understanding of accessibility as a checklist or tick box exercise through the notion of "Critical Access Studies" (Hamraie, 2017). Theoretically this work builds on "Crip" and feminist perspectives (Guffey, 2018; Kafer, 2013; McRuer, 2006) that critique the dominant narratives of design for their prioritization of straight, white, male perspectives. Within this context Touretteshero became increasingly interested in new forms of built environment design and management that would support a diversity of bodies and minds to take up space meaningfully and equitably together. In 2017, with funding from Arts Council England's Change Makers programme, Touretteshero and BAC started a new partnership and the Relaxed Venue project formally began.

4. The first relaxed venue

The first step in creating the Relaxed Venue methodology was to develop a bespoke design process through which the methodology could be devised, tested and delivered. Considerations were given to existing design structures such as the Design Council's double diamond (designcouncil.org.uk) and Stanford Institute of Design's d.School's numerous "tools for taking action" (dschool.stanford.edu). Although these existing structures have undoubtably guided a plethora of designers and non-designers through inclusive processes, they have also been critiqued for perpetuating acts of colonialism and displacing "alternative" practices, knowledges, and world views (Akama, Hagen, & Whaanga-Schollum, 2019). The structure through which the Relaxed Venue methodology was developed therefore needed to draw on knowledge from outside of the built environment and design spheres, in particular from the Disability Justice movement and people such as Mia Mingus (leavingevidence.wordpress.com), Alice Wong (Wong, 2020; Disability Visibility, disabilityvisibilityproject.com), collectives such as Sins Invalid (sinsinvalid.org) and the 10 Principles of Disability Justice (Berne, Morales, Langstaff, & Invalid, 2018). Drawing on these disabled-led practices a new

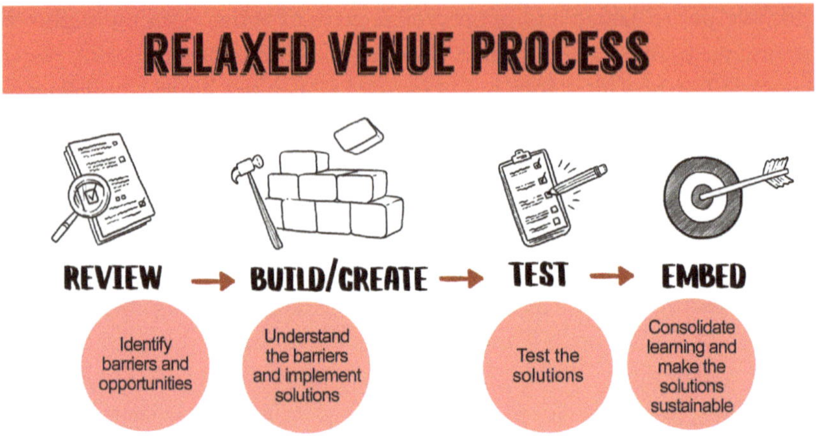

Fig. 1 Relaxed venue—process.

framework for designing a Relaxed Venue was developed, with four key phases: review, build, test and embed (Fig. 1).

4.1 Phase 1—Review

This phase focused on getting to know BAC as an organization and identifying barriers and opportunities. The Touretteshero team worked with staff from across BAC to undertake an accessibility review—identifying barriers and opportunities from across the organization's systems, spaces and services. These were consolidated into a "barrier report" which was then reviewed with the senior leadership team. Key themes within the report included signage, thresholds and doors, circulation, chill-out spaces, resting points, lighting and sound. The barriers were categorized into four areas: statutory, short term, medium term and long term.

4.1.1 Statutory barriers

Statutory barriers are those which fall within the existing legal context of The Equality Act 2010 (legislation.gov.uk). This includes things such as having ramps at a compliant gradient and door frames at a compliant width. Statutory barriers are the things that a venue is required to do by law and in the context of the partnership between Touretteshero and BAC should therefore be dealt with outside of the Relaxed Venue project.

4.1.2 Short term barriers

Short term barriers are accessibility concerns that once identified can be removed immediately—for example the placement of sanitary bins in accessible toilets which are often positioned in a way that stops a wheelchair user from being able to safely make a side transfer onto the toilet. One solution to this is using high contrast tape to indicate where on the floor the bin should be located and talking to the cleaning staff about this change.

4.1.3 Medium term barriers

Medium term barriers are areas of change that require the allocation of specific time and money—for example redesigning a website to make it more accessible. Removing these barriers requires discussion, planning and a strategy. These are the barriers that were prioritized during the development of the Relaxed Venue methodology at BAC and formed the basis of the build, test and embed phases of the project.

4.1.4 Long term barriers

Long term barriers are those which need to be addressed strategically by a venue over a longer time scale and/or with additional funding—for example identifying space for and installing a Changing Places toilet facility (changing-places.org).

4.2 Phase 2—Build

During the second phase of the process a series of multidisciplinary non-hierarchical teams were created, bringing together staff from across BAC (from front of house to senior producers and the artistic director). Each team worked with Touretteshero to consider a distinct category from within the barrier report, for example "signage" or "recruitment." Each group was assigned a clear remit of work and budget and subsequently worked with Touretteshero to understand the barrier and imagine, implement and review solutions. During the build phases six primary areas of change were identified: digital, structural, creative, community, physical and emotional (Fig. 2). Giving equal attention to structural and creative, physical and emotional changes was an important framing early on in the Relaxed Venue project. This approach builds on the notion of "architecture in action" described as:

> *"An architecture that is in and among the actions of living, of activism, of social practices ... that is engaged with the politics of the place in which one lives, an architecture that questions the rules of current architectural and urban practices to promote new ways of working."*
>
> **Petrescu & Trogal, 2017: 11**

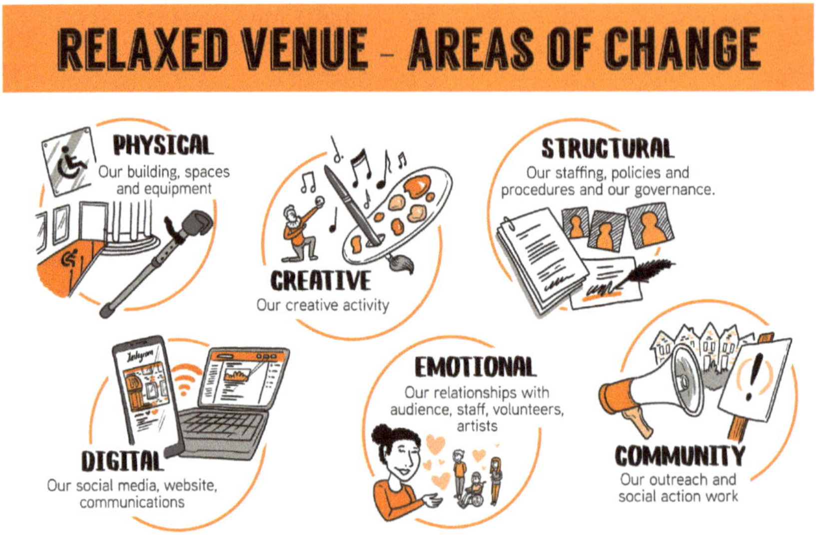

Fig. 2 Relaxed venue—primary areas of change.

4.3 Phase 3—Test

During this phase the solutions for barriers implemented in the build phase were tested through a mixed methods approach. This involved consultation in person and via surveys with key stakeholders such as audiences and staff. To ensure the project was informed by additional lived and professional expertise a dynamic network of critical friends was established to support the design process. This network included disabled and neurodivergent people, accessibility consultants and disabled-led organizations including Shape Arts (shapearts.org.uk), VocalEyes (vocaleyes.co.uk) and Attitude is Everything (attitudeiseverything.org.uk). Acknowledging the problematic histories of disabled people's knowledge being undervalued and their labor being unpaid (Brown & Leigh, 2018; Inckle, 2018; McRuer, 2006), the network of critical friends was paid for their time and an additional access budget was provided to support their involvement. Once the consultation with stakeholders was complete and the multidisciplinary teams were confident that the design solutions were working there were four key moments of "stress testing" the building at different scales.

4.3.1 Stress testing the building

Stress testing the building in the context of the Relaxed Venue project was a process that aimed to strategically bring higher numbers of disabled and neurodivergent people into the BAC building to work in different ways for sustained periods of time. Similar to the concepts of "lived experience iteration" in Inclusive Design (Langdon, Clarkson, & Robinson, 2010) and "experience-based co-design" in participatory research (Donetto, Pierri, Tsianakas, & Robert, 2015), the intention of the stress tests was to examine the usefulness of the designed solutions by bringing people with a range of lived experiences and access requirements into real-world public scenarios. Feedback after each stress test was integrated into the design process and iterated over an 18-month period. The first stress test was having two disabled-led companies performing shows in the BAC building at the same time. The second was a residency in which three disabled people and disabled-led companies were simultaneously making new work in the building. The third was a large-scale creative event where 600 people (including approximately 175 disabled children and young people) came to BAC over 2 days (see Fig. 3). The final stress test was a two-week relaxed festival showcasing the work of disabled artists with a large number of disabled audience members attending and using the building over a sustained period of time. 40 artists contributed to the relaxed festival of which 85% identified as disabled and/or neurodivergent. The stress testing process enabled issues to be highlighted that had not been considered in previous phases of the project, or by BAC historically—for example, questioning why a fire risk assessment states that a maximum of 5 wheelchair users

Fig. 3 Brewing in Battersea event. *Photo by: Kevin Moran Photography.*

can be upstairs in the building at any given time. Policies such as this show the assumptions made about who will or won't access an environment, creating a politics of space and place (Certoma, Clewer, & Elsey, 2012) in which disabled and neurodivergent people continue to be labeled in the problematic category of "extreme users." The stress testing process demonstrated assumptions and rules that weren't previously understood as barriers because they related to scenarios that had historically never happened.

4.4 Phase 4—Embed

The goal of the final phase was to collate the learning from the three previous phases and find strategies to make the designed solutions as consistent and sustainable as possible. At this stage in the process there were multiple accessible solutions in place that had become part of BAC's standard practice. This included consistent subtitles on social media and promotional content, the permanent opening of a dedicated chill-out space for staff and visitors and the redesigned accessible seating and tables in the cafe. However, there were also elements of the work that were not completed within the timeframe. This resulted in some key recommendations for BAC and Touretteshero moving forward. These outstanding elements emphasized the reality that accessibility is not an end-goal but rather something that changes and evolves over time. During the embed phase, BAC staff, the Touretteshero team and key stakeholders such as audience members and artists worked together to review key achievements and challenges from the Relaxed Venue project. These were categorized within the six primary areas of change: digital, structural, creative, community, physical and emotional.

A key component of the embed phase was to recruit an Inclusive Practice Manager to take the project forward after the four phases had been completed. Whilst the expectation at BAC was that the responsibility for accessibility and inclusivity is shared amongst the entire staff team, having someone on the senior management team with the necessary skills and experience to support, reinforce and advocate for this work was an enormous asset.

5. Moments of change

There were a number of key moments of change within the development of the Relaxed Venue methodology; some were tangible examples of change and others were moments when change was visible. These moments are outlined below.

5.1 Dealbreakers and stopping the process

One of the reasons that 2017 felt like the right time to start the Relaxed Venue project, was that BAC was in the midst of a large-scale capital project and the rebuilding and reopening of their largest performance space The Grand Hall after a devastating fire in 2015. These elements meant that for the first time in the building's history BAC had been transformed from a previously non-accessible town hall to a fully wheelchair accessible environment. A lot of that access was dependent on a single passenger lift that was also used to move props and performance sets around the building. The newly installed lift did not work during the early stages of the Relaxed Venue project. Without a working lift Jess (who uses a wheelchair) could not access her desk on the top floor of the building or invite other disabled people who needed the lift to work with her at BAC.

> *"My feeling at that moment was that disabled people have waited long enough to access this type of opportunity, it's important that we don't compromise by limiting my (and other disabled people's) work to a separate working space on the ground floor. Being able to access the building equally and without stress was a dealbreaker, it wasn't possible, so we stopped the project."*
>
> ***Jess Thom***

This moment in the process highlighted that although the aim of the Relaxed Venue methodology is to move beyond the established perspectives of what an accessible environment is, this process is only possible when certain dealbreaker accessibility considerations have been made.

5.2 Powerful ideas and training

As part of the Relaxed Venue project, Touretteshero delivered training to every member of staff at BAC, from bar staff to producers and the senior leadership team. As well as discussions of the different models of disability and disability equality, these training sessions introduced staff to powerful disabled-led thinking from across disability arts, activism and culture—for example the concept of "forced intimacy," a term coined by Mia Mingus to "refer to the common, daily experience of disabled people being expected to share personal parts of ourselves to survive in an ableist world" (leavingevidence.wordpress.com). This training helped to create a shared understanding and vocabulary amongst all BAC staff and increased opportunities for staff to think and learn collectively about how to take an anti-ableist approach to their different areas of work.

5.3 The scratch hub

During the development of the Relaxed Venue methodology, BAC was also designing a co-working environment in the basement of the building called "The Scratch Hub." Touretteshero were asked to consult on accessibility concerns during the Scratch Hub design process. Three members of the Touretteshero team and the Artistic Director of BAC used early proposals and drawings of the Hub to inform a speculative design (Dunne & Raby, 2013) process. The team created different use scenarios in which a diverse group of people would navigate and work together in the space. This process acknowledged what Tharp and Tharp (2018) describe as a "relentless normativity" where disabled, divergent and non-normative experiences remain absent in the imaginaries of many contemporary designers. During this process Touretteshero created a series of critical questions for the architects. These questioned were designed to encourage the architects to imagine a diversity of bodies and minds undertaking a diversity of activities within the Scratch Hub space. The questions included, can two wheelchair users make a cup of tea in the kitchen and then sit and have a meeting together? Can a wheelchair user get something from the storage cupboard? Can a visually impaired member of staff let themselves into the Scratch Hub and get to their desk independently? The answer to all of these questions when initially asked was "no" and therefore the plans were redesigned to make the space more accessible and to ensure that multiple disabled and neurodivergent people could access and work in all of the Hub spaces.

5.4 Leveraging change inside and out

During the Relaxed Venue project there was an increasing disconnect between the inclusive changes being made inside BAC and the physical environment directly outside the building. Outside of BAC during the project were a series of uneven paving stones which posed a barrier and risk to wheelchair users and other disabled people. However, these paving stones were not within BAC's control.

> *"The flagstones outside were a growing issue; I had even damaged my wheelchair on them and it felt like the inside and outside of BAC were two different worlds at that time."*
>
> ***Jess Thom***

Using the Relaxed Venue process as leverage, Touretteshero and BAC successfully lobbied Wandsworth Council who agreed to remove the paving

Fig. 4 Chill-out Space at Battersea Arts Centre, designed by Rhiannon Armstrong.

stones and repave the street outside to make it flat and accessible. This process highlighted how improving the accessibility of a specific space can be used as a catalyst for change and have positive repercussions for neighboring places and spaces.

5.5 Chill-out room

Part of the Relaxed Venue project was creating a permanent chill-out space at BAC for visitors and staff (see Fig. 4). The space was designed by interdisciplinary disabled artist Rhiannon Armstrong and includes handmade weighted blankets, cushions and mats, low-level indirect lighting, the colors of which can be controlled by the user, and a range of different accessible seats and resting areas.

Initially BAC had concerns about losing the small meeting room identified to become the chill-out space but through conversations with Touretteshero, BAC realized that they were not losing a meeting room but instead gaining a chill-out space. When the space opened it had an immediate positive impact on organizations wanting to hire space at BAC as a dedicated chill-out space was not something that many other commercially available environments were offering. Shortly after the chill-out space opened, artists and staff members were using the space regularly in different ways and the feedback was extremely positive. Not only was the space designed to support anyone for whom a chill-out space is an access requirement, it also sent a message of love and solidarity to other disabled and neurodivergent people. As Hamraie notes "from a doorframe's negative space to the height of shelves and cabinets, inhabitants" bodies are simultaneously imagined, hidden, and produced by the design of the built worlds' (Hamraie (2017): 19).

> *"The chill-out space at BAC is essentially a love letter to disabled people in the shape of a room. I think this is a radical way of approaching what an accessible environment might be and the message it sends to disabled and neurodivergent people is one of love and care. I would like to challenge architects to think about how their buildings might send other love letters to disabled and neurodivergent people in the future."*
>
> **Jess Thom**

6. Three relaxed commitments

In reflecting on the development of the methodology, we hope to create more opportunities for architects, built environment specialists and researchers across multiple disciplines to identify and remove barriers in the built environment in new and innovative ways. During the process of creating the methodology at BAC, three key commitments emerged: "No New Barriers" "Equality of Experience" and "Reduce Fuss" (Fig. 5). These commitments are significant as they propose a way of designing and managing an environment that centers disabled and neurodivergent people's experiences and challenges the limiting parameters set out in accessibility legislation, regulations and policy.

6.1 No new barriers

This commitment accepts that you may be working in an environment in which you inherit disabling barriers and that designing an accessible space may be at odds with maintaining heritage features. However, this doesn't dictate that the environment can never be accessible or that change isn't possible, it means that extra care and attention need to be given to accessible solutions. A commitment to no new barriers means considering access at every stage of any future project and to creating no new disabling barriers within the systems, spaces and structures of the future.

> *"If every architect, committed to no new barriers from tomorrow, in everything that they made, then very quickly you would radically change the landscape of our built environment".*
>
> **Jess Thom**

Fig. 5 Three relaxed commitments.

Importantly, barriers in this context are not just the physical and structural components of a space such as stairs and toilets, but also sensory, attitudinal and communicative barriers that may be produced and reenforced in the design of an environment. The six areas of change defined within the Relaxed Venue project—digital, structural, creative, community, physical and emotional—are a useful starting point for thinking about barriers beyond physical and visual concerns. Physical and visual barriers to the built environment continue to dominate accessibility legislation and policy, driving an approach to "designing disability" (Guffey, 2018) that leaves the requirements of many neurodivergent people behind and fails to account for the reality of intersectional lived experiences. As Hamraie (2017) notes, how we approach the intersections of disability with race, gender, class, and aging *matter* for questions of belonging and justice in the built environments of the future.

6.2 Equality of experience

Equality of experience underlines the importance of giving equal attention and thought to the different elements of how someone might experience an environment. This does not mean that the goal is for everyone to experience exactly the same thing. Instead the goal is to value different experiences equally and to ensure that the way in which an environment is designed does not create exclusionary and unequal ways of being and doing for disabled or neurodivergent people. Equality of experience is an attempt to situate the ideals of "cross-disability solidarity" and "collective access" established in the 10 Principles of Disability Justice (Berne et al., 2018) within the context of the built environment. Therefore, designing towards an equality of experience means challenging ableist and normative ways of doing which isolate different communities, valuing the experiences of a diversity of bodies and minds and creating environments in which communities can co-exist equally through processes of collective liberation.

6.3 Reduce fuss

Reduce Fuss acknowledges the accumulative impact that experiencing barriers has on individual and community experiences of the built environment. By committing to reducing fuss around access requirements a venue is ensuring that requirements are understood, embedded and met as seamlessly as possible.

> *"One of the things that is pervasive as a disabled neurodivergent person is that if you explain some of the barriers you experience people think you're being picky and pernickety. For example, having to wait before entering a building because the staff don't know where the key is for the stairlift. By seeing this as picky what is missed is the accumulative impact of experiencing these things and the messages that this sends out about who is or isn't valued and who is expected within a space."*
>
> ***Jess Thom***

Designing for reduced fuss therefore means not only considering the structural barriers within a built environment but also the psycho-emotional barriers—those which disabled and neurodivergent people experience through processes of internalized oppression produced and emphasized by ableist ideologies and practices in design (Reeve, 2012; Thomas, 1999).

The three relaxed commitments may seem simple but we believe they have the potential to reframe how accessibility is conceptualized within a design process and to challenge normative assumptions about who can and will be present in the designed worlds of the future.

7. Conclusion

Taking a relaxed approach or working to become a Relaxed Venue does not mean sitting back and going with the flow or passively accepting that things cannot change. On the contrary it means thinking actively, critically and creatively about taking up space together. It means acknowledging the legacies of resistance from the disabled and neurodivergent people and movements of the past. The Relaxed Venue methodology begins to challenge a depoliticised understanding of accessibility as a check-list or tick-box exercise, an end goal that can be achieved and completed.

But the work has only just begun. The decisions made about what to design, how to design it and who will use it are not value-neutral but bring to life different visions and possibilities of worlds that can exclude and oppress or nurture and liberate. We hope in sharing our experiences of developing the Relaxed Venue methodology with you we can sound a call to action for new movements and environments that are designed to support a diversity of bodies and minds to take up space meaningfully, ethically and equitably together.

References

Akama, Y., Hagen, P., & Whaanga-Schollum, D. (2019). Problematizing replicable design to practice respectful, reciprocal, and relational co-designing with indigenous people. *Design and Culture, 11*(1), 59–84. https://doi.org/10.1080/17547075.2019.1571306.

Attitude is Everything. www.attitudeiseverything.org.uk.

Battersea Arts Centre. www.bac.org.uk.

Berne, P., Morales, A. L., Langstaff, D., & Invalid, S. (2018). Ten principles of disability justice. *WSQ: Women's Studies Quarterly, 46*(1), 227–230. https://doi.org/10.1353/wsq.2018.0003.

Brown, N., & Leigh, J. (2018). Ableism in academia: Where are the disabled and ill academics? *Disability & Society, 33*(6), 985–989. https://doi.org/10.1080/09687599.2018.1455627.

Certoma, C., Clewer, N., & Elsey, D. (Eds.). (2012). *The politics of space and place.* Newcastle: Cambridge Scholars Publishing.

Changing Places. www.changing-places.org.

Crow, L. (1996). Including all of our lives: Renewing the social model of disability. In J. Morris (Ed.), *Encounters with strangers: Feminism and disability.* London: Women's Press.

Denshire, S. (2013). Autoethnography. *Sociopedia, 12*(1), 221–236. https://doi.org/10.1177/205684601351.

Design Council. www.designcouncil.org.uk.

Disability Visibility. www.disabilityvisibilityproject.com.

Donetto, S., Pierri, P., Tsianakas, V., & Robert, G. (2015). Experience-based co-design and healthcare improvement: Realizing participatory Design in the Public Sector. *The Design Journal, 18*(2), 227–248. https://doi.org/10.2752/175630615X14212498964312.

Dunne, A., & Raby, F. (2013). *Speculative everything: Design, fiction, and social dreaming.* Cambridge, Massachusetts: MIT.

Garland-Thomson, R. (1997). *Extraordinary bodies: Figuring disability in American culture and literature.* NY: Columbia UP.

Guffey, E. (2018). *Designing disability: Symbols, space, and society.* NY: Bloomsbury Academic.

Hamraie, A. (2013). Designing collective access: A feminist disability theory of universal design. *Disability Studies Quarterly, 33*(4). Available: http://dsq-sds.org/article/view/3871/3411. [Accessed 01.04.22].

Hamraie, A. (2016). Universal design and the problem of "post-disability" ideology. *Design and Culture, 8*(3), 285–309.

Hamraie, A. (2017). *Building access: Universal design and the politics of disability.* Minneapolis, MN: University of Minnesota Press.

Hyde, R. (2017). Ways to be public. In D. Petrescu, & K. Trogal (Eds.), *The social (re)production of architecture: Politics, values and actions in contemporary practice.* London: Routledge.

Imrie, R. (2010). Disability, embodiment, and the meaning of home. In *Toward enabling geographies: 'Disabled' bodies and minds in society and space.* Burlington, VT: Ashgate.

Inckle, K. (2018). Unreasonable adjustments: The additional unpaid labour of academics with disabilities. *Disability & Society, 33*(8), 1372–1376. https://doi.org/10.1080/09687599.2018.1480263.

Kafer, A. (2013). *Feminist queer Crip*. Bloomington, IN: Indiana UP.

Langdon, P., Clarkson, J., & Robinson, P. (Eds.). (2010). *Designing inclusive futures*. NY: Springer.

McNabb C. H. (2016). Relaxed Performance and Shakespeare. Touretteshero, blog. Available: https://www.touretteshero.com/2016/07/11/relaxed-performance-and-shakespeare/ [Accessed 17.08.2022].

McRuer, R. (2006). *Crip theory: Cultural signs of queerness and disability*. NY: NYU Press.

Mia Mingus. www.leavingevidence.wordpress.com.

Oliver, M. (1990). *The politics of disablement*. Basingstoke: Macmillan.

Oliver, M. (2013). The social model of disability: Thirty years on. *Disability & Society., 28*(7), 1024–1026. https://doi.org/10.1080/09687599.2013.818773.

Petrescu, D., & Trogal, K. (Eds.). (2017). *The social (re)production of architecture: Politics, values and actions in contemporary practice*. London: Routledge.

Reeve, D. (2012). Psycho-emotional Disablism. In N. Watson, A. Roulstone, & C. Thomas (Eds.), *Routledge handbook of disability studies, edited by*. London: Routledge.

Rosqvist, H. B., Chown, N., & Stenning, A. (2020). *Neurodiversity studies: A new critical paradigm*. London: Routledge.

Sedgman, K. (2018). *Theatre etiquette, behaviour policing, and the live performance experience*.

Shape Arts. www.shapearts.org.uk.

Simpson, H. (2018). Tics in the theatre: The quiet audience, the relaxed performance, and the Neurodivergent spectator. *Theatre Topics, 28*(3), 227–238. https://doi.org/10.1353/tt.2018.0046.

Sins Invalid. www.sinsinvalid.org.

Tharp, B.M, and Tharp, S, M. (2018). Discursive design: Critical, speculative, and alternative things. Cambridge, Massachusetts: MIT.

Thom, J. (2011). *Wall of sound. Touretteshero, blog*. Available: https://www.touretteshero.com/2011/05/26/wall-of-sound/ (accessed 01.04.22).

Thom, J. (2016). *Relaxed performances—the FAQs. Touretteshero, blog*. Available: https://www.touretteshero.com/2016/03/16/relaxed-performances-the-faqs (accessed 01.04. 2022).

Thomas, C. (1999). *Female forms: Experiencing and understanding disability*. Buckingham: Open University Press.

VocalEyes. www.vocaleyes.co.uk.

Wong, A. (2020). *Disability visibility: First person stories from the twenty-first century*. NY: Vintage Books.

CHAPTER SIXTEEN

Chronic pain & chronic illness: A Crash Course in Cloudspotting

Raquel Meseguer Zafe*
Founder, Unchartered Collective and PMStudio Resident, Watershed, Bristol, United Kingdom
*Corresponding author: e-mail address: uncharteredcollective@gmail.com

Contents

In 2016 I (accidentally) set off a security alert by lying down on the sixth floor of the Southbank Centre, London. I was 9 years into my chronic pain experience, and had just begun to define my need to rest as something other than shameful. I had begun to notice the poetry that is (sometimes) possible in the pause and to see public rest as a practice. I had begun to challenge myself to find the most exquisite places to lie down and was excited by the idea of mapping cities in this way.

Nine years may seem like a long time. I spent many years in what Alison Kafer calls "curative time" (2013, p. 27) pursuing a return to my pre-injury capacity—as was expected of me, but I admit I was hell bent on it. Then I spent time in what I call my "wilderness years," trying to figure out how to live well with pain, shedding ideas of what my life should look like in my 30s. This was time I spent with the writings of Lorde (1996) and Hedva (2016). Time I was the absent family member. Time I spent in bed,

Developments in Neuroethics and Bioethics, Volume 5
ISSN 2589-2959
https://doi.org/10.1016/bs.dnb.2022.08.001

wondering how I might have agency from that site (Meseguer Zafe, 2022). For me, it has been by enacting the (at times) awkward practice of lying down in public that enables me to be in the world: I lie down on trains so I can see my family and travel for work; I lie down in galleries and theaters so I can have art in my life. Put simply, if I don't lie down in public, my world can get very, very small.

On that day in 2016, I had found a quiet spot up on the sixth floor of the Southbank Centre. I had rolled out my little mat. I had deliberately placed my shoes next to one another tidily to signal intention rather than misbehaving. I had my earphones in. I was listening to music and giving in to the floor, which helps to reset my pain levels, when a security guard tapped me on the shoulder and told me that I wasn't allowed to lie down. He couldn't tell me why it was okay for the young man a few meters away to sit on the floor with his laptop, but not okay for me to lie down. All he could say was that he'd been "alerted." After telling him about my condition his team conferred over radio and eventually made the first aid room couch available to me for 40 min. Which was a solution of sorts. But I had gone to an art center, and I didn't want to be hidden away in a windowless room that smelt of TCP. It made it clear to me that the horizontal body is not welcome in public, that the horizontal body is seen at worst as a threat, and at best as an inconvenience.

1. Public interior and crip spaces

The kick-back I get when I rest in public might be to do with a cultural squeamishness about private needs showing up in public spaces. In her article *Place of Safety* (2021), Khensani de Klerk insists on a collective and public "home." Writing as a black woman in Cape Town she envisions "*a public interior, where a sense of home is not exclusively private and where access to the mountain and the ocean create an atmosphere of well-being, host gathering, and allow for solitude without fear.*" De Klerk is writing about the natural resources of Cape Town, but her call for collective and equitable access to the assets of a place, be they green spaces, or theaters, or galleries, resonates with

"The world is simply not designed for me." *Graphic Design: Peter Moffat.*

me. As does her notion of a "public interior," which might be at the heart of the spaces I have been attempting to create in my theater work and advocacy.

Since that day in 2016, every space I have created has attempted to challenge the etiquette of our public space, to make horizontal bodies more visible, to re-frame and elevate acts of rest and to convey something of the lived experience of a 'invisible' disability like chronic pain. Informed by the work of disabled artists and scholars that have gone before me, like Liz Crow and her piece "Bedding Out" and Joanna Hedva's "Sick Woman Theory," I think of this as a *crip* practice and of creating *crip* spaces: pockets of non-normative time and space, where bodyminds (Price, 2014) are welcome to rest. Crip is a term reclaimed in disability scholarship and disability justice movements. It denotes a desire to "jolt people out of their everyday understanding of bodies and minds, of normalcy and deviance" (Kafer, 2013, p. 14). Below, I will discuss one *crip* resting space in detail, and then address the qualities and textures of the spaces I would like to see in the world.

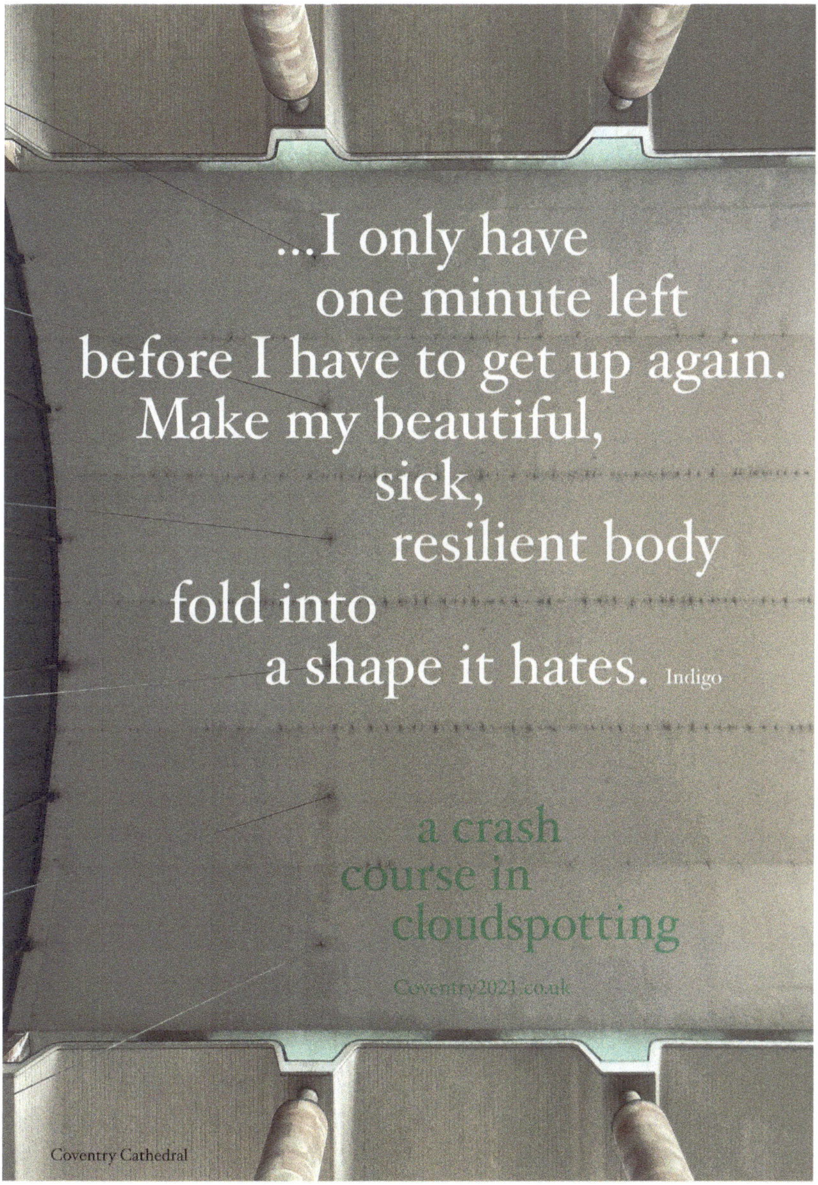

"I only have 1 min left before I have to get up again. Make my beautiful, sick, resilient body fold back into a shape it hates." *Graphic Design: Peter Moffat.*

2. A Crash Course in Cloudspotting

After triggering that security alert, I began to collect other people's stories about their attempts to rest in public. So far I have gathered nearly 300 stories from people who experience chronic pain, chronic fatigue, and other disabilities and neurodivergence. I have found these stories to be frank and truthful telegrams about what it is like to navigate the world in a body that is hurt or fatigued. Some are pithy one liners, others are longer meditations on how public rest has been integrated into people's lives, or double lives:

> *"...as a child, I didn't know what was wrong, and was very invested (well into adulthood, actually) on hiding my difficulties and maintaining the illusion of 'normal.' One of my main tools for coping was Toilet Floors.*
>
> *Many, many times a day I would embrace the cool tiles, my body pressed against torn scraps of toilet paper, my head next to ambiguous pools of water. I clung to the foulness, because it was still the better option than finding myself horizontal in front of others. I lay there, and for a small precious moment, I was permitted to inhabit my reality. I lay there, and I longed for the end of the day and home and the luxury to lie when I needed to; the luxury not to have to lie, about needing to lie."*
>
> ***Extract from Toilet Floors by A. Berry***

These resting stories are the narrative thread of an audio installation and live performance called *A Crash Course in Cloudspotting*. The piece challenges unquestioned assumptions of vertical culture, and invites an audience of 10 to engage in a collective act of rest: to lie down together in public and listen to an audio journey where the resting stories duet with an ever morphing soundscape. The installation itself is created by a circle of overlapping 3 m high fabric panels, suspended from the ceiling and floating slightly above the ground. Inside the installation are eight 5 in. thick mats, topped in midnight blue surge wool, 2 rostra both roughly the size of a small single bed, and a wheelchair space. It is a soft and ever-shifting space.

A central aim of the piece is to make hidden acts of rest visible, tangible and present for audiences. To do this we use light as a medium for presence and have developed technology that allows 48 remote participants to communicate their rest with the audience via an App. When you press 'start resting' on the App, your designated hexagon on the screen lights up, and a light in the physical installation switches on. We later developed the technology to simultaneously trigger a sound loop, so that these acts of rest create the soundtrack our audience encounter when entering the space. The lighting and music are therefore the live and truthful communication of our participants' rest.

Image for *A Restful City* App. *Photo credit Tom Metcalfe.*

Something we hadn't anticipated in the making of this piece, is the extent to which lying down changes our sense of self and can create a different site and mode of spectatorship. In this 'restful spectatorship' audience were guided into a more open and vulnerable mode, in which stories seemed to land differently in the body. Afterwards, audiences spoke of their felt sense of the situations described, and of embedding the stories in their own map of 'the city.' For example, one story takes place in a local cinema where a participant-author has negotiated the possibility to lie down with the cinema management. But when they do so, an usher puts black and yellow hazard warning tape around their body. Several people talked about how vividly they pictured this happening at their local cinema.

3. Poetry in the pause

I would argue that rest enables a different mode of being and of experiencing the world. It might be said to awaken a more poetic orientation. For me, this includes dropping into a deeper sense of time; attending to the moment with softness and care; a steadiness that comes with seeing things as they are; the everyday detail and beauty that can come into sharp focus; the epic in the everyday.

Poster for Cloudspotting. "Because she hurts she zooms out, right, right out." *Graphic Design: Peter Moffat.*

To collectively pause can also enable a different mode of togetherness. In summer 2021 I co-curated a *Rest Improvised* event at Arnolfini Gallery Bristol UK, with Gaylene Gould, Creative Director of The Space to Come. The audience were invited to lie down in the main gallery at twilight and to choose a painting to rest with. They were guided by my voice and by Jamie McCarthy's music, as we improvised in response to the Frank Bowling exhibition *Land of Many Waters*. When the event finished the room was hushed and still. First, people found their friends. There were long hugs, and some tears. Then people talked about the painting they had chosen to rest with, and the thoughts and memories that had been evoked. They said things like *"I have never spent that long with an artwork. Why have I never done that?"* The event undid the norms of a contemporary art gallery, and created the possibility for a different experience of art and togetherness.

If we agree in the poetry in the pause, could we set out to design the most beautiful and exquisite resting spaces? Spaces that aren't hidden away, but that are proud and central parts of our cityscapes? Spaces that celebrate difference, and challenge cultural attitudes toward people who present differently in public? In this city people could lie on warm wooden structures to experience the city skyline from a new perspective or come together to marvel at a full moon. Bus stops have hammocks. Libraries have chaise-longs and plush red carpets that adults (as well as children) are free to roll around on. A gallery exhibit hangs on the ceiling beneath tatami mats and cushions. Theaters have a front row of bean bags. Cafes have low benches and tables for horizontal conversations. These kinds of spaces would enable a different way of being together and knowing one another.

To create true resting spaces takes care. It requires humility and empathy to think about how bodies feel in a space. First and foremost the space would need to be safe, warm and inviting. Many people rest best in quiet, tranquil spaces, but others rest best when given the permission to make noise and to make movement. Many people prefer dimly lit spaces that feel somewhat cocooned. There can be a tension between the safety we associate with bright, open, busy spaces and these preferences. And indeed in the difference between seemingly opposing preferences. But this can be overcome with commitment and leadership. For example, Jess Thom's Relaxed Venue Methodology made Battersea Arts Centre the first Relaxed Venue in the World. This extends an invitation to audiences to be relaxed about the noise or the silence, the stillness or the movement they need to be most comfortable. The methodology also includes the first permanent Resting Space in the UK, designed by neurodivergent artist Rhiannon Armstrong. Opened in 2019, this space has an array of cushions and mats to choose from, and has a door which can be closed if deep rest is needed.

A good place to begin this enquiry would be to rest in public yourself—to spend some time exploring how your body feels in different spaces, where and why you are able to rest, and where and why you are not. This would develop an embodied approach to designing our public spaces.

Poster for Cloudspotting "I wonder if you will stay awhile." *Graphic Design: Peter Moffat.*

4. Crip wisdom & Crip futures

In her book *Queer, Feminist, Crip,* Kafer writes damningly about how we write disability out of our futures (2013). We focus on cure and prevention, and posit the absence of disability as *progress.* In this deficit model of disability my condition is framed as a tragic burden to society and something to be fixed, something to be "conquered" (see for example, Borsook "A Future Without Chronic Pain" 2012, which has three separate mentions of "conquering" chronic pain). Some neuroscience further pathologizes my condition by describing the maladaptive changes in the structure of the brain and nervous system.

Shorters (2022) talks eloquently about the dangers of deficit framing communities. Although often with good intentions, deficit framing erodes the dignity of those communities:

"...if I'm a Black person, I'm taught to believe that I must deficit-frame my people. I must dramatize the disparity, right? And then the behavior is to then stigmatize my people so that I can attract resources, right? Well, this culture of denigration for dollars means that, yes, you'll attract the resources, but you do so by writing your population into the public consciousness as inferior, as ineffective, as pathological."

Shorters advocates for what he calls 'asset framing', which places the aspiration of a person or community forefront in your mind, as opposed to the cognitive trap of deficit framing. In more colloquial terms, I have a colleague who will only describe their disabilities if you also listen to *all* their abilities. And in my own work I insist on dignity while I tread the line between acknowledging the difficulties of my condition (I do not shy away from that), and articulating the particular ways I experience the world and the crip wisdom I have gained. This crip wisdom centers on spaciousness; kindness; a poetic orientation to the world; creative ways to move with pain; an understanding that things take the time they take and of my lack of control; and the ability to define my own markers of success. This wisdom has (eventually) meant I live my life with chronic pain, far better than I lived my life before chronic pain.

5. The resting spaces network and restful cities

Since 2017 I have been advocating for a Resting Spaces Network, and seeking to map cities for rest friendly spaces. While this could be someone's favorite bench, pew in a church, sofa in a cafe, or a department store where the staff are relaxed about people lying on the beds, ideally this would be a network of galleries, theaters and other maintained public spaces who would declare themselves rest friendly. In this vision, no one would ever be more than 15 min away from a place where they are welcome to rest, in their city center.

Change comes slowly. *A Crash Coure in Cloudspotting* aims to affect (or exquisitely infect) the culture of the hosting theater or gallery. Each time we set up at a new venue, we hope some of the things put in place for the installation and performance, remain after we have left. In my hometown of Bristol I have also worked with local venues to create horizontal events. For example, Watershed now runs Horizontal Cinema, where you can request a bean bag or a yoga mat and pillow for any screening of any film.

As part of *A Crash Course Cloudspotting* at Coventry UK City of Culture 2021, we ran an event called Walk + Talk + Rest with local charity Grapevine, who work with people experiencing isolation in Coventry and Warwickshire. At the event 40 people gathered to lie down in the main square, the Cathedral and the Herbert Art Gallery where my story and the story of several members of Coventry Youth Activists were read out. Attended by the Mayor of Coventry, we issued a challenge to Coventry to become the first Restful City in the UK. Since then, Bristol City Council have funded my project *Toward a Restful City* which will run Horizontal Events and begin to map Bristol for rest friendly spaces. My aim is how possible it is to shift the culture of a city, through these projects and challenges.

6. Conclusion

In this chapter, I hope to have conveyed something of all we miss when we disdain rest and the horizontal, and to have set curious minds off on adventures into more embodied public design. To seal the point, I will leave you with a final resting story:

> *I lay down in the Guggenheim in Bilbao while my husband went off to explore other galleries. When he came back I was surrounded by members of the public who had copied me and lain down to appreciate the architecture of the building from a new perspective.*
>
> ***A New Perspective, Anonymous***

References

Borsook, D. (June 2012). *A future without chronic pain, Cerebrum.* https://www.ncbi.nlm.nih.gov/pmc/articles/PMC3574803/.

Gould, G. *The Space to Come.* https://www.thespacetocome.com.

Hedva, J. (2016). Sick Woman Theory. *Mask Magazine.*

Kafer, A. (2013). *Queer, Feminist, Crip.* Indiana University Press.

Lorde, A. (1996). *Audrey Lorde Compendium.* Pandora.

Meseguer Zafe, R. (2022). The potential and poetics of rest. In B. Wihstutz, & E. Backhausen (Eds.), *Out of TIme? Temporatlity in Disability Performance.* Routledge.

Price, M. (2014). *The Bodymind Problem and the Possibilities of Pain. Hypatia.* https://cpb-us-w2.wpmucdn.com/u.osu.edu/dist/3/41686/files/2017/01/Price-The-Bodymind-Problem-vlbk4h.pdf.

Shorters, T. (February 2022). *A cognitive skill to magnify humanity, on being.* https://onbeing.org/programs/trabian-shorters-a-cognitive-skill-to-magnify-humanity/.

Further reading

BAC. *Relaxed Venue Methodology.* https://bac.org.uk/relaxed-venue/.

Meseguer Zafe, R. *A Crash Course in Cloudspotting archive.* https://cloudspotting.live/archive.

Renel, W., & Thom, J. (2022). Relax and resist: Reflections on the Touretteshero Relaxed Venue methodology. In J. Gardner, & J. Boys (Eds.), *vol. 5. Developments in Neuroethics and Bioethics* Elsevier. 347–364.

MIX

Papier | Fördert gute Waldnutzung

FSC® C083411

Kolibri 360 GmbH
Ferdinand-Jühlke-Straße 7
99095 Erfurt, Deutschland
produktsicherheit@kolibri360.de